Akribeia

History of Metaphysics: Ancient, Medieval, Modern

Edited by

VOLUME 6

The titles published in this series are listed at *brill.com/hmtp*

Akribeia

Certainty and Ontology of Mathematics in Alessandro Piccolomini's De certitudine mathematicarum

By

Álvaro José Campillo Bo

BRILL

LEIDEN | BOSTON

This work has been completed with the support of the Research Irish Research Council Research Grant (GOIPG/2019/1194) and the 2024 Zétesis Postdoctoral Scholarship (Accademia *Vivarium Novum* and Fondazione Roma).

IRISH RESEARCH COUNCIL
An Chomhairle um Thaighde in Éirinn

FONDAZIONE ROMA

ACCADEMIA
VIVARIUMNOVUM

Library of Congress Cataloging-in-Publication Data

Names: Campillo Bo, Álvaro José, author.
Title: Akribeia : certainty and ontology of mathematics in Alessandro Piccolomini's De certitudine mathematicarum / Álvaro José Campillo Bo.

Description: Leiden ; Boston : Brill, 2025. | Series: History of metaphysics : ancient, medieval, modern, 2666-9307 ; volume 6 | Revision of author's thesis (Ph.D.)—University College Dublin. | Includes bibliographical references. | Identifiers: LCCN 2024053148 (print) | LCCN 2024053149 (ebook) | ISBN 9789004724136 (hardback) | ISBN 9789004724143 (ebook)
Subjects: LCSH: Piccolomini, Alessandro, 1508–1578. De certitudine mathematicarum. | Mathematics—Philosophy—Early works to 1800. | Certainty—Early works to 1800. | Exact (Philosophy)—Early works to 1800. | Mathematics, Greek—Early works to 1800. | Philosophy, Ancient.
Classification: LCC QA8.4 .C28 2025 (print) | LCC QA8.4 (ebook) | DDC 510.1—dc23/eng/20241226
LC record available at https://lccn.loc.gov/2024053148
LC ebook record available at https://lccn.loc.gov/2024053149

Typeface for the Latin, Greek, and Cyrillic scripts: "Brill". See and download: brill.com/brill-typeface.

ISSN 2666-9307
ISBN 978-90-04-72413-6 (hardback)
ISBN 978-90-04-72414-3 (e-book)
DOI 10.1163/9789004724143

This book is printed on acid-free paper and produced in a sustainable manner.

A Vicente e Isabel.

Contents

Acknowledgements

I wish to extend my heartfelt gratitude primarily to my Ph.D. supervisor, Dragos Calma. His constant assistance, unwavering encouragement, and scholarly insights have guided me throughout the research journey culminating in this book. Without him, and the trust he once put in me, this book would simply not exist. I deem myself truly privileged to have had the chance to learn from him the seriousness of philosophical and historical inquiry combined with the most exemplary benevolence and kindness.

I wish to acknowledge Vincenzo De Risi, Svetla Slaveva-Griffin, and Robert Goulding for their assistance. Their advice and encouraging comments have significantly contributed to the refinement of this work.

I also want to acknowledge the support from Evan King, Louis Jansen, Giuseppe Vitale, Matthew Vanderkwaak, Odile Gilon, Iulia Székely, and Elizabeth Curry.

I am particularly indebted to Accademia Vivarium Novum, the institution where I gained an understanding of classical languages. Its nurturing environment significantly enriched my educational experience and provided me with the linguistic skills leading to this research. Because of their generosity and inspiring cultural work, I am deeply thankful to all its members and students.

With affectionate appreciation, I also bear in mind the uplifting support given to me by Kostiantyn Riabtsev, Aidan McKay, Aanandavardhan Nyaupane, and Luca Sicurello.

I wish to extend my thanks also to Alfonso García Marqués. His wise counsel and deep love for the ancient world have been an inspiring light on my journey.

In closing, I want to show my utmost gratitude to Antonio Pérez Ramos who, through his exceptional erudition and careful tutelage, acquainted me with Proclus' philosophy of mathematics for the very first time.

Figures and Tables

Figures

Tables

PART 1

Introduction

> Et tamen in rebus quoque apertis noscere possis,
> si non advertas animum, proinde esse quasi omni
> tempore semotum fuerit longeque remotum.
> Cur igitur mirumst, animus si cetera perdit
> praeter quam quibus est in rebus deditus ipse?
> Deinde adopinamur de signis maxima parvis
> ac nos in fraudem induimus frustraminis ipsi.
>
> LUCRETIUS, *De rerum natura*, IV, vv. 810–817

∵

1.1 Mathematical Akribology: a Perennial Question

Is mathematics certain?

The present world around us, wherein we employ laptops and aeroplanes conceived through aeronautics, where we manage our finances and engage in resource oversight based on statistical data, appears to confirm our profound reliance on mathematics. If an activity where mathematics is involved goes wrong, we blame human beings for their bad calculations, not mathematics itself. Fully understood and aptly exercised, we hold, mathematics never lies, it does not fail. Once something has been rightly concluded in mathematics, we accept that this is the case, and we hold it to be absolutely certain.

But why is mathematics certain?

How can we explain that mathematics is certain to such a degree as to gain our immediate trust?

Such questions invite us to engage in what I shall call "akribology", i.e., in the philosophical discourse that studies and accounts for the traits of exactness, epistemic transparency, and certainty.

The certainty of mathematics is even more intriguing as its answer is not as evident as it may initially appear. What is it in mathematical reasoning that presents itself as transparent and unequivocal, thereby fostering spontaneous trustworthiness? And what exactly prompts this phenomenon? Perhaps it is

 | DOI:10.1163/9789004724143_002

because of the type of procedures mathematics adopts, which we consider a paradigm of rigorous science. Perhaps such precision and irrefutability stem from the type of realities mathematics deals with. As known, the akribology of mathematics has given rise to doubts amongst both great philosophers and scientists – and none of these doubts is new.

In this book, I will deal in depth with a critical episode in the history of the early modern akribology of mathematics. This is the intellectual challenge posed by the Sienese philosopher Alessandro Piccolomini (1508–1578) through his opusculum *De certitudine mathematicarum disciplinarum* (Rome, 1547) (= *De certitudine*).

Pondering some of the key questions regarding the akribology of mathematics of his time, Piccolomini came to a daunting conclusion: surely mathematics is the most certain human discourse, indeed *so certain* that its demonstrations cannot possibly be scientific. Why so? Because the transparency and exactness of mathematics are due to the unreality of its object of study: mathematical items are not unqualifiedly real things (or substances, in traditional metaphysical terms), but simplified fictions produced by our minds. This is the reason why mathematics is so easily and undoubtedly known: we enact these diaphanous quantitative "beings" in our minds, fictions so simple and transparent that they could not possibly exist outside of our imagination. But can real science be about something unreal? Can something unreal be held by genuine causative connections? What do we learn about reality through mathematics? By asking such questions, Piccolomini's opusculum confronted his readers with serious doubts regarding the scientificity of mathematical disciplines. So much so that his little book posed the gravest threat to the mathematical community of his time, causing a wide controversy during the critical centuries of the so-called scientific revolution: the highly debated "question about the certainty of mathematics".

But what motivated Piccolomini to argue such a thing? Was his view unusual or extreme in the 16th century? And how is it that his argument was effective as to be discussed years later in foreign universities?

In the following chapters, I will present a technical and historically contextualised analysis of the most relevant passages of Piccolomini's text, along with an examination of its sources. Thus, I will focus also on retracing the impact of a second, pivotal text on which Piccolomini greatly relied in writing *De certitudine*: Proclus of Lycia's (412–485) *Commentary on the First book of Euclid's Elements* (= *In Euclidem*), a work indissolubly linked to the early modern history and philosophy of mathematics, to the history of modern science and, hence, to the intellectual history of the West.

1.2 Piccolomini's Intellectual Background to *De certitudine*

Alessandro Piccolomini, of Sienese origin, was steeped in the literary humanist tradition, mastering ancient Greek and gaining deep familiarity with the classical Latin sources. Renowned from a young age for his linguistic and literary prowess within Siena's erudite circles, he went on to receive a distinguished philosophical and scientific education at the University of Padua from 1538 to 1542.[1] There, Piccolomini followed the lectures of the Aristotelian and humanist Vincenzo Maggi (1498–1564),[2] the renowned Averroist Marco Antonio "Genua" De' Passeri (1491–1563), and the mathematician Federico Delfino (1477–1547).[3] In 1539, Bernardino Tomitano (1517–1576) was appointed to the professorship of logic, where he may have had Piccolomini amongst his students.[4]

Maggi was a member of the intellectual society of scholars called the *Accademia degli Infiammati* of which Piccolomini would become not only a member but also one of its "princes" (a position of general coordinator re-elected every four months).[5] This erudite group was founded in 1540 by the bishops Ugolino Martelli (1519–1592) and Leone Orsini (1512–1574), the humanist Benedetto Varchi (1502–1565), and the Venetian nobleman and accomplished mathematician (also a student of Delfino) Daniele Barbaro (1514–1570). Piccolomini, like the other *Infiammati*, was committed to translating Greek and Latin works into vernacular Italian to enhance the literary culture of a broader audience. During these years, he translated into Italian book VI of Vergil's *Aeneis*, published in the collective publication *I sei primi libri dell'Eneide di Virgilio tradotti* (Venice, 1540), some passages from Ovid's *Metamorphoses*, and Xenophon's *Oeconomicus* (Venice, 1540).

Notably, Piccolomini would later produce the first-ever Italian work on Aristotelian natural philosophy: *La prima parte della filosofia naturale* (Rome, 1551) and *La seconda parte della filosofia naturale* (Rome, 1554). He also produced the first vernacular book on Aristotelian logic with *L'instrumento della*

1 A. Cotugno, "Alessandro Piccolomini", in M. Sgarbi (ed.), *Encyclopedia of Renaissance Philosophy* (Cham: Springer International Publishing Switzerland, 2022).

2 S. Di Mare, "Vincenzo Maggi", in M. Sgarbi (ed.), *Encyclopedia of Renaissance Philosophy.*

3 B. Baldi, *Le vite de' Matematici. Edizione annotata e commentata della parte medievale e rinascimentale a cura di Elio Nenci* (Milan: Franco Angeli, 1998), pp. 524–525.

4 M. T. Girardi, "Bernardino Tomitano", in M. Sgarbi (ed.), *Encyclopedia of Renaissance Philosophy.*

5 Girardi, "Accademia degli Infiammati", in M. Sgarbi (ed.), *Encyclopedia of Renaissance Philosophy.*

Filosofia (Rome, 1551),[6] of the *Rhetorica* with *I tre libri della Retorica d'Aristotele* (Venice, 1571), and also of the *Poetica* with *Il libro della Poetica d'Aristotele* (Venice, 1572).[7] Additionally, Piccolomini wrote several Italian treatises on natural philosophy and astronomy. Some of them are *Della sfera del mondo* (Venice, 1540), *Delle stelle fisse* (Venice, 1543), or *Della grandezza della Terra et dell'Acqua* (Venice, 1561).

Piccolomini's Paduan teacher, Marco Antonio De' Passeri, was recognised as one of the authorities on (Pseudo-)Simplicius' commentary on *De anima*.[8] He influenced a great number of students at Padua through a particularly long teaching career: he taught at the local university from 1517 to 1563. He was a leading figure amongst the so-called "Simplician Averroists". This label refers to a group of philosophers who read Pseudo-Simplicius' teachings in his commentary on *De anima* as aligning with the doctrines of Averroes' *Magnum commentarium*. Pseudo-Simplicius' commentary was made widely available to the readership first through its 1527 Greek edition by Francesco Torresano (fl. 1529–1546) and in its 1543 Latin translation by another of De' Passeri's students, the Hellenist Giovanni Faseolo (1517–1571).[9] From this interpretation tendency a new, late species of Paduan Averroism, leaning towards Neoplatonic positions, emerged. This new form of eclectic Averroism will be central in accounting for Piccolomini's positions. As we shall see, the Averroist reading of Pseudo-Simplicius is, allegedly, one of the direct sources determining Piccolomini's philosophy of mathematics.[10]

6 See M. Sgarbi, *The Italian Mind: Vernacular Logic in Renaissance Italy (1540–1551)* (Leiden: Brill, 2014), pp. 175–205.

7 L. Bianchi, "Per una storia dell'aristotelismo 'volgare' nel Rinascimento: problemi e prospettive di ricerca", in *Bruniana & Campanelliana* 15/2 (2009), pp. 367–385.

8 In this work I will follow Steel and Bossier, who have convincingly argued that the commentary traditionally attributed to Simplicius should be counted amongst the works of Priscian of Lydia. Thus, I will refer to this author as Pseudo-Simplicius, "Simplicius", or Priscian interchangeably. See C. Steel, *The Changing Self: A Study on the Soul in Later Neoplatonism: Iamblichus, Damascius and Priscianus* (Brussels: Paleis der Academiën, 1978), *passim*; G. Gabor, *Colloquium 1 The Authorship of the Pseudo-Simplician Neoplatonic Commentary on the De Anima. Proceedings of the Boston Area Colloquium of Ancient Philosophy* (Leiden: Brill, 2020), pp. 1–22. Regarding Genua, see L. Spruit, "Marco Antonio Genua", in M. Sgarbi (ed.), *Encyclopedia of Renaissance Philosophy*; L. Spruit, *Species Intelligibilis: From Perception to Knowledge: 2. Renaissance Controversies, Later Scholasticism, and the Elimina* (Leiden: Brill, 1995), pp. 159–163; B. Nardi, *Saggi sull'Aristotelismo padovano dal secolo XIV al XVI* (Florence: Sansone, 1958), pp. 365–443.

9 See Simplicius, *Commentaria Simplicii profundissimi et acutissimi philosophi in treis* [sic] *libros De anima Aristotelis*, transl. Giovanni Faseolo (Venice: Hieronymus Scotus, 1543).

10 For Genua's denial of mathematical causation see Marco Antonio Genua De' Passeri, *In tres libros Aristotelis De Anima* (Venice: Damianus Zenarus, 1576), I, p. 9. This passage will be commented upon in section 3.3.3.

During his third year at the University of Padua, Piccolomini developed an interest in the teachings of the Bolognese professor Ludovico Boccadiferro (1482–1545).[11] As we learn from his correspondence,[12] Piccolomini was familiar with Boccadiferro's doctrines even before personally attending his lectures. In a letter dated June 9th, 1541, Piccolomini requested his fellow *Infiammato* Varchi, who had attended the lectures at the Bolognese *studium* in 1540–41, to send him the transcriptions of Boccadiferro's teachings.[13] Despite this connection, Piccolomini was unable to secure detailed transcriptions of the lectures he sought, leading him to leave Padua and pursue Boccadiferro's teachings the following year in Bologna.

Another of the intellectual figures influencing Piccolomini between 1538 and 1543 was Don Diego Hurtado de Mendoza (1503–1575), the Spanish imperial ambassador to the Republic of Venice from 1539 to 1546. An advocate for the technical usefulness of mathematics, a competent Hellenist and friend of numerous humanists such as Pietro Bembo, Mendoza owned a collection of around 300 Greek manuscripts, most of which can be found in the Biblioteca de El Escorial's Greek index.[14] Mendoza offered his help and support to some other members of the *Infiammati* in Padua, in whose projects of vernacular translation and divulgation he was also interested: he helped Varchi with his translation of Livy and provided Maggi with a manuscript for his edition of Aristotle's *Poetica*.[15]

Mendoza shared this fondness for the application of mathematics to machinery with Niccolò Fontana Tartaglia (*c.*1500–1557), of whom he became both a mathematical disciple and a patron. Mendoza's interests in mechanics were of extraordinary magnitude and might have been increased by the interest in

11 Spruit, "Ludovico Boccadiferro", in M. Sgarbi (ed.), *Encyclopedia of Renaissance Philosophy*.

12 *Lettere al Varchi* 66, 67 and 68, published in and annex to F. Cerrata's *Alessandro Piccolomini: letterato e filosofo senese del Cinquecento* (Siena: Accademia degli Intronati, 1960), pp. 264–272 and in Cerrata, "An Account of the Early Life of the Accademia degli Infiammati in the letters of Alessandro Piccolomini", in *The Romanic Review* 48 (1957), pp. 249–264.

13 In letter 71 we read: *Non mi sovien altro che dirvi, salvo che mi voliate bene e mi avisiate alcuna volta degli studii di Bologna e come il Bocca di Ferro vi sodisfa.* In letter 67: *Harò molto caro M. Benedetto, poi che così vi lodate del Bocca di Ferro, che duraste di tanta fatiga di far che io havesse 2 o 3 lettioni de quelle che fatte tenete più belle, che gli habbia lette in questi giorni, nel secondo del mese di giugno, non vi si scordi.*

14 P. L. Rose, "The New Science of Tartaglia and Benedetti", in *The Italian Renaissance of Mathematics: Studies on Humanists and Mathematicians from Petrarch to Galileo* (Genève: Librarie Droz, 1975), pp. 151–158, see p. 154. A study of Mendoza's manuscripts can be found in A. Hobson, *Renaissance Book Collecting: Jean Grolier and Diego Hurtado de Mendoza, their Books and Bindings* (Cambridge: Cambridge University Press, 1999).

15 Hobson, *Renaissance Book Collecting*, pp. 79–80.

the technical utility of mathematical literature shared by several Venetian patricians and an ever-growing number of technician-mathematicians at their service in Venice's arsenals.

At some point between 1545–46, Mendoza acquired a Greek manuscript of Pseudo-Aristotle's *Quaestiones mechanicae* from Bessarion's collection and translated it into Spanish when attending the Council of Trent.[16] Tartaglia made Mendoza one of the interlocutors in his dialogue, *Quesiti e Inventioni diverse* (Venice, 1538). In books VII and VIII of this work, Mendoza's persona voices the mechanics of Pseudo-Aristotle: the argument concerns statics, namely, problems on the balance, the lever, and a discussion about the laws of equilibrium or, as Tartaglia puts it, the "science of weights" or *scientia dei pesi*. Tartaglia amends these views to align them with his own perspectives.

It was under the influence of Mendoza that Piccolomini produced his Latin paraphrase of the *Quaestiones mechanicae*, to which his opusculum *De certitudine mathematicarum* was later added as an attachment. We know that Piccolomini had finished his Latin paraphrase of the mechanics before the first day of 1540. The evidence is found in Piccolomini's vernacular work *De la institutione di tutta la vita del'huomo nato nobile* (Venice, 1542), a 16th-century bestseller that, in its 1560 version, was translated both into Spanish and French and widely used as a textbook on moral education. The dedicatory letter of the first edition was written on the first of 1540 at Padua. In book III, chap. 17, Piccolomini addresses the influential Sienese poet Laudomia Forteguerri (1515–1555), to whom the work is dedicated. After praising the practical and mechanical arts and their role in the education of the noble spirit, he adds:

> Aristotle himself wrote about it [mechanics] in a brief but beautiful book which, due to significant inaccuracies and corruptions, is very difficult and has not been clarified by anyone; I have paraphrased it persuaded by the very illustrious Don Diego Mendoza, currently Ambassador of His Majesty to the Venetian lords.[17]

---

16 A. G. Keller, "Machines as Mathematical Instruments", in L. B. Cormack, S. A. Walton, J. A. Schuster (eds), *Mathematical Practitioners and the Transformation of Natural Knowledge in Early Modern Europe* (Cham: Springer, 2017), pp. 115–130, see p. 120, footnote.

17 Unless otherwise specified, all the translations are of my own. See Alessandro Piccolomini, *De la institutione di tutta la vita del'huomo nato nobile* (Venice: Hieronymus Scotus, 1545), III, 17, p. 72: *Aristotele istesso n'ha scritto in un suo breve ma bellisimo libro, sopra il quale, per essere per le grandi scorrettioni e corrutioni di testi, che vi sono, difficilissimo e da nissuno alluminato, ho fatto io una Parafrasi à persuasione del multo Illust. Signor Don Diego Mendozio, al presente Ambasciatore di sua Maiestà appresso i signori Venetiani.*

This text shows that Piccolomini completed his paraphrase, encouraged by Mendoza, when he was a student in Padua. Nonetheless, it would be published seven years later, in 1547 (see below).

After attending Boccadiferro's lectures in 1543, Piccolomini would travel back to his native Siena and remain there for three years. Back in his hometown, Piccolomini undertook some responsibilities as an attendant of the archbishop of Siena. He was an active member of the Sienese society *Accademia degli Intronati*, an erudite community also committed to bringing classical works to a wider audience through vernacular translation. There is evidence that he was appointed to teach moral philosophy for at least one year (by 1545) at the local university.[18] In 1546, Piccolomini left Siena and travelled to Rome, where he was to remain for the following seven years serving as a man of letters in the Roman curia.

Coincidentally, Diego Hurtado de Mendoza was assigned to Rome and Siena to develop his diplomatic functions from 1547 to 1552, which means that Piccolomini and the erudite nobleman would converse again. Indeed, Piccolomini's successive years are highly linked to the Spanish noble house of Mendoza: most likely recommended by the imperial ambassador, the Sienese was taken under the protection of Cardinal Francisco de Mendoza y Bobadilla (1508–1566) and his younger brother, bishop Fernando de Mendoza y Bobadilla (fl. *c.*1533). To the former Piccolomini dedicated his 1551 *L'instrumento della Filosofia*, while he honoured the latter in the dedicatory of *De certitudine*.

Thus, the project of publishing a widely readable Latin version of the Aristotelian mechanical problems, suggested by the imperial ambassador years before, was fulfilled by the 15th of March of 1546 in Rome under the patronage of bishop Fernando de Mendoza, and in his own palace (*Romae, ex aedibus ipsis tuis*).[19] To the Latin paraphrase, Piccolomini added his opusculum *De certitudine* and a *Praefatio* to the paraphrase, finished by March of 1546.[20] Since the *Praefatio* contains a mention of his "little book on mathematical certainty" and invokes Proclus,[21] it can be inferred that Piccolomini wrote *De certitudine*

18 See Cerrata, *Alessandro Piccolomini*, pp. 52–55.

19 Alessandro Piccolomini, *De certitudine mathematicarum scientiarum* (Rome: Antonio Blado, 1547), f. LXXI.

20 Biblioteca Nazionale Centrale di Firenze, II, IV, 214, f. 1, l.12: *Romae, quinto idus martias, MDXLVI*.

21 Alessandro Piccolomini, *In mechanicas quaestiones Aristotelis paraphrasis* (Rome: Antonio Blado, 1547), f. III: *Quo fit ut alia sit causa qua in primo certitudinis gradu mathematicae scientiae reponantur, sicut nos abunde in libello quem fecimus De certitudine mathematicarum scientiarum opinionem hanc nostram manifestissimis rationibus declaravimus. Cuius quidem opinionis ansam praebuit Proclus in I Elementorum.*

between 1540 and March 1546. After that, he composed the *Praefatio* to the paraphrase. The three texts were finally published in 1547 and reprinted in 1565 (Venice, Traianus Curtius).

In sum, the following two doxographical keys should be taken into consideration when interpreting Piccolomini's *De certitudine*:

(1) Piccolomini's university education took place in the 16th-century Paduan and Bolognese context, influenced by the teachings of masters such as the Simplician Averroist Genua De' Passeri and the natural philosopher Ludovico Boccadiferro.

(2) *De certitudine mathematicarum* was produced as an addition to Piccolomini's paraphrase of the Pseudo-Aristotle's *Quaestiones mechanicae*, whose Latin paraphrase the Sienese had finished seven years before by suggestion of Diego de Mendoza in Padua, whom he would meet again in Rome.

1.3 *De certitudine Mathematicarum*: Themes and Hermeneutical Keys

The complete title of Piccolomini's opusculum is *Commentarium de certitudine mathematicarum disciplinarum in quo de resolutione, diffinitione et demonstratione necnon de materia et fine logicae facultatis quam plura continentur ad rem ipsam, tum mathematicam, tum logicam maxime pertinentia*. It consists of twelve chapters in addition to a *Praefatio* and a *Peroratio*. The titles of the chapters are as follows:[22]

Chapter I, On the purpose or end of the logical faculty.
Chapter II, On the subject-matter of the same logical faculty.
Chapter III, On the division and resolution of logic and its parts, both in discovery and judgement.

22 Alessandro Piccolomini, *De certitudine, Index capitum*: *Capitulum I, De scopo sive fine logicae facultatis*; *Capitulum II, De materia eiusdem logicae facultatis*; *Capitulum III, De divisione, et resolutione logica ac de partibus eiusdem logicae, tum inveniendi, tum iudicandi*; *Capitulum IV, De demonstratione et eius partibus, sive speciebus*; *Capitulum V, De quibusdam assumptis ad declarandum medium demonstrationis potissimae*; *Capitulum VI, De medio demonstrationis potissimae*; *Capitulum VII, Quae praelibanda sunt ad mathematicas disciplinas pertinentia et primum de materia ipsarum*; *Capitulum VIII, An sit finis mathematicis disciplinis et de earum utilitate*; *Capitulum IX, De problematibus et theorematibus mathematicis et eorum partibus*; *Capitulum X, De resolutione et compositione mathematica; Capitulum XI: An certitudo mathematica ex vi demonstrationum potissimarum oriri dicenda sit; Capitulum XII: Quae tandem sit vera causa cur mathematicae disciplinae in primo gradu certitudinis ab Averroe positae fuerint.*

Chapter IV, On demonstration and its parts or species.
Chapter V, On certain assumptions for explaining the middle-term of unqualified demonstration.
Chapter VI, On the middle-term of unqualified demonstration.
Chapter VII, What must be briefly mentioned regarding the mathematical disciplines and first about their subject-matter.
Chapter VIII, Whether there is an end to the mathematical disciplines and on their utility.
Chapter IX, On mathematical problems and theorems and their parts.
Chapter X, On mathematical resolution and composition.
Chapter XI: Whether mathematical certainty is to be said to arise from the force of unqualified demonstrations.
Chapter XII: What, in the end, is the true cause why mathematical disciplines were posited by Averroes in the first degree of certainty.

De certitudine's chapters can be arranged in accordance with four broad topics. Chapters I, II, and III work as an introduction to logic and its methodologies, whereas chapters IV, V, VI, and X discuss the traits of a would-be paradigmatic scientific demonstration and its instantiation in mathematics. Chapters VII, VIII, and IX present Piccolomini's philosophy of mathematics, including mathematical ontology and epistemology. Chapters XI and XII contain a discussion on mathematical akribology, demonstrative theory, and the ontology of mathematics.

As acknowledged by both historians of philosophy and science, Piccolomini's *De certitudine* triggered one of the most critical debates in the early modern philosophy of mathematics: the question about the certainty of mathematics.[23]

23 D. Levitin, *Ancient Wisdom in the Age of the New Science* (Cambridge: Cambridge University Press, 2015), part 4.5, pp. 313–325; J. Longeway, "The Place of *Demonstratio Potissima* in some 16th-Century Accounts of Mathematics", in J. Biard (ed.), *Raison et démonstration: Les commentaires médiévaux sur le Seconds analytiques* (Turnhout: Brepols, 2015), pp. 223–251; D. Cozzoli, "Alessandro Piccolomini and the certitude of mathematics"; A. De Pace, *Le matematiche e il mondo: Ricerche su un dibattito in Italia nella seconda metà del Cinquecento* (Milan: Francoangeli, 1993); T. Schöttler, "From Causes to Relations: The Emergence of a Non-aristotelian Concept of Geometrical Proof out of the *Quaestio De Certitudine Mathematicarum*", in *Society and Politics* 6/2 (2012), pp. 29–47; P. Mancosu, *Philosophy of Mathematics and Mathematical Practice in the Seventeenth Century* (Oxford: Oxford University Press, 1999), pp. 13–35; P. Mancosu, "Aristotelian logic and Euclidean mathematics: Seventeenth-century developments of the quaestio de certitudine mathematicarum", in *Studies in History and Philosophy of Science Part* 23/2 (1992), pp. 241–265; G. Ferrari, "Dimostrazione Matematiche e Conoscenza Scientifica in Alessandro Piccolomini", in H. Burns, F. P. Di Teodoro, G. Bacci (eds), *Saggi di Letteratura architettonica da Vitruvio a Winckelmann III* (Firenze: Olschki, 2010), pp. 215–233;

Amongst many others, Francesco Barozzi (1537–1604), Pietro Catena (1501–1576), Petrus Ramus (1515–1572), the Jesuits Benito Pereyra (1535–1610), Christopher Clavius (1538–1612), Marcin Śmiglecki (1564–1618), and Giuseppe Biancani (1566–1624), the Spanish sceptic Francisco Sánchez (1551–1623), Pierre Gassendi (1592–1655), the Englishmen Thomas Hobbes (1588–1679), John Wallis (1616–1713), Isaac Barrow (1630–1677), and even the Neapolitan Giambattista Vico (1668–1744) partook in this controversy.

Amongst the figures just indexed, Francesco Barozzi deserves special attention.[24] Barozzi was a nobleman of Venetian origin, 29 years younger than Piccolomini, a disciple of the Cretan Andreas Donos (1476–1550), and of the – by then elder – De' Passeri at the University of Padua. Barozzi was friends with Daniele Barbaro (another of Piccolomini's acquaintances). A highly proficient Hellenist and mathematician, Barozzi was the owner of a rich collection of Greek manuscripts, today partially preserved at the Bodleian Library. Perhaps spurred by Barbaro, he actively and publicly opposed Piccolomini's views since

---

H. E. Velilla-Jiménez, "El debate sobre la certeza de las matemáticas en la filosofía natural de los siglos XVI y XVII (*De quaestio* [sic] *de certitudine mathematicarum*)", in *Estudios de Filosofía* 3 (2018), pp. 59–93; P. Palmieri, "On *scientia* and *regressus*", in H. Lagerlund, B. Hill (eds), *The Routledge Companion to Sixteenth Century Philosophy* (New York: Routledge, 2017), pp. 319–349; S. Higashi, *Penser les mathématiques au XVIe siècle* (Paris: Editions Classiques Garnier, 2018); D. Rabouin, "Le rôle de Proclus dans les débats sur la 'mathématique universelle' à la Renaissance", in A. Lernould (ed.), *Études sur le Commentaire de Proclus au premier livre des Éléments d'Euclide* (Saint-Paul Fribourg: Presses Universitaires du Septentrion, 2007), pp. 217–223; G. C. Giacobbe, "Epigoni nel seicento della "Questio de certitudine mathematicarum": Giuseppe Biancani", in *Physis* 18 (1976), pp. 5–40; Giacobbe, "La riflessione metamatematica di Pietro Catena", in *Physis* 15/2 (1973), pp. 178–196; Giacobbe, "La Quaestio de certitudine mathematicarum all'interno della scuola padovana", *Atti del convegno di storia della logica* (1974), pp. 95–112; Giacobbe, "Un gesuita progressista nella "Quaestio de certitudine mathematicarum" rinascimentale: Benito Pereyra", in *Physis* 19 (1977), pp. 51–86; L. M. Carolino, "Mathematics and the late aristotelian theory of science: The '*Quaestio de Certitudine Mathematicarum*' in seventeenth-century Portuguese universities", in V. Navarro, W. Eamon (eds), *Beyond the black legend: Spain and the Scientific Revolution* (Valencia: Universidad de Valencia, 2007), pp. 399–412; N. Jardine, "Epistemology of the sciences", in C. B. Schmitt, Q. Skinner, E. Kessler (eds), *The Cambridge History of Renaissance Philosophy* (Cambridge: Cambridge University Press, 1988), pp. 685–711; J. de Garay, "Proclus' Reception in the Sixteenth Century: Commentary on the First Book of Euclid's *Elements*", in E. Anagnostou, K. Parry (eds), *Later Platonists and their Heirs among Christians, Jews, and Muslims* (Leiden Boston: Brill, 2023), pp. 459–479.

24 M. Cosci, "Francesco Barozzi", in M. Sgarbi (ed.), *Encyclopedia of Renaissance Philosophy*; Rose, *A Venetian Patron and Mathematician of the Sixteenth Century: Francesco Barozzi: 1537–1604* (Pisa: Giardini Editori, 1977); B. Boncompagni, "Intorno alla vita ed ai lavori di Francesco Barozzi", in *Bullettino di bibliografia e di storia delle scienze matematiche e fisiche* XVII (1884), pp. 831–837.

1559 at least in three ways: (i) by writing and publishing in 1560 a refutation of *De certitudine*; (ii) by addressing the question of the certainty of mathematics in his teaching at the University of Padua in the course 1559–1560; and (iii) by providing the first ever printed Latin edition of Proclus' *In Euclidem* (1560). Through his Latin edition, the 23-year-old Barozzi arguably became one of the main authorities regarding Proclus' text, its manuscript grounding, and its hermeneutics – a precious weapon in his dispute against Piccolomini's reading of the text. Although Barozzi's answer is of the greatest importance in the history of the 16th-century reception of Proclus' commentary and of Piccolomini's challenge, I will mention it only occasionally in this book, without offering an analysis of it.[25]

1.4 State of the Art

In recent scholarship, Piccolomini's *De certitudine* has been understood mostly in the light of this controversy. The modern philosophical understanding of the debate was defined by Giacobbe with his series of papers on the question of the certainty of mathematics, and by De Pace through her 1993 seminal monograph on 16th-century mathematics and physics. Other recent studies addressing Piccolomini's opusculum are those of Mancosu on 16th-century mathematics, including the spreading of the question about mathematical certainty, and on syllogistic and mathematical proof; Cozzoli's paper on Piccolomini's philosophy of science and its reception in scholarship; Longeway's paper on Piccolomini's *demonstratio potissima*; and the excellent monograph by Higashi on the 16th-century philosophy of mathematics. Felgner has included Piccolomini's *De certitudine* and some of the authors in the subsequent debate on mathematical proof in his monograph on the history of theorem demonstration, emphasising the constructivist aspects of Piccolomini's ontology of mathematics.[26]

---

25 An in-depth scrutiny of Barozzi's counterattack to Piccolomini will be the object of a future study of mine.

26 Giacobbe, "Il *Commentarium De Certitudine*"; Giacobbe, "Francesco Barozzi"; P. Freguglia, *Ars analytica. Matematica e methodus nella seconda metà del Cinquecento* (Busto Arsizio: Bramante Editore, 1988); De Pace, *Le matematiche e il mondo*; Mancosu, *Philosophy of Mathematics and Mathematical Practice*; P. Mancosu, M. Mugnai, *Syllogistic Logic and Mathematical Proof* (Oxford: Oxford University Press, 2023), chapter 3; M. Hellyer, *Catholic Physics: Jesuit Natural Philosophy in Early Modern Germany* (South Bend, Indiana: University of Notre Dame, 2005); D. Cozzoli, "Alessandro Piccolomini and the certitude of mathematics"; Longeway, "The Place of *Demonstratio Potissima*"; J. Biard, "La certitude

All the accounts of Piccolomini's philosophy of mathematics in *De certitudine* agree on the undeniable influence of Proclus' *In Euclidem* on it. After Aristotle, who is the most frequently cited authority with 37 mentions in *De certitudine*, Proclus is referenced by name 32 times in the opusculum, followed by Averroes with 13 mentions. Proclus is cited once in the *Praefatio* and once in the text of the paraphrase (see Table 1 in the Appendix for a transcription of all references).

In his studies on the early modern reception of *In Euclidem*, Claessens has shown that it was valued not only because of the authority which Proclus held as a Greek commentator but also for the philosophical and historical insights it provided.[27] It was also adopted as a source of mathematical knowledge and arguments on the usefulness of mathematics. It was used in all these ways in *De certitudine*: Piccolomini not only used it as a source concerning the philosophy of mathematics, but he also drew out mathematical technicalities (such as the parts of the demonstration or the distinction between theorems and problems) and wrote an encomium to mathematical usefulness following Proclus.[28]

De certitudine has been read, using De Pace's phrase, as an "Aristotelian critique of mathematics as a science."[29] This emphasis on the insurmountable opposition between Aristotelian natural philosophy and the quantification of natural phenomena, the clash between the Aristotelian theory of science and mathematical practice, and the question thereby raised concerning the

des mathématiques et ses fondements selon Piccolomini" in M. F. Piéjus, M. Residori (eds), *Alessandro Piccolomini (1508–1579), un siennois à la croisée des genres et des savoirs* (Paris: Sorbonne PSN, 2012), pp. 247–258; Higashi, *Penser les mathématiques*; U. Felgner, *Philosophy of Mathematics in Antiquity and in Modern Times* (Springer: Birkhäuser Cham, 2023), chap. 7.

27 See G. Claessens, "Clavius, Proclus, and the Limits of Interpretation: Snapshot-idealization versus Projectionism"; Claessens, "Sixteenth- and Seventeenth Century Views on Mathematical Imagination", in Ch. Lüthy, C. Swan, P. J. J. M. Bakker, C. Zittel (eds), *Image, Imagination, and Cognition: Medieval and Early Modern Theory and Practice* (Leiden Boston: Brill, 2018), pp. 208–226; Claessens, "Reception and the Textuality of History: Ramus and Kepler on Proclus' History and Philosophy of Geometry", *Texts, Transmissions, Receptions: Modern Approaches to Narratives* (Boston: Brill, 2015), pp. 281–294; see also Claessens, "Kepler's Platonism in the *Harmonice mundi*", in *De mundi recentioribus phænomenis: Cosmologie et science dans l'Europe des Temps modernes, xve–xviie siècles. Essais en l'honneur de Miguel Ángel Granada* (Brepols, 2022), pp. 303–317; Claessens, "The Drawing Board of Imagination: Federico Commandino and John Philoponus", *Journal of the History of Ideas* 76/4 (2015), pp. 499–515.

28 Alessandro Piccolomini, *De certitudine*, f. LXXXVI; f. LXXXXVIIII; ff. LXXXXVIIII–LXXXXVIIIIv; f. C.; f. LXXXXVIIIv.

29 De Pace, *Le matematiche e il mondo*, chap. 1, title: *La critica Aristotelica della matematica come scienza.*

epistemic status of mathematical demonstration has guided the reception of Piccolomini's *De certitudine* ever since Barozzi's rebuttal.

Thus, our received reading of *De certitudine* holds that Piccolomini undertook a theoretical attack against mathematical means of demonstration. The core of this attack is that mathematical demonstrations cannot instantiate a *demonstratio potissima*, i.e., the paradigmatic form of syllogistic demonstration which is ranked as the one having the highest scientific status. This is so because of three reasons:

Firstly, (1) when mathematical proofs are syllogistically formalised, the middle-term that connects the entire proof does not fulfil the conditions that, according to Piccolomini, are those of the paradigmatic demonstration: the middle-term is not the definition of the passion to be concluded of the subject. This claim leans on the preceding Latin debates on the syllogistic arrangement of mathematical proof.

Secondly, (2) Piccolomini's attack denies that any kind of causality is involved in mathematical proofs. Piccolomini posits that mathematical demonstrations are in some sense explanative devices, but they are not (and cannot possibly be) *causally* explanative. Therefore, mathematical proofs not only do not rank amongst the most scientific kind of logical arrangement, but the very scientific status of mathematics, in an epistemic framework where "knowing is knowing through causes", is thrown into grave doubt. Mathematics, as the science of quantity, cannot speak of any essential aspects of the natural world: quantity amongst natural entities is categorised as a mere accident. And a discourse about accidents cannot reveal anything relative to the essence of natural substances.

Thirdly (3), according to Piccolomini, genuine science consists in obtaining the definitions of substances. However, mathematical objects are not substances. Even worse, mathematical objects are so epistemically transparent (or certain) that their definitions are known by us from the very beginning and previous to any demonstration. This is shown by the arrangement of Euclid's *Elementa*, which provides the reader with all its definitions from the very beginning. Thus, if doing science is finding previously unknown definitions of substances, mathematics is not a science, as it fails to fulfil both requirements.

It is worth noting that, despite his overall argumentation against the scientificity of mathematical demonstration, Piccolomini occasionally refers to mathematics as *scientia* – albeit in a deliberately improper sense – and also as *disciplina*. Piccolomini's followers and epigones will make far more belligerent use of the arguments contained in *De certitudine* (see Pereyra's quotation below).

As seen, questionable as its scientific status is shown to be, Piccolomini gladly accepts the (by then proverbial) claim of Averroes that mathematics possesses the highest degree of certainty.[30] If that is so, mathematical certainty cannot stem from the procedures of mathematical proof since, as said, mathematical proofs are hardly real demonstrations. Whence does it come, then? Piccolomini answers that mathematical certainty originates in the special ontological status of mathematical items. In sustaining this, he adopted a particular mathematical ontology: we are told that *all* mathematical disciplines deal with phantastic[31] entities which cannot be instantiated in the sensible world, or that have no counterparts in it: bi-dimensional or part-less entities, perfect angles, or pure numbers cannot be found in sensible matter.

It is to be noted that neither this debate nor the positions on this question are new. Traces of a similar discussion can be already seen in late ancient philosophers such as Themistius, Simplicius, Philoponus, and crucially, Proclus. Its presence can be also found in the 13th- and 14th-century Latin tradition: following Averroes' elaborations on the akribology of mathematics vis-à-vis that of natural philosophy, authors such as Albert the Great, Thomas Aquinas, or John Buridan are amongst the better-known predecessors to the Renaissance discussions. As usual in their cultural framework, their elaborations were developed as *passim* notes within the commentary tradition or, as in the case of Buridan, inspired by some passages in *Aristoteles Latinus*. Perhaps the most influential amongst the sections stirring the Latin debates on mathematical certainty are Aristotle's comments in *Metaphysica* II, 3, 995a (from whose Latin translation the very notion of *certitudo mathematica* came into Latin) as well as Averroes' elaboration on this precise section in his *Tafsīr* (تفسير) or *Expositio*. Other *loci* are *Metaphysica*, I, 2, 982a, l. 19–28; VI, 1, 1025b, l. 3–7; *Ethica Nichomachea* I, 3,1094b, l. 24; I, 1098a, l. 27; *De caelo*, 306a, l. 27; *De anima*, I, 402a, l. 2; I, 419a, l. 16–21. In all these *loci*, the certainty (ἀκριβές and derived forms) of mathematical disciplines is brought out by Aristotle and suggested as an epistemological paradigm which is, however, unachievable in all disciplines.

Rose has provided evidence that the question of mathematical certainty, as linked to the speculation on mechanics and their gnoseological status, had 16th-century Italian predecessors before Piccolomini. A prominent case is

30 Averroes, *Commentaria in libros Metaphysicorum Aristotelis*, in *Aristotelis opera cum Averrois commentariis* (Venice: Giunti, 1562–1574), vol. 8, lib. II, p. 35v, l. 16–27.

31 I will use the transliterations "phantasy", "phantastic", and related terms to translate the Greek φαντασία and its cognates. This choice is motivated by the fact that some Latin authors distinguished between *imaginatio* and *phantasia* as different faculties (see section 2.2.4). Such nuances in meaning would be obscured if both terms were translated simply as "imagination". Additionally, the choice of one term over another will prove significant in examining Proclus' growing influence on Piccolomini after 1540.

that of Leonardo Da Vinci (1452–1519) who discussed the question of mathematical certainty related to mathematical applicability to sensible items.[32] In the university context, Marco Antonio Zimara (*c.*1470–1532), Agostino Nifo (1473–1545), and Ludovico Boccadiferro (1482–1545) can also be counted as northern Italian predecessors in this question and, as I shall show, direct influences on Piccolomini's views.[33]

As mentioned above, Barozzi answered Piccolomini's challenge defending the scientific status of mathematical proofs 13 years later, in his *Opusculum in quo una oratio, et duae quaestiones: altera de certitudine, et altera de medietate mathematicarum continentur* (Padua, 1560) by discussing the interpretation of some critical sections of Proclus' commentary and opposing Piccolomini's reading of them. Moreover, scholars have pointed out that Piccolomini's attack on the scientific status of mathematics was so devastating that mathematicians all around Europe felt the utmost necessity of either answering or surrendering to his challenge. That the Jesuit order assumed this controversy as an internal debate (also related to the role and time to be given to mathematics in the curriculum of the *ratio studiorum*) immensely contributed to its dissemination and greatly increased the number of intellectuals (both within and outside the order) partaking in it.[34] To offer just one example of how arguments derived from *De certitudine* were provocatively weaponised in the subsequent debates, teaching at the *Collegium Romanum* in the 1560s, the Jesuit Pereyra argued as follows:

> My opinion is that the mathematical disciplines are not properly sciences, and I am led to this opinion by other reasons, but especially by this one argument. To know is to understand a thing through the cause through which it exists, and science is the result of demonstration. Demonstration, however, (I am speaking of the most perfect kind of demonstration) must consist of things that are *per se* and proper to what is being demonstrated, while those things that are accidental and common are excluded from perfect demonstrations. But the mathematician neither considers the essence of quantity, nor does he treat its properties

32 Rose, "Certitudo Mathematicarum from Leonardo to Galileo", in *Leonardo nella scienza e nella tecnica: Atti del Simposio Internazionale di storia della Scienza* (Firenze Vinci: Giunti, 1969), pp. 43–49.

33 Higashi, *Penser les mathématiques*, pp. 91–120.

34 For the impact and diffusion of the question within the Jesuit order see B. Machado Mota, *O estatuto da Matemática em Portugal nos Séculos XVI e XVII* (Lisboa: Fundação Calouste Gulbenkian, 2011), part 2; see also Velilla-Jiménez, "Francisco Sánchez and the Quaestio de certitudine mathematicarum: A Sceptical approach", in *Endeavour* 46/4 (2022), pp. 1–10.

> as they arise from such an essence, nor does he explain them by their proper causes for which they belong to quantity, nor does he make his demonstrations from proper and *per se* predicates, but from common and accidental ones. Therefore, mathematics is not properly a science. The major premise of this syllogism does not require proof, for it is clearly derived from what is written by Aristotle in *Analytica posteriora* I. The confirmation of the minor premise is drawn from what Plato writes in *Respublica* VII, where he says that mathematicians dream about quantity and, in their demonstrations, proceed not scientifically but based on some assumptions. Therefore, he [Plato] does not wish to call their [the mathematicians'] teachings intellection (*intelligentiam*) or science, but just discursive thought (*cogitationem*). Proclus, in book I of his commentaries on Euclid, writes extensively in agreement with this view.[35]

Another 17th-century remarkable episode of this controversy is that of Isaac Barrow (1630–1677), who adopted Proclus' text (usually quoting it directly in Greek) as one of his sources in his *Lectiones mathematicae XXIII* published in Cambridge (1664). Taking the question from Gassendi, Barrow faced the challenge of defending the scientific status of mathematical demonstrations.[36] Newton himself, Barrow's disciple, showed interest in the akribology of mathematics and its role in securing certain knowledge through its application to natural phenomena.[37]

---

35 Pereyra's lectures at the *Collegium Romanum* were published in 1576. See the original Latin in Benito Pereyra, *De communibus omnium rerum naturalium principiis et affectionibus libri quindecim* (Paris: Sonnius, 1576), pp. 24–25: *Mea opinio est, Mathematicas disciplinas non esse proprie scientias, in quam opinionem adducor tum aliis, tum hoc uno maxime argumento. Scire est rem per caussam cognoscere propter quam res est, et scientia est demonstrationis effectus, demonstratio autem (loquor de perfectissimo demonstrationis genere) constare debet ex his quae sunt per se et propria eius quod demonstratur, quae vero sunt per accidens et communia excluduntur a perfectis demonstrationibus. Sed mathematicus neque considerat essentiam quantitatis, neque affectiones eius tractat prout manant ex tali essentia, neque declarat eas per proprias caussas, propter quam insunt quantitati, neque conficit demonstrationes suas ex praedicatis propriis et per se, sed ex communibus, et per accidens, ergo doctrina mathematica non est proprie scientia. Maior huius syllogismi non eget probatione, etenim aperte elicitur ex his quae scripta sunt ab Aristotele in I Posteriorum analyticorum. Confirmatio minoris ducitur ex his quae scribit Plato in VII libro De republica dicens mathematicos somniare circa quantitatem, et in tractandis suis demonstrationibus non scientifice sed ex quibusdam suppositionibus procedere. Quamobrem non vult doctrinam eorum appellare intelligentiam aut scientiam, sed tantum cogitationem, in quam sententiam multa scribit Proclus in I libro suorum Commentariorum in Euclidem.*

36 P. Mancosu, "Aristotelian logic and Euclidean mathematics", pp. 258–264.

37 D. Levitin, *The Kingdom of Darkness: Bayle, Newton, and the Emancipation of the European Mind from Philosophy* (Cambridge New York: Cambridge University Press, 2022),

1.5 Reassessing Piccolomini's Text

The goal of this book is to offer a historical and philosophical analysis of *De certitudine* paying special attention to Piccolomini's sources and motivations.

Nevertheless, to do so, I will have to defy (or rather, complete) some of the views commonly held in scholarship regarding the goals and the scope of Piccolomini's opusculum. As stated, *De certitudine* contains an attack on the scientific status of mathematical demonstration. Higashi has recently provided great insight into the details of the philosophy of mathematics and logical doctrines contained in *De certitudine* as well as regarding its debts to the northern Italian (Paduan) intellectual milieu.[38] My point is not that this account is wrong, but, rather, that some details can still be added. Dwelling on the progress made by Higashi, I shall propose some additions and innovations which I consider crucial for better grasping Piccolomini's work.

I claim that relevant details for understanding the scope of *De certitudine* have been overlooked precisely because of the great impact and success it achieved. Barozzi's answer to Piccolomini's reading of Proclus and the intervention of the Jesuits Clavius and Pereyra in the subsequent debate rendered this, at first somewhat parochial dispute, into a debate of much larger proportions. The success of *De certitudine* had an accidental hermeneutic effect: Piccolomini's argumentation was read in isolation from its original intellectual context. I argue that the inheritance of the controversy on mathematical certainty (doubtlessly of the utmost importance) has kept out of sight some of the goals and presuppositions of *De certitudine*. In other words, the momentum and scope of the questions that *De certitudine* raised caused many of its contextual details to be overlooked whereas the general attention has been focused on Piccolomini's most controversial and provocative arguments.

In the next pages, I will index the five hermeneutical keys that have been generally overlooked in reading *De certitudine* as a whole:

---

pp. 60–80; Levitin, *Ancient Wisdom in the Age of the New Science*, especially part 4.5; I. Stewart, "Mathematics as philosophy: Barrow and Proclus", in *Dionysius* 18 (2000), pp. 151–181; M. S. Mahoney, "Barrow's mathematics: between ancients and moderns", in M. Feingold (ed.), *Before Newton: The Life and Times of Isaac Barrow* (Cambridge New York: Cambridge University Press, 2021) (Transformations: Studies in the History of Science and Technology), pp. 179–249; N. Guicciardini, *Isaac Newton on Mathematical Certainty and Method* (Cambridge MA: Massachusetts Institute of Technology Press, 2011).

38 Higashi, *Penser les mathématiques*, especially pp. 123–276; See also Higashi, "Giorgio Valla et Alessandro Piccolomini. Quelques aspects de la réception de Proclus ê la Renaissance", in *Bulletin of Liberal Arts Education Center* 27 (2007), pp. 31–52.

1.5.1 *Anti-Mertonian Tradition in Italy*

First is the influence of a previous northern Italian trend in the philosophy of mathematics strongly opposed to the doctrines of the so-called *Calculatores Oxonienses* or Mertonians. The intellectual approach of the Mertonians of introducing quantification in the resolution of some problems in natural philosophy (what I shall call from now on "the Mertonian approach") was imported into Italy in the 14th century. In many cases, the Italo-Mertonian tradition assumed what I have called a "full-correspondence" reading of mathematical statements: the claim that quantifications describe and represent non-mathematical realities. From this persuasion, it follows that the Italian Calculators that quantified natural phenomena offer a legitimate portrait of some non-mathematical entities and that the resulting middle science of *latitudines* (as it was called) is able to account, through quantification, for non-mathematical phenomena.

Some Italian philosophers would develop ever-increasing hostile answers against this way of thinking. The core of the anti-Mertonian thought is that the use of mathematics in dealing with sensible entities is an epistemically unsound move. Mathematical items are imaginative productions which do not capture or reflect the reality of natural things, let alone reveal their essences. Arbitrary mathematical imagination may produce, for instance, qualitative degrees which are impossible in the real world (such as infinite warmth). Thus, against the full-correspondence reading of the epigones of the Calculators, the anti-Mertonian thinkers leaned towards non-correspondence views of mathematics. As I shall defend with arguments and texts, Piccolomini belonged to this trend of thought.

1.5.2 *Philosophy of Mechanics*

The second point is that *De certitudine* was not an isolated work, but it was conceived and published as an attachment to the Latin paraphrase of the Pseudo-Aristotelian *Quaestiones mechanicae*. While highly relevant passages on the philosophy of mathematics can be found in the paraphrase and its *Praefatio*, they have never been connected to the contents of *De certitudine*. As an examination of these sections reveals, *De certitudine* was also intended to propose a philosophy of mathematics that would elevate mechanics, which Piccolomini portrayed as a branch of contemplative philosophy, to a more esteemed intellectual position.

Nonetheless, this elevation of a discipline that hitherto had lacked academic and intellectual prestige involved a danger. By declaring that mechanics is a form of contemplative philosophy and offering to the readership its first whole Latin adaptation, Piccolomini could be understood as the promoter of a subalternate science which, through the mathematisation of kinetical processes, attempts to explain the movement of sensible entities. In other words, mechanics could be understood as a quantitative emulation of natural philosophy in the Mertonian manner.

Piccolomini was aware of this danger and took pains to avoid such a reading of the discipline he was attempting to foster. Thus, chapter VI in the first question of his Latin paraphrase of the *Quaestiones mechanicae* contains an openly hostile piece of anti-Mertonian philosophy which intends to show that mechanical propositions and principles do not hold when applied to non-mathematical (i.e., sensible) entities: the mathematical principles ruling the balance are falsified by real balances, which do not behave exactly as they should according to the principles of the mathematical balance. Thus, Piccolomini concludes that mechanical propositions and, generally, mathematical truths only hold within the abstract realm of mathematics.

The Sienese revisited these criticisms in the final chapter (XII) of *De certitudine* and yet again in the *Praefatio*, once he fully developed his mathematical ontology in the opusculum. I will demonstrate how reading these passages from the *Praefatio* and the Latin paraphrase, in conjunction with *De certitudine*, can provide a more comprehensive understanding of Piccolomini's theory and objectives.

1.5.3 *Demonstrative Regress*

The third point relates to the long-held Paduan debates on *regressus demonstrativus* and their relation to the so-called *demonstratio potissima*. To my knowledge, Palmieri and Longeway are the only scholars who have detected a link between *regressus* and *demonstratio potissima* in *De certitudine*.[39] I will elaborate on this proposal and provide both textual and logical evidence that Piccolomini's *demonstratio potissima* is a reciprocal intertwining of two syllogisms. Nevertheless, Piccolomini's *regressus* stands as a distinctive example of this logical structure. The common arrangement of *regressus* as developed in Padua ever since Peter of Abano (*c.*1257–1316) involved (at least) three steps: (1) a demonstration *quia* (effect-to-cause); (2) an intermediate, non-deductive step in which the nature of the cause is scrutinised; and (3) a demonstration *propter quid* (cause-to-effect) – under certain conditions called *potissima* by authors such as Nifo or Jacopo Zabarella (1533–1589). Piccolomini's *regressus* as presented in *De certitudine* lacks step 2, the most controversial and discussed of the three because of its vagueness and non-apodictic character. Hence, the

39 Palmieri, "On *scientia* and *regressus*"; Longeway, "The Place of *Demonstratio Potissima*".

arrangement of the highest sort of demonstration according to Piccolomini is that of two interconnected, reciprocal syllogisms *quia* and *propter quid*.

1.5.4 *"Galenic" and "Anti-Galenic" Analysis*

Another of the overlooked themes in *De certitudine* is the reception of the methodology of mathematical analysis and its relation to *regressus* and *demonstratio potissima*. This ancient methodology consisted of reducing a proposition-to-be-solved in such a way that it could be derived from or constructed upon (in the case of problems) already known propositions or principles. In some cases, the analysis of a mathematical proposition could lead to a new to-be-demonstrated proposition as a requirement for the proof under analysis or even to a hitherto non-stated axiom. In a second synthetic (σύνθεσις) stage, the studied proposition is deduced from the previous propositions to which it has been reduced. Thus, analysis-synthesis was held both as a heuristic and justificatory methodology in its different steps.[40]

In the 16th century, the terms *resolutio* and *compositio* became part of the technicalities present in the debates inspired by Galen's *viae*. Northern Italian philosophers educated in this tradition were inclined to read mentions to analysis-synthesis of recovered Greek sources in light of Galen's *Ars parva*, i.e., as logical devices going from signs to their causes (analysis) and then from the found causes back to the signs (synthesis). In this way, Galen's *via resolutiva* was interpreted by some philosophers as corresponding to Aristotle's demonstration *quia*, and the *via compositiva* was understood as the *propter quid*. This is what I will call the "Galenic" reading of analysis and synthesis. Ever since Peter of Abano (see above), such hermeneutics paved the way for the development of the doctrine of demonstrative regress.

The influence and presence of analysis in *De certitudine* is remarkable. Piccolomini devotes several paragraphs to discussing what *resolutio* (the Latin term for ἀνάλυσις) is, and how it works in a demonstration. In chapter III of *De Certitudine* Piccolomini describes different types of analysis following Eustratius of Nicaea's (*c.*1055–*c.*1120) commentary on *Analytica posteriora* and Proclus' *In Euclidem*. As far as I am aware, this section has received little to no attention in scholarship. Importantly, Piccolomini adds to Eustratius' list the *resolutio mathematica*. Against the commonly held Galenic interpretation, he insists that analysis must be distinguished from *demonstratio quia* as a

---

40 J. Hintikka, U. Remes, *The Method of Analysis: Its Geometrical Origin and Its General Significance* (Dordrecht Boston: Springer, 1974), especially pp. 7–22; and also Hintikka, "Method of Analysis: A Paradigm of Mathematical Reasoning?", in *History and Philosophy of Logic* 33/1 (2012), pp. 49–67.

scientific procedure: it should not be understood as a part of *regressus* leading to a *demonstratio potissima*.

In Piccolomini's "anti-Galenic" view, analysis is a checking (syllogistic) tool to be applied to proofs after they are produced to ensure their validity. Thus, Piccolomini holds that analysis is neither a heuristic tool nor a form of demonstration. Moreover, under the guide of Proclus, Piccolomini offers in chapter x the, arguably, first preserved case of a *resolutio mathematica* in syllogistic form: taking Euclid's proof of the first problem in book I, the Sienese inversely retraces all the steps of the demonstration back to its principles, either self-evident or indemonstrable truths. I will examine Piccolomini's views on analysis and illustrate them in the wider context of his logical theory.

1.5.5 *"Simplician" Averroism*

The fifth and last hermeneutical key to be emphasised is that Piccolomini was undeniably influenced by the so-called "Simplician" Averroism. By this, I mean that (unsurprisingly) many features of the 16th-century Paduan elaborations on Averroes' reading of Aristotle's works by using Pseudo-Simplicius' (i.e., Priscian of Lydia's) commentary on *De anima* have a decisive presence in *De certitudine*.

The four most important points in which Averroes' influence is noticeable in the opusculum are (i) Piccolomini's justification of his model of *demonstratio potissima* as a form of *regressus* – in which two quotations from Averroes are adopted as the main authority; (ii) Piccolomini's theory of *principia activa*, used in his refusal of mathematical efficient causation, explicitly attributed to Averroes; (iii) Piccolomini's position regarding the co-eternity and inherence of quantity to prime-matter, also of Averroistic origin in *De substantia orbis*; (iv) an implicit "Simplician" Averroistic psychology as the theoretical framework for his interpretation of Proclus' *In Euclidem*.

The particular species of Averroism influencing Piccolomini is that which followed the lead of the Paduan master Genua De' Passeri and his interpretation of Pseudo-Simplicius' commentary on *De anima* as one of the most authoritative philosophical guides for rightly grasping Aristotle. Genua's persuasion was that Averroes' interpretation of Aristotle's *De anima* was in fundamental agreement with (and hence could be enriched through) Pseudo-Simplicius' psychological positions.

That Piccolomini was educated in the circles where this species of Platonising Averroism flourished is one of the keys to better grasping his interpretation of Proclus in *De certitudine*. "Simplicius" is invoked 10 times within *De certitudine*, out of which 2 times are direct allusions to the commentary on *De anima* within an argument on the akribology of mathematics. Moreover, Pseudo-Simplicius'

In De anima was the source of some views on the philosophy of mathematics which, arguably via Genua, were destined to play a major role in Piccolomini's work, namely (a) that there is no causation in mathematical objects, (b) that mathematical objects cannot be classified as substances, and (c) that, hence, mathematics cannot be held as real contemplative philosophy. I will sustain that Pseudo-Simplicius sets the direction in which Piccolomini interpreted Proclus' *In Euclidem* and that his theories are behind Piccolomini's view that mathematical demonstrations are not *potissimae*.

In general, Alessandro Piccolomini's *De certitudine mathematicarum* is a work full of beautiful paradoxes and hidden contextual presuppositions. The Sienese uses Proclus' *In Euclidem*, a work which puts high value on mathematical knowledge, to put into question the demonstrative status of mathematics. At the same time, against the *Simpliciani* at Padua, he portrays mechanics, a discipline held intellectually uninteresting, as a fully contemplative philosophy. Also, the so-called subalternate sciences neither have natural objects as their subject nor do they deal with truths regarding the physical world. Additionally, no example of the much-debated *demonstratio potissima* is formalised by Piccolomini; that this precise logical arrangement was a case of *regressus* should have been so clear to Piccolomini's intended readership that he barely felt the necessity of explaining it. The philosophy of mathematics which Piccolomini opposes, the Mertonian theory, was so widely discussed that it is referenced only by key terms known to the readership. Many of Piccolomini's arguments concerning the disputes amongst the "Simplician" Averroists are only elusively introduced. Finally, the meaning of the main concept of the opusculum, *certitudo*, is never explained.

Piccolomini's production on themes related to the philosophy of mathematics and logic extends beyond the limits of the paraphrase and the 1547 opusculum: they are also present in his vernacular works. Some questions on the philosophy of mathematics are addressed in *Prima* and *Seconda parte della filosofia naturale* (Venice, 1551, 1554, respectively), an exposition of Aristotle's organon is conveyed in his *Instrumento della filosofia* (Rome, 1552), and some details in a rather conventional way of understanding subalternation are found in *La prima parte delle theoriche* (Siena, 1558).[41] Nevertheless, the philosophy of mathematics and the theory of demonstration expressed in Piccolomini's

41 Alessandro Piccolomini, *Prima parte della filosofia naturale* (Venice: Vincenzo Valgrisi, 1551); Alessandro Piccolomini, *Seconda parte della filosofia naturale* (Venice: Vincenzo Valgrisi, 1554); Alessandro Piccolomini, *Instrumento della filosofia* (Venice: Giovanmaria Bonelli, 1552); Alessandro Piccolomini, *La prima parte de le theoriche* (Venice: Giordano Ziletti, 1558).

vernacular works differ in many respects from Piccolomini's views in the *De certitudine*.[42] Overall, one has the impression that, after *De certitudine*, Piccolomini expressed more conventional opinions regarding the philosophy of mathematics. As Sgarbi has shown, in a 1541 letter to the poet Pietro Aretino (1492–1556), Piccolomini expressed that it was the primary goal in his philosophical work in Italian to proceed "according to the ways of the Peripatetics, not by translating, but by expanding where necessary, in such a way, however, that I do not deviate from the original opinion of Aristotle first and foremost, and then from his most important Greek interpreters".[43] Thus, by Piccolomini's own admission, his aim in his Italian philosophical production was to present the Peripatetic natural and moral philosophy, while ostensibly refraining, as far as possible, from elaborating on his personal views or using (Neo-)Platonic sources.

Because of these reasons, I saw it fitting to interpret Piccolomini's philosophy in *De certitudine* without appealing to his later vernacular production to attempt to make sense of his claims. Thus, this study shall be circumscribed to Piccolomini's 1547 publication, including the Latin paraphrase of the *Quaestiones*, its *Praefatio*, and *De certitudine*. As I will show, even in these texts some hints of a fluctuating, ongoing philosophy can be retraced.

1.6 The Map of This Book

This work is divided into four parts.

Part 1, the one the reader is now about to finish, has been an introduction to the main characters, philosophical and historical themes, and hermeneutics of the study.

Part 2 deals with Piccolomini's sources and intellectual context. This part shall provide the necessary historical and philosophical background for later addressing the core arguments of *De certitudine*. The themes introduced in this part are: the philosophy of mathematics of Piccolomini's Neoplatonic

42 On the differences between the views on scientific demonstration put in *De certitudine* and those in Piccolomini's later vernacular production see Cozzoli, "Alessandro Piccolomini and the certitude of mathematics". See also the excellent account of Piccolomini's vernacular logic and natural philosophy in Sgarbi, *The Italian Mind*, pp. 195–212. I will address the discrepancies between Piccolomini's ontology of mathematics in *De certitudine* and the views expressed in his *Prima* and *Seconda parte della filosofia naturale* in section 3.5.2.

43 Quoted in Sgarbi, *The Italian Mind*, p. 181: [...] *secondo la via dei peripatetici; non traducendo, ma ampliando dove bisogna, di maniera però che io non mi parti dal parer primariamente di Aristotile e di poi dei primi suoi greci espositori.*

sources (2.1), including mathematical analysis (2.1.2), the ontology of mathematics (2.1.3), common mathematics (2.1.4), and Priscian on mathematical insubstantiality and causal idleness (2.1.5); a survey on the ontology and epistemology of *phantasia* (2.2); an inquiry on the history of the syntagm *certitudo mathematica* and its meaning in Piccolomini's work (2.3); an account of the Mertonian, anti-Mertonian, and Ludovico Boccadiferro's downgrading of mathematical sciences (2.4); some details on Pseudo-Aristotle's *Quaestiones mechanicae* and the philosophy of mechanics in Piccolomini's early Latin paraphrase and the *Praefatio* (2.5); and an introduction to the Latin debates surrounding the notion of *demonstratio potissima* dwelling on *De certitudine*'s 13th- to 16th-century sources (2.5).

Part 3 consists of an analysis of Piccolomini's opusculum by putting at work all the themes featured in Part 2. These chapters will tackle the following issues: Piccolomini's own model of the *demonstratio potissima* (3.1); his take on mathematical analysis and its epistemological consequences (3.2 and 3.3); Piccolomini's denial of mathematical causation and his "Simplician" hermeneutics of Proclus' *In Euclidem* (3.4); *De certitudine*'s philosophy of mathematics, its theory of *abstractio*, its akribology of mathematics, and their philosophical consequences (3.5); Piccolomini's anti-Mertonian philosophy as recast in the last chapter of *De certitudine* (3.6); and, lastly, the presence and purpose of common mathematics in the opusculum (3.7).

Part 4 provides a summary of the results reached through the present study and some conclusive remarks.

PART 2

Piccolomini's Sources and Context

2.1 Proclus in the Latin Context and Neoplatonic Themes in *De certitudine*

Proclus, the Platonic Successor or "Diadochus", stands as a towering figure amongst the great philosophers of antiquity. He crafted one of the most comprehensive and systematic renditions of late Neoplatonic philosophy through his works and commentaries.[1] His impact extended across Islamic, Byzantine, Latin Medieval, Renaissance, and Early Modern philosophy, and he can be ranked amongst the most influential of the thinkers of late antiquity.[2]

The legacy of Proclus in the Latin tradition can be considered to have occurred in two manners: directly and indirectly.

Regarding the indirect spread of Proclus' ideas, two highly influential sources, not attributed to Proclus, which, nevertheless, conveyed metaphysical ideas inspired by his philosophy were the *corpus Dionysiacum* and the *Liber de causis*. The writings attributed to Dionysius the Areopagite, undoubtedly indebted to Proclus, reached apostolic status within the Latin world, influencing philosophical and theological speculation from John Scot Eriugena (815–877) to

1 L. Siorvanes, "Proclus' life, works, and education of the soul", in S. Gersh (ed.), *Interpreting Proclus: From Antiquity to the Renaissance* (Cambridge: Cambridge University Press, 2014), pp. 33–56; A. Sheppard, "Proclus as exegete", in S. Gersh (ed.), *Interpreting Proclus*, pp. 57–79; R. Chlup, *Proclus, An Introduction* (Cambridge: Cambridge University Press, 2012), Epilogue: "Proclus' Legacy"; section I. 3, pp. 32–44.

2 P. Adamson, F. Karfík, "Proclus' Legacy", in P. D'Hoine, M. Martijn (eds), *All From One: A Guide to Proclus* (Oxford: Oxford University Press, 2016), pp. 290–332; R. Chlup, *Proclus, An Introduction*, pp. 279–295; S. Gersh, "One Thousand Years of Proclus", in S. Gersh (ed.), *Interpreting Proclus*, pp. 1–29; Claessens, "Proclus in the Renaissance", in M. Sgarbi (ed.), *Encyclopedia of Renaissance Philosophy*; J. A. Aertsen, "Ontology and Henology in Medieval Philosophy (Thomas Aquinas, Master Eckhart and Berthold of Moosburg)", in E. P. Bos, P. A. Meijer (eds), *On Proclus and His Influence in Medieval Philosophy* (Leiden: New York, and Cologne, 1992), pp. 120–140; P. O. Kristeller, "Proclus as a Reader of Plato and Plotinus, and His Influence in the Middle Ages and in the Renaissance", in J. Pépin, H. D. Saffrey (eds), *Proclus lecteur et interprète des anciens, Actes du colloque international du Centre National de la Recherche Scientifique, 2–4 octobre 1985* (Paris: Centre national de la recherche scientifique, 1978), pp. 191–211; R. Imbach, "Le (néo-)platonisme médiéval: Proclus latin et l'école dominicaine allemande", in *Revue de théologie et de philosophie* 110 (1978), pp. 427–448.

 | DOI:10.1163/9789004724143_003

Nicholas of Cusa (1401–1464), Marsilio Ficino (1433–1499), and onwards.[3] For its part, the *Liber de causis*, a propositionally arranged treatise partly based on Proclus' *Elementatio theologica*, which, nonetheless, possesses philosophical and theoretical originality of its own, can be justly numbered amongst the most relevant philosophical treaties in the 13th century. Most likely produced in al-Kindī's (*c.*801–873) circle under the Arabic name *Aristotle's Book of the Pure Good* (كتاب إيضاح لأرسطوطالس في الخير المحض, *Kitāb Īḍāḥ al-Arisṭūṭālis fī l-ḫayr al-maḥḍ*), it was translated into Latin in 12th-century Spain.[4] Commonly attributed to Aristotle or to al-Fārābī (870–950), the *Liber de causis* had become a well-known source of metaphysical insight in the Latin context by the 13th century. This contributed to the spread of Proclean-like philosophical tenets amongst the Latin *intellectualité*.[5] The *Liber de causis* was placed in the

3 M. Vlad, "L'Être premie entre Proclus et Denys l'Aréopagite", in D. Calma (ed.), *Reading Proclus and the Book of Causes: On Causes and the Noetic Triad* (Leiden Boston: Brill, 2022), Vol. 3, pp. 89–116; J. M. Dillon, "Dionysius the Areopagite", in S. Gersh (ed.), *Interpreting Proclus*, pp. 111–124; Gersh, "One Thousand Years of Proclus", in S. Gersh (ed.), *Interpreting Proclus*, pp. 4–5; Saffrey, "Les Néoplatoniciens et les Oracles chaldaïques", in *Revue des études augustiniennes* 27 (1981), pp. 209–225; I. P. Sheldon-Williams, "Henads and Angels: Proclus and the Pseudo-Dionysius", in *Studia Patristica* 11 (1972), pp. 65–71; P. Megna, "Per Ficino e Proclo", in F. Baussi, V. Fera (eds), *Laurentia laurus: Scritti offerti a Mario Martelli* (Messina: Centro Interdipartimentale di Studi Umanistici, 2004), pp. 313–362.

4 D. N. Hasse, "Three Double Translations from Arabic into Latin by Gerard of Cremona and Dominicus Gundisalvi", in D. Calma (ed.), *Reading Proclus and the Book of Causes: Translations and Acculturations* (Leiden Boston: Brill, 2021), Vol. 2, pp. 247–276; J. Janseens, "Doubles traductions et omissions: une approche critique en vue d'une édition de la traduction latine du *Liber de causis*", in D. Calma (ed.), *Reading Proclus and the Book of Causes*, Vol. 2, pp. 275–316.

5 D. Calma, "Du néoplatonisme au réalisme et retour, parcours latins du *Liber de causis* aux XIIIe–XVIe siècles", in *Bulletin de philosophie médiévale* 54 (2012), pp. 217–276; Calma, "Introductio to Reading Proclus and the Book of Causes: Notes on the Western Scholarly Networks and Debates", in D. Calma (ed.), *Reading Proclus and the Book of Causes: Western Scholarly Networks and Debates* (Leiden Boston: Brill, 2019) Vol. 1, pp. 18–46; I. Caiazzo, "La première réception du *Liber de causis* en Occident (XIIe–XIIIe siècles)", in D. Calma (ed.), *Reading Proclus and the Book of Causes*, Vol. 1, pp. 18–46; E. Wakelnig, "Les Chapitres sur les thèmes métaphysiques d'al-ʿĀmirī et l'anonyme Kitāb al-ḥaraka: deux interprétations du *Liber de causis* en arabe", in D. Calma (ed.), *Reading Proclus and the Book of Causes*, Vol. 2, pp. 183–198; R. C. Taylor, "Contextualizing the Kalām fī maḥḍ al-khair / *Liber de causis*", in D. Calma (ed.), *Reading Proclus and the Book of Causes*, Vol. 2, pp. 211–232; J. Rachak, "La présence du *Liber de causis* dans l'œuvre d'Ibn Bāǧǧa et Pseudo-Ibn Bāǧǧa: un philosophe péripatéticien du XIe/XIIe siècle de l'Occident islamique", in D. Calma (ed.), *Reading Proclus and the Book of Causes*, Vol. 2, pp. 233–245; F. Retucci, "Liber de causis in Thomas of York", in D. Calma (ed.), *Reading Proclus and the Book of Causes*, Vol. 1, pp. 47–70; L. Miolo, "Le *Liber de causis* et l'*Elementatio theologica* dans deux bibliothèques anglaises: Merton College (Oxford) et Peterhouse (Cambridge)", in D. Calma (ed.), *Reading Proclus and the Book of Causes*, Vol. 1, pp. 71–120; O. Weijersm, "Les gloses sur le *Liber de causis* dans les manuscrits parisiens", in D. Calma (ed.), *Reading Proclus and the Book of Causes*, Vol. 1, pp. 121–151;

statues of the University of Paris by 1255 together with Aristotle's *Metaphysica* (*vetus* and *nova*),[6] and commented upon by highly influential figures including (amongst others) Albert the Great (1200–1280), Roger Bacon (1219–1292), Thomas Aquinas (1225–1274), Siger of Brabant (1240–1280), Giles of Rome (1243–1316), and later authors such as Heymericus De Campo (1395–1460).[7] As Calma has shown, the influence of the *Liber de causis* is far greater than it was previously thought, for it was commented on and taught even after William of Moerbeke's (1215–1286) Latin rendering of Proclus' *Elementatio theologica* in 1268.[8] Indeed, philosophers in the Latin context found in the *Liber de causis* doctrines absent from both Aristotle and the *Elementatio theologica* which were valued on their own.[9] Another case of indirect influence in the Latin world is that of Boethius (*ob.* 524), who used Proclus' *Tria opuscula* and *In Timaeum* in writing his *De consolatione philosophiae* (book III, poem 9).

Regarding the direct influence of Proclus in the Latin tradition, his writings were relatively scarce up to the mid-13th century. Moerbeke also translated the *Tria opuscula*, *In Parmenidem* and some passages of *In Timaeum*. After him, some figures in the Italian context such as Ambrose Traversari (1386–1439), Pietro Balbi (1399–1479), and George of Trebizond (1395–1486) rendered some of Proclus' works into Latin.[10] By the mid 15th century, the list of Proclus'

K. Krause, H. Anzulewicz, "From Content to Method: the *Liber de causis* in Albert the Great", in D. Calma (ed.), *Reading Proclus and the Book of Causes*, Vol. 1, pp. 152–180; M. E. Malgieri, "Citing the Book of Causes, IV: Henry of Ghent and His (?) Questions on the Metaphysics", in D. Calma (ed.), *Reading Proclus and the Book of Causes*, Vol. 1, pp. 181–209; J. Casteigt, "Le chapitre xi[xii] du Liber de causis commenté par Albert le Grand: l'unité de la cause et de son effet comme immanence réciproque", in D. Calma (ed.), *Reading Proclus and the Book of Causes*, Vol. 3, pp. 476–502; J. M. Counet, "Duns Scot et le *Liber de causis*", in D. Calma (ed.), *Reading Proclus and the Book of Causes*, Vol. 1, pp. 210–251.

6 C. D'Ancona, "The *Liber de causis*", in S. Gersh (ed.), *Interpreting Proclus*, pp. 137–161.

7 Calma, "Metaphysics as a Way of Life: Heymericus de Campo on Universals and the "Inner Man"", in *Vivarium* 58/4 (2020), pp. 305–334; M. Meliado, *Sapienza peripatetica: Eimerico di Campo e i percorsi del tardo albertismo* (Münster: Aschendorff, 2018), pp. 177–179; Calma, Imbach, "Heymeric de Campo, auteur d'un traité de métaphysique. Étude et édition partielle du Colliget principiorum", in *Archives d'Histoire Doctrinale et Littéraire du Moyen Age* 80 (2013), pp. 277–412.

8 Taylor, "'Abd al-Laṭīf al-Baghdādī's Epitome of the Kalām fī maḥḍ al-ḫayr (*Liber de Causis*)", in M. E. Marmura (ed.), *Islamic Theology and Philosophy* (New York: State University of New York Press, 1984), pp. 236–248; D. Piché, "Causa prima et esse dans le *Liber de causis* selon Thomas d'Aquin et Siger de Brabant", in *Dialogue* 38 (1999), pp. 75–97.

9 Calma, "The Exegetical Tradition of Medieval Neoplatonism. Considerations on a Recently Discovered Corpus of Texts", in D. Calma (ed.), *Neoplatonism in the Middle Ages, Vol. I: New Commentaries on* Liber de causis *(ca. 1250–1350)* (Turhout: Brepolis, 2016), pp. 11–52.

10 Saffrey, "Pietro Balbi et la première traduction latine de la Théologie platonicienne de Proclus", in P. Cockshaw, M. C. Garand, P. Jodogne (eds), *Miscellanea codicologica F. Masai dicata* (Gand: E. Story-Scientia, 1979), pp. 425–437.

works available to the Latin readership included the *Theologia Platonica*. This translation movement resulted in notable landmarks in the spreading of Proclus' philosophy. Within the German Dominican tradition, thinkers such as Dietrich of Freiberg (1250–1310) and Berthold of Moosburg (*ob.* 1361) began to elevate Platonic metaphysics, as they thought it was epitomised in Proclus' works, above that of Aristotle.[11] The influential presence of Proclus can also be retraced in Nicholas of Cusa, beginning with his *De coniecturis*.[12]

Other works of Proclus became accessible in Greek by the 15th century: Ficino had access to Proclus' *In Alcibiadem* (which he partially translated), the *Elementatio physica*, *In Rempublicam*, *In Cratylum*, and the passages known as the work Περὶ τῆς ἱερατικῆς τέχνης (translated by Ficino as *De sacrificio et magia*).[13] A partial Latin version of Proclus' *In Timaeum* was published by Niccolò Leonico Tomeo (1456–1531) in 1525.[14] By the mid-15th century, the *Hypotyposis astronomicarum positionum* and *In Euclidem* (hardly a century before Piccolomini) were also recovered to the Latin West (see below, section 2.1.1). The influence of Proclus' rediscovered works has been also attested in 16th-century philosophers such as Francesco Patrizi (1529–1597), who translated and published the *Elementatio theologica* (arguably the most popular of Proclus' works) and *Elementatio physica* (1583), and Jacopo Mazzoni (1548–1598).[15]

---

11 E. King, *Supersapientia: Berthold of Moosburg and the Divine Science of the Platonists* (Leiden Boston: Brill, 2021); King, "*Sapiens modernus*: The Reception of Dietrich of Freiberg in Berthold of Moosburg", in D. Calma, E. King (eds), *The Renewal of Medieval Metaphysics Berthold of Moosburg's Expositio on Proclus' Elements of Theology* (Leiden Boston: Brill, 2021), pp. 243–281; King, "Eriugenism in Berthold of Moosburg's *Expositio super Elementationem theologicam Procli*", in D. Calma (ed.), *Reading Proclus and the Book of Causes*, Vol. 1, pp. 394–437; M. Führer, Gersh, "Dietrich of Freiberg and Berthold of Moosburg", in S. Gersh (ed.), *Interpreting Proclus*, pp. 299–317; R. Imbach, "Au-delà de la métaphysique: Notule sur l'importance du Commentaire de Berthold de Moosburg OP sur les *Éléments de théologie*", in D. Calma (ed.), *Reading Proclus and the Book of Causes*, Vol. 1, pp. 376–393.

12 Gersh, "Nicholas of Cusa", in S. Gersh (ed.), *Interpreting Proclus*, pp. 318–352.

13 D. J. J. Robichaud, *Plato's Persona: Marsilio Ficino, Renaissance Humanism, and Platonic Traditions* (Philadelphia: University of Pennsylvania Press, 2018), pp. 142–143, and p. 167; Robichaud, "Fragments of Marsilio Ficino's Translations and Use of Proclus' *Elements of Theology* and *Elements of Physics*: Evidence and Study", in *Vivarium* 54/1 (2016), pp. 46–107; M. J. B. Allen, "Marsilio Ficino", in S. Gersh (ed.), *Interpreting Proclus*, pp. 353–379.

14 Niccolò Leonico Tomeo, *Opuscula nuper in lucem edita* (Venice: Vitalis, 1525). See also E. Gamba, "Leonico Tomeo, Niccolò", in M. Sgarbi (ed.), *Encyclopedia of Renaissance Philosophy*.

15 T. Leinkauf, "Francesco Patrizi", in S. Gersh (ed.), *Interpreting Proclus*, pp. 380–402; Leinkauf, *Il neoplatonismo di Francesco Patrizi come presupposto della sua critica ad Aristotele* (Florence: La Nuova Italia Editrice, 1990); Claessens, "Sixteenth- and Seventeenth-Century Views on Mathematical Imagination"; De Pace, *Le matematiche e il mondo*, chap. 4.

It is to be stressed that Proclus' identity and his exact chronology were unclear to some of his 16th-century readers, and, most likely, also to Piccolomini. One illustrative case has been made by Goulding:[16] Raffaele Maffei's (1451–1522) 1506 encyclopaedic work *Commentariorum urbanorum 38 libri*. This work contained one of the most extensive and widely read collections of biographies of philosophers and writers. Proclus is one of the figures listed in the index.

Maffei mostly based his biographies on the Byzantine encyclopaedia *Suda*. Following the information concerning Proclus found there, Maffei stated that Proclus was a disciple of Syrianus (*ob.* 437) and the teacher of Marinus of Neapolis (450–500), that he commented on Homer, Hesiod, and Plato, that, after Porphyry (*ob.* 301), Proclus wrote against the Christians (Maffei says that Proclus "barked against us", *latravit*), and that his views were criticised by John Philoponus (490–570). Nevertheless, confusingly, Maffei adds that Proclus was the preceptor of the Roman emperor Marcus Aurelius (121–180) and that he was made consul by the imperial power.[17] This last point was taken from the life of Marcus Aurelius in the *Historia Augusta*, wrongly attributed by Maffei to Aelius Spartianus: the life of Marcus Aurelius was written by Iulius Capitolinus. The line reads as follows:

> He [Marcus Aurelius] also had as teachers of literature the Greek Alexander from Cotiaeum, and, for Latin literature, Trosius Aper, Polio, and Eutychius Proculus from Sicca.[18]

As seen, a philosopher named Eutychius Proculus, coming from the city of Sicca in North Africa, is mentioned as one of the preceptors of the emperor-philosopher. It seems that Maffei erroneously assimilated Proclus to this

16 R. Goulding, *Defending Hypatia: Ramus, Savile, and the Renaissance Rediscovery of Mathematical History* (Dordrecht: Springer, 2010), pp. 165–168.

17 Raffaele Maffei, *Commentariorum urbanorum 38 libri* (Rome: J. Besicken, 1502), Lib. XVIII, "Proclus", p. 143: *Proclus Lycius discipulus Syriani, philosophus Platonicus, praefuit scholae Atheniensi, cuius discipulus et successor Marinus Neapolitanus fuit. Scripsit plura in philosophia et grammatica, commentarios in totum Homerum, in Hesiodi Erga et Hemeras, in Rempublicam Platonis. Preterea contra Christianos Epicheremata XVIII. Hic est Proclus qui post Porphyrium secundus contra nos latravit. Adversus quem Ioannes cognomento grammaticus scripsit apologiam, dictitans eum, quamquam in rebus Graecanicis magnum, stultum tamen esse et in hac parte indoctum, auctor Suidas. Is est cuius hodie commentarios in Platonem habemus. Praeceptor etiam Marci Antoni* [*i.e., Marci Aureli*], *quem ad consulatum usque provexit, ut auctor Spartianus.* Note that Maffei does not index Euclid amongst the authors on which Proclus commented.

18 *Historia Augusta*, "Marcus Aurelius", II, 3. See also D. Steinbauer, "Proculus", in H. Cancik, H. Schneider (eds), *Brill's New Pauly: Encyclopedia of the Ancient World* (Leiden: Brill, 2007), xi. 926: *Usus praeterea grammaticis Graeco Alexandro Cotiaensi, Latinis Trosio Apro et Polione et Eutychio Proculo Siccensi.*

Eutychius Proculus, sending Proclus Diadochus back to the 2nd century AD in his chronology. As Goulding reports, this was an extended mistake even committed by the accomplished Barozzi, who followed Maffei *ad litteram* as his source regarding Proclus' chronology.[19] In sum, the labyrinths and confusions of doxography of the 15th- to 16th-century made Proclus predate Plotinus!

2.1.1 *Proclus'* In Euclidem *and Its Reception in* De certitudine

2.1.1.1 Proclus, Philosopher of Mathematics

Proclus' *In Euclidem* is arguably amongst the most influential and innovative sources of Piccolomini's *De certitudine* as it is the only surviving ancient Greek commentary on Euclid's work. Of it, two prologues and two books commenting on the first book of Euclid's *Elementa* came to the Latin world in the 15th century.[20]

---

19 Goulding, *Defending Hypatia*, p. 165 and p. 168. See also Francesco Barozzi, *Procli Diadochi Lycii philosophi platonici ac matematici probatissimi in primum Euclidis Elementorum librum commentariorum ad universam mathematicam disciplinam principium eruditionis tradentium libri IV* (Padua: Gratiosus Perchacinus, 1560), *Praefatio ad lectorem*, p. 2r: […] *alter M. Antoninus, a quo etiam (ut refert Spartianus) ad consulatum usque provectus fuit.*

20 Proclus, *In primum Euclidis Elementorum librum commentarii*, ed. G. Friedlein (Leipzig: Teubner, 1874); Proclus, *A Commentary on the First Book of Euclid's Elements*, ed. G. R. Morrow (New Jersey: Princeton University Press, 1970); Proclus, *Commentary on the First Book Of Euclid's Elements*, ed. Th. Taylor (London: Printed for the Author, 1792), re-printed Prometheus Trust 2006; Proclus, *Commento al I libro degli «Elementi» di Euclide*, ed. M. Timpanaro Cardini (Pisa: Giardini Editori, 1978). For an analysis of the philosophical themes of Proclus' text see D. J. O'Meara, "Mathematics and the sciences", in P. D'Hoine, M. Martijn (eds), *All From One*, pp. 167–182; O'Meara, "Geometry and the Divine in Proclus", in T. Koetsier and L. Bergmans (eds), *Mathematics and the Divine* (Amsterdam: Elsevier, 2005), pp. 133–145; O'Meara, *Pythagoras Revived: Mathematics and Philosophy in late Antiquity* (Oxford: Oxford University Press, 1989); J. L. Cleary, "Proclus' Philosophy of Mathematics", in G. Bechtle, D. J. O'Meara (eds), *La philosophie des mathématiques de l'Antiquité tardive Actes du colloque international Fribourg, Suisse, 24–26 septembre* (Saint-Paul Fribourg: Editions Universitaires Fribourg Suisse, 1998), pp. 85–102; G. MacIsaac, "The Soul and Discursive Reason in the Philosophy of Proclus", Ph.D. (South Bend, Indiana: University of Notre Dame, 2001); MacIsaac, "Projection and Time in Proclus", in J. Inglis (ed.), *Medieval Philosophy and the Classical Tradition in Islam, Judaism and Christianity* (London: Curzon Press, 2009), pp. 83–105; MacIsaac, "Geometrical First Principles in Proclus' Commentary on the First Book of Euclid's Elements", in *Phronesis* 59 (2014), pp. 44–98; MacIsaac, "Νόησις, dialectique et mathématique chez Proclus", in. A. Lernould (ed.), *Études sur le Commentaire de Proclus*, pp. 125–138; C. Helmig, *Forms and Concepts: Concept Formation in the Platonic Tradition* (Berlin: De Gruyter, 2012); Helmig, "Proclus' Criticism of Aristotle's Theory of Concept Formation in *Analytica posteriora* II.19", in F. A. J. de Haas, M. Leunissen, M. Martijn (eds), *Interpreting Aristotle's Posterior Analytics in Late Antiquity and Beyond* (Leiden: Brill, 2010), pp. 27–55; Helmig, "Iamblichus, Proclus, and Philoponus on Parts, Faculties and *ousiai* of Soul and the Notion of Life", in K. Corcilius, D. Perler (eds), *Partition of the Soul in Ancient, Medieval and Early Modern*

The two prologues are of the greatest philosophical and historical relevance.[21] The first prologue elaborates on the philosophy of mathematics, including discussions on the ontological status of mathematical objects and their intermediary position between Intellect and Soul (chapters I, II, V, and VI), the possibility of a common mathematical science (chapters III, IV, and VII), on mathematical usefulness in theology, physics, politics, ethics, rhetoric, and banausic arts (chapters VIII, IX, and X), discussions on the division and

---

Philosophy (Berlin: De Gruyter, 2014), p. 149–178; Steel, "Breathing Thought: Proclus on the Innate Knowledge of the Soul", in J. J. Cleary (ed.), *The Perennial Tradition of Neoplatonism* (Leuven: Leuven University Press, 1997), pp. 293–309; I. Mueller, "Mathematics and Philosophy in Proclus' *Commentary on Book I of Euclid's Elements*", in *Proclus, lecteur et interprète des anciens: Actes du Colloque international du CNRS, Paris, 2–4 octobre 1985*, pp. 305–318; Mueller, "Aristotle's Doctrine of Abstraction in the Commentators", in R. Sorabji (ed.), *Aristotle Transformed: the Ancient Commentators and their Influence* (Ithaca NY: Duckworth, 1990), pp. 463–80; Mueller, "Foreword", in Proclus, G. R. Morrow, *Proclus: A Commentary on the First Book of Euclid's Elements*, pp. ix–xxxi; Mueller, "Aristotelian Objections and Post-Aristotelian Responses to Plato's Elemental Theory", in C. Horn, J. Wilberding (eds), *Neoplatonism and the Philosophy of Nature* (Oxford: Oxford University Press, 2012), pp. 129–146; J. Opsomer, "Mathematical Explanation and the Philosophy of Nature, Late Ancient Philosophy: Astronomy and the Theory of the Elements", in *Documenti e studi sulla tradizione filosofica medievale* 23 (2012), pp. 65–106; Opsomer, "Syrianus, Proclus, and Damascius", in F. Sheffield and J. Warren (eds), *Routledge Companion to Ancient Philosophy* (New York London: Routledge, 2013), pp. 626–642; O. Harari, "Methexis and Geometrical Reasoning in Proclus' *Commentary on Euclid's Elements*", in *Oxford Studies in Ancient Philosophy* 30 (2006), pp. 361–389; Harari, "Proclus' Account of Explanatory Demonstrations in Mathematics and its Context", in *Archiv für Geschichte der Philosophie* 90 (2008), pp. 137–164; Harari, "Les preuves démonstratives dans le Commentaire de Proclus au premier livre des *Éléments* d'Euclide", in A. Lernould (ed.) *Études sur le Commentaire de Proclus*, pp. 197–216; Martijn, "Proclus' geometrical method", in P. Remes, S. Slaveva-Griffin (eds), *The Routledge Handbook of Neoplatonism* (Abingdon: Routledge, 2014), pp. 145–159; D. Nikulin, *Matter, Imagination and Geometry: Ontology, Natural Philosophy and Mathematics in Plotinus, Proclus and Descartes* (Ashgate: Aldershot, 2002); Nikulin, "Imagination et mathématiques chez Proclus", in A. Lernould (ed.), *Études sur le Commentaire de Proclus*, pp. 139–160; M. Schmitz, *Euclids Geometrie und ihre mathematiktheoretische Grundlegung in der neuplatonischen Philosophie des Proklos* (Würzburg: Königshausen und Neumann, 1997); L. Siorvanes, *Proclus: Neoplatonic Philosophy and Science* (New Haven: Yale University Press, Edinburgh University Press, 1996); E. Évrard, "FANTASIA chez Proclus", in M. Fattori, M. Bianchi (eds), *Phantasia – imaginatio, V. Colloqui internazionale Roma 9–11 gennaio 1986* (Rome: Edizioni dell'Ateneo Roma, 1988), pp. 57–68.

21 On the composition and structure of the two prologues, see A. Lernould, "Introduction" and "Analyse des Prologus I et II de l'*In Euclidem*", in A. Lernould (ed.), *Études sur le Commentaire de Proclus*, pp. 11–25; O'Meara, "L'eloge des sciences mathématiques dans la philosophi de l'Antiquité tardive", in A. Lernould (ed.), *Études sur le Commentaire de Proclus* pp. 57–67; O'Meara, *Pythagoras Revived*, pp. 156–176; A. Bernard, "L'arrière-plan rhétorique de la thórie de l'activité mathématique chez Proclus", in A. Lernould (ed.), *Études sur le Commentaire de Proclus*, pp. 67–88.

specificity of mathematical disciplines (chapters XII and XIII), mathematical methodologies and demonstrations (chapter XI and XIV), and some speculations on the origin and meaning of the term "μαθηματικά" (chapter XV).

The second prologue, approaching the subject of the commentary, deals with geometry. It includes discussions on the ontological status of geometrical objects, their production, and their relation to common mathematics (chapter I), the methods of geometry (chapter II), the usefulness of geometry (chapter III), the history of geometry culminating in Euclid (chapters IV and V), and on the goal, nature, and arrangement of Euclid's *Elementa geometrica* (VI, VII, VIII), and particularly, of book I (IX).

The two books of *In Euclidem* provide a detailed account of the axioms and propositions of Euclid's I, including theological, philosophical, logical, mathematical, and terminological discussions. Notably, the commentary also contains Proclus' original mathematical contributions, such as his influential (attempt at a) proof of the fifth postulate as if it were a theorem.[22]

As expounded by O'Meara,[23] in composing *In Euclidem*, Proclus drew on the earlier tradition of commentaries and discussion on Euclid's *Elementa*, adopting insights from (now mostly lost) works by Porphyry, Pappus of Alexandria (*ob.* 350), Plutarch of Athens (350–430), and Syrianus of Alexandria (the last two authors direct acquaintances of his). Syrianus' *In Metaphysica* has also been shown as a valuable source of Proclus' *In Euclidem*: it contains counterarguments targeting Aristotle's criticisms of the Platonic and Pythagorean views on the ontological status of mathematical objects, the theory of the unfolding or projection of mathematical entities onto phantasy,[24] and criticisms of Aristotle's theory of mathematical abstraction.[25] Proclus also made use of

---

22 V. De Risi, *Leibniz on the Parallel Postulate and the Foundations of Geometry* (Cham: Springer, 2010), chap. 2.1; De Risi, "The development of Euclidean axiomatics", in *Archive for History of Exact Sciences* 70 (2016), pp. 591–676.

23 O'Meara, "Mathematics and the sciences", especially pp. 169–171; O'Meara, *Pythagoras Revived*, pp. 142–175.

24 See ex. gr. Syrianus, *In Metaphysica*, M, f. 91, l. 29–34 and f. 95, l. 29–36.

25 A. Longo, "*In Euclidem*, Prologue I, chapitre 6: Ce que Proclus doit à son maître Syrianus dans les arguments avancés contre la thèse de l'existence "postérieure" des objets mathématiques", in A. Lernould (ed.), *Études sur le Commentaire de Proclus*, pp. 45–54; Mueller, "Syrianus and the Concept of Mathematical Number", in G. Bechtle, D.J. O'Meara (eds), *La philosophie des mathématiques de l'Antiquité tardive*, pp. 71–84; S. Klitenic Wear, *The Teachings of Syrianus on Plato's Timaeus and Parmenides* (Leiden Boston: Brill, 2011). See also the Introduction to Syrianus, *On Aristotle Metaphysics 3–4*, transl. J. M. Dillon, D.J. O'Meara, ed. R. Sorabji (London: Bloomsbury Academic, 2008); Syrianus, *On Aristotle Metaphysics 13–14*, transl. J. M. Dillon, D. J. O'Meara, ed. R. Sorabji (London: Bloomsbury Academic, 2006).

Nicomachus of Gerasa's *Introductio arithmetica* and of book III of Iamblichus' (245–325) *De communi mathematica scientia*, as is attested by the account of ὅλη μαθηματική in the first prologue of *In Euclidem*. Other of Proclus' sources are Eudemus' (*c.*370–300 BC) *Historia geometrica*, Geminus' (1st century BC) lost work Φιλοκαλία, and writings of Apollonius of Perga (*c.*240–190 BC) and of Archimedes (*c.*287–212 BC).

2.1.1.2 The Latin Reception

Proclus' *In Euclidem* was a later rediscovery of the Latin world. It was recovered in the 15th century, arguably brought to Italy by Giovanni Aurispa (1376–1459) and Bessarion (1403–1472). Recent scholarship has emphasised the remarkable role played by *In Euclidem* in 16th- and 17th-century mathematical and philosophical culture and explored the ways its contents were rediscovered and spread.[26] Published by Simon Grynaeus in 1533 in the same volume as the *editio princeps* of Euclid's *Elementa*,[27] *In Euclidem* was entirely translated into Latin at least three times in the 16th century: by Bartolomeo Zamberti (1473–1543), whose translation is preserved at Munich, Bayerische Staatsbibliothek, Clm 6, fol. 1r–154v, by Giovanni Battista Gabia (*c.*1500–1590), in Milan, Ambrosiana Library, P. 51 sup., and by Barozzi. Another complete Latin translation of Proclus' commentary following Grynaeus' text was finished in 1638 by the Spaniard Vicente Mariner (*ob.* 1642). It is preserved at Madrid, Biblioteca Nacional de España, 9871, fol. 45–299.[28]

Of the three complete translations produced in the 16th century, only that of Barozzi knew a printed edition in 1560–13 years after *De certitudine*. Therefore,

---

26 Rose, "Humanist Culture and Renaissance Mathematics: The Italian Libraries of the Quattrocento", in *Studies in the Renaissance* 20 (1973), pp. 46–105; Á. J. Campillo Bo, "The Forgotten Gifts of Hermes: The Latin Reception of Proclus' *Commentary on Euclid's Elements*", in *Mediterranea, International Journal on the Transfer of Knowledge* 8 (2023), pp. 193–278; P. Adamson, F. Karfík, "Proclus' Legacy", especially s.15.2; Claessens, "Proclus in the Renaissance", pp. 1–5; Claessens, "Het denken verbeeld. De vroegmoderne receptie (1533–1650) van Proclus' *Commentaar op het eerste boek van Euclides' Elementen*", Ph.D. (Leuven: Katholieke Universiteit Leuven, 2011); Rabouin, "Le rôle de Proclus dans les débats sur la 'mathématique universelle'"; M. O. Helbing, "La fortune des commentaires de Proclus sur le premier livre des Éléments d'Euclide à l'époque Galilée", in G. Bechtle, D. J. O'Meara (eds), *La philosophie des mathématiques de l'Antiquité tardive*, pp. 173–193; and L. Maierù, "La diffusione di Proclo, commentatore di Euclide, nel Cinquecento", in *11° Annuario del Liceo Scientifico "B. G. Scorza"* (Soveria Mannelli: Calabria Letteraria Editrice, 1999), pp. 49–68.

27 Euclid, Proclus, *Στοιχείων βιβλία ιε΄ ἐκ τοῦ Θέωνος συνουσιῶν. εἰς τοῦ αὐτοῦ τὸ πρῶτον ἐξηγημάτων Πρόκλου βιβλίον* ed. Simon Grynaeus (Basel: I. Hervagius, 1533).

28 Campillo Bo, "From Poetics to Mathematics: Vicente Mariner's Latin Translation of Proclus' *In Euclidem*", in *Noctua* 11/2 (2024), pp. 258–294.

the spreading of Proclus' *In Euclidem* through the printing-press occurred mainly through these two editions: the deficient Greek text of Grynaeus (which takes as its source a single manuscript, Oxford, Corpus Christi College Library, 97), and the Latin version of Barozzi who, unsatisfied with the text of Grynaeus' edition, undertook by himself the endeavour of reconstructing Proclus' Greek text by collating several manuscripts.[29] He used the resulting, refined Greek version as the basis for his published Latin translation.

It must be underscored that, through Grynaeus' edition, every scholar acquiring the first ever printed edition of the Greek text of Euclid's *Elementa* would *ipso facto* also have access to Proclus' commentary. As attested by the page-number quotations of Proclus' text in *De certitudine*, this was the case for Piccolomini as well: he seemingly used the Greek text of Grynaeus' edition for composing his opusculum. In his study, Higashi has suggested that Piccolomini might have had access to Grynaeus' edition through his patron Mendoza, who owned the book as preserved in his collection.[30]

Nevertheless, it might be doubted whether the Sienese had constant access to the commentary while writing *De certitudine* because of the numerous misquotations in page numbers. Ex. gr., on f. C, Piccolomini cites p. 46, 47, 48 and 61 of Proclus' commentary in support of his opinions concerning the differences between theorems and problems, and the way they are related to prior demonstrations. These pages do not coincide with the sections in which Proclus provides his commentaries on any of those topics in Grynaeus' edition: p. 61, however, matches the commentary on proposition II (III, 223), in which Proclus opens with some references to how problems and theorems can be constructed through different diagrams which point to the same truth. Yet again, on f. CIX Piccolomini refers to Proclus' text and provides the page number.[31] This Greek line is not found on p. 47 of Grynaeus' edition but on p. 57. The mistakes in the references of Proclus' commentary are too numerous

29 According to Rose's research in Rose, *A Venetian Patron*, pp. 127–128, those are: Milan, Ambrosiana A. 164 sup., Bologna, Biblioteca Universitaria, 2293 (fol. 1–149v). Rose proposes that the manuscripts which Barozzi obtained from Andrea Doni and Fabrizio Garzoni might be Vat. Barb. Gr. 101, Vat., Barb. Gr. 145, Estense Library, α T.9.11 (Puntoni 46), and Venice, Marciana VI. 11 (coll. 1409). Oxford, Bodleian Library, Barocci 161, coming from Barozzi's own library, preserved handwritten scholia and corrections arguably in Barozzi's hand.

30 Higashi, *Penser les mathématiques*, p. 129.

31 Alessandro Piccolomini, *De certitudine*, f. CIX: Quod idem asseverat Proclus *In Primum Elementorum* pagina 47: Πανταχοῦ [μὲν] γὰρ τὸ ἀναγκαῖον ἔχουσιν οἱ γεωμετρικοί λόγοι διὰ τὴν ὑποκειμένην ὕλην.

to be attributed to a manual error in the printing types. Hence, it might be possible that Piccolomini had no constant access to Grynaeus' *editio princeps* while composing his opusculum and was thus misquoting or confusing the pages either from memory or from his notes. Nevertheless, the lacunae in Piccolomini's Greek quotations fit those in the text of Grynaeus, and his attention to quoting page numbers suggests that the 1533 edition was most likely his primary (if not sole) source.[32]

Regarding the philosophical relevance of its Latin rediscovery, it is crucial to note that many of the themes discussed in Proclus' text and used by authors such as Giorgio Valla (1447–1500), Petrus Ramus (1515–1572), Jerónimo Muñoz (1520–1591),[33] Conrad Dasypodius (1532–1600), Federico Borromeo (1564–1631), Johannes Kepler (1571–1630), and the Jesuits Francisco Suárez (1548–1617)[34] and Athanasius Kircher (1602–1680)[35] found resonance in early modern philosophy. Some of the topics explored in Proclus' commentary that gained significant prominence in the 16th and 17th centuries were: the legitimacy of mathematical methods in comprehending the natural world, the concept of innate mathematical knowledge, the intermediary status of mathematics between physics and metaphysics, the notion of a *mathesis universalis*, the criteria in classifying the ever-expanding array of mathematical disciplines, and the history of mathematics.[36]

---

32 The two Greek quotations in *De certitudine* from Proclus are found in f. CIX (as just mentioned) and on f. CVII–CVIIV: τήν γε μὲν αἰτίαν καὶ τὸ διὰ τί πολλοῖς ἔδοξεν ἡ γεωμετρία μὴ θεωρεῖν. [καὶ] ταύτης γάρ ἐστι καὶ ὁ Ἀμφίνομος τῆς δόξης Ἀριστοτέλους κατάρξαντος. *Quid amplius quaerimus pro hac sententia?* Both readings coincide in their lacunae with those of Grynaeus: in the former, the particle μὲν is absent, whereas in latter the initial καὶ lacks in the second sentence.

33 On Muñoz's reception and mathematical and philosophical use of Proclus' *In Euclidem*, see Campillo Bo, "Jerónimo Muñoz's Reception of Proclus' *In Euclidem*: Philosophy of Mathematics and an Attempt to Prove the Parallel Postulate", in *Early Science and Medicine* (2023), pp. 631–658.

34 See Frascisco Suárez, *Metaphysicae Disputationes*, I, III, *Quomodo metaphysica prima principia confirmet ac tueatur*.

35 See Athanasius Kircher, *Musurgia universalis, sive Ars magna Consoni et Dissoni, in X libros digesta* (Rome: Francisci Corbelletti, 1650), Vol. I, book X, pp. 446–447. I wish to express my gratitude to Professor Stephen Gersh for directing my attention to the use of Proclus' *In Euclidem* in Kircher's theory of music.

36 For an excellent account of the early modern rediscovery of the history of mathematics and the role played by Proclus' commentary in it, see R. Goulding, *Defending Hypatia*.

Some philosophers, such as Jerónimo Muñoz,[37] Francesco De' Vieri (1524–1591),[38] or Francesco Patrizi,[39] showed some interest in incorporating the teachings found in Proclus' *In Euclidem* into a wider (Neo-)Platonic framework. Nonetheless, this seems to have been rarely the case. This is a momentous detail in accounting for the reception of *In Euclidem* in the 16th century: it was mostly read as an isolated source of mathematical knowledge, history of mathematics, and philosophy of mathematics; no great efforts were made by its most influential readers to grasp its contents as part of Proclus' complete philosophical views. This was also the case for Piccolomini, who circumscribed himself to the sole Proclean source of *In Euclidem* in writing *De certitudine*, without ever appealing to any other writing of the Diadochus.

2.1.1.3 Proclus in Piccolomini's *De certitudine*

Proclus' *In Euclidem* is, as said, a momentous theoretical and argumentative resource pervading Piccolomini's account of mechanics, geometry, and the philosophy of mathematics. Already mentioned in the *Praefatio* to the paraphrase, Proclus reappears in *De certitudine* as an authority who (according to Piccolomini's reading) anticipated his views that (i) mathematical demonstrations cannot instantiate a *demonstratio potissima*, and (ii) that mathematical certainty stems from the special ontological status of mathematical objects:

> Indeed, I decided to repress this opinion of mine within myself, although it appeared most convincing to me and was supported by numerous reasons. I did so, however, to avoid seeming as if I had uttered nearly a paradox (*quasi* παράδοξον) until I found Proclus himself to affirm

37 Navarro-Brotons has shown that Muñoz was familiar with other works of Proclus, such as *In Alcibiadem* and *In Timaeum*. See V. Navarro-Brotons, *Jerónimo Muñoz: Matemáticas, cosmología y humanismo en la época del Renacimiento* (Valencia: Universidad de Valencia, 2019); Navarro-Brotons, *Disciplinas, saberes y prácticas: Filosofía natural, matemáticas y astronomía en la sociedad española de la época moderna* (Valencia: Universidad de Valencia, 2014); Navarro-Brotons, E. Rodríguez Galdeano, "The Cultivation of Astronomy in Spanish Universities in the Latter Half of the 16th Century", in M. Feingold, Navarro-Brotons (eds), *Universities and Science in the Early Modern Period* (New York London: Springer), 2006, pp. 83–98.

38 See S. Fellina, *Platone allo Studium Fiorentino-Pisano (1576–1635). L'insegnamento di Francesco de' Vieri, Jacopo Mazzoni, Carlo Tomasi, Cosimo Boscagli, Girolamo Bardi* (Mantova: Scripta, 2019), p. 67.

39 For an account of Francesco Patrizi's reception of Proclus see Leinkauf, "Francesco Patrizi".

> the same, and being greatly delighted that I found such a rich witness, I subsequently asserted this very thing loudly and repeatedly.[40]

Because of this argument, Piccolomini's *De certitudine* became one of the first works in the Latin world in which Proclus' *In Euclidem* was used (and interpreted) as a main authority in a philosophical discussion. Moreover, in the passage above, Piccolomini confesses that he had always been of the persuasion that mathematical proofs were not scientifically perfect. He likely refers to some ideas inspired by his education as a "Simplician" Averroist (see 2.1.5 and 3.4.1). He also reveals that, after he came across Proclus' *In Euclidem*, he was no longer afraid to express his opinion. Since then, we are told, backed by the Greek commentator's authority, Piccolomini felt entitled to proclaim publicly his reservations concerning the scientific status of mathematical demonstrations. Piccolomini's avowed change of attitude after reading Proclus (if we are to believe him) will be useful in explaining some of his theoretical shifts from the paraphrase to the composition of *De certitudine*.

Cozzoli has pointed out that the two main themes of Piccolomini's *De certitudine* can be jointly found in one particular section of Proclus' *In Euclidem* where he comments on the first proposition in book I (= I.1).[41] It reads as follows:

> What is called "proof" we shall find sometimes has the properties of a demonstration in being able to establish what is sought using definitions as middle terms, and this is the perfect form of demonstration (ἀποδείξεως τελειότης), but sometimes it attempts to prove through signs. This point should not be overlooked. Although geometrical propositions always derive their necessity from the matter under investigation, they do not always reach their results through demonstrative methods (διὰ τῶν ἀποδεικτικῶν μεθόδων). For example, when from the fact that the exterior angle of a triangle is equal to the two opposite interior angles it is shown that the sum of the interior angles of a triangle is equal to two right angles, how can this be called a demonstration based on the cause? Is not the middle-term used here only a sign?[42]

---

40 Alessandro Piccolomini, *De certitudine*, f. LXXII: *Verumenimvero hanc meam sententiam, quamvis mihi constantissime probaretur ac quampluribus rationibus fulciretur, in me ipso tamen, ne quasi παράδοξον a me dictum videretur, eousque comprimendam duxi, donec Proclum ipsum hoc idem cognoscens, maxima animi laetitia affectus, testem tam locupletem nanctus id ipsum dehinc clara voce frequenter asserui* [...].

41 Cozzoli, "Alessandro Piccolomini and the certitude of mathematics", p. 166.

42 Proclus, *In Euclidem*, p. 206, l. 12–23. For Proclus' *In Euclidem*, I will always propose the English translation of G. R. Morrow, slightly modifying it when required. See

As we saw above and shall explore in the next sections, this precise passage is partially quoted in Greek by Piccolomini.[43] The two main topics which stand as the doctrinal core of *De certitudine* are present here, namely:

(a) The deficiency of mathematical demonstrations: in Proclus' case, proposition 32 of Euclid's first book (= I.32) concludes only, imperfectly, through "evidence" or "signs".

(b) The origin of mathematical certainty (here expressed as "necessity") is bestowed upon the discipline by the ontological status of mathematical items rather than by mathematical demonstrative methods.

Regarding (a), the term demonstration by "sign" or "token" (σημεῖον and τεκμήριον respectively) indicates a kind of proof from non-causal evidence. This notion was elaborated by Aristotle (and the subsequent tradition): it was classified as one of the weakest forms of demonstration.[44] Aristotle's example in *Poetica*, 16, 1454b, l. 20 ff. is when Odysseus' nurse recognises him through a scar: Odysseus' scar is not what makes him Odysseus; rather, it is a contingent mark that carries no explanatory power over his identity or essence. Proclus declares that Euclid's proof of I.32 is a demonstration of this sort, as it concludes through an accidental sign.

In (b) Proclus establishes that, although some mathematical demonstrations are logically flawed, all of them are certain and necessary, therefore the epistemic superiority of mathematics (necessity, irrefutability, precision) must stem from the special nature of its objects, not from its demonstrative methods.

Proclus expresses his reservations about mathematical demonstration by invoking the controversial proof of proposition I.32. This will be the theorem adopted by Piccolomini as a paradigm of the inappropriateness of mathematical demonstration – and, of even greater importance, the one frequently used by Aristotle in *Analytica posteriora* as a case of scientific knowledge (see 2.6.2). I will attempt to show that Piccolomini's particular reading of this passage from Proclus' *In Euclidem* (and many other *loci*) as undermining the scientific

---

A Commentary on the First Book of Euclid's Elements, p. 161: τὴν δὲ λεγομένην ἀπόδειξιν ὅτε μὲν καὶ τὰ ἴδια τῆς ἀποδείξεως ἔχουσαν εὑρήσομεν ἀπὸ τῶν ὁρισμῶν μέσων τὸ ζητούμενον δεικνύουσαν – αὕτη γὰρ ἀποδείξεως τελειότης – ὅτε δὲ ἐκ τεκμηρίων ἐπιχειροῦσαν. καὶ δεῖ μὴ λανθάνειν. πανταχοῦ μὲν γὰρ τὸ ἀναγκαῖον ἔχουσιν οἱ γεωμετρικοὶ λόγοι διὰ τὴν ὑποκειμένην ὕλην, οὐ πανταχοῦ δὲ περαίνονται διὰ τῶν ἀποδεικτικῶν μεθόδων. ὅταν γὰρ διὰ τοῦ τὴν ἐκτὸς τοῦ τριγώνου γωνίαν ἴσην εἶναι δύο ταῖς ἐντὸς καὶ ἀπεναντίας δεικνύηται τὸ τρίγωνον ἴσας ἔχον τὰς ἐντὸς τρεῖς γωνίας δυσὶν ὀρθαῖς, πῶς ἀπ᾽αἰτίας ἡ ἀπόδειξις αὕτη, πῶς δὲ οὐχὶ τεκμήριόν ἐστι τὸ μέσον;.

43 For the Greek quotation see Alessandro Piccolomini, *De certitudine*, f. CIX.

44 See also Aristotle, *Analytica priora*, II, 27. For a history of this concept see J. Allen, *Inference from Signs: Ancient Debates about the Nature of Evidence* (Oxford: Clarendon Press, 2001).

status of mathematical demonstration can be accounted for by appealing to another Greek text under the influence of which the Sienese read Proclus, namely Pseudo-Simplicius' *In De anima* (see 2.1.5., 2.3.3, 3.4.3, and 3.5.2).

Regarding the presence of the Diadochus in other parts of the 1547 publication, there is only mention of Proclus in Piccolomini's 1540 Latin paraphrase of the *Quaestiones mechanicae*. Proclus is called upon to affirm that mathematics can demonstrate the same conclusion through different but *true* causes.[45] This claim, unless tremendously qualified, is incompatible with Piccolomini's denial of mathematical causation in *De certitudine*. This might be a hint of Piccolomini's ongoing philosophical development during the 40s, culminating in his unhesitant denial of mathematical causation.

In the next sections, I shall introduce Proclus' philosophy of mathematics, paying special attention to three themes whose influence determined the philosophical and argumentative structure of both *De certitudine* and the *Praefatio* to the Latin paraphrase. I ultimately intend to answer the questions (1) Which parts of Proclus' commentary did Piccolomini use in composing the core of *De certitudine*? and (2) How did he read them?

The three themes from *In Euclidem* which I will treat in the next sections are (a) Proclus' account of mathematical analysis, (b) his ontology of mathematics, and (c) Proclus' notion of a common mathematical science.

2.1.2 *Mathematical Analysis*

One of the themes which had the greatest influence on the logical theories proposed in *De certitudine* is the methodology known as mathematical analysis. In Proclus' *In Euclidem* the abstract term ἀνάλυσις appears 21 times, whereas the verbal form ἀναλύω is found 7 times, and the adjectival form ἀναλυτικός/ή/όν 2 times: a total of 30 times.[46] On its part, σύνθεσις (in a logical and/or metaphysical sense – for I do not count its occurrences when it means "addition" or the "composition" of lines) appears a total of 16 times.[47]

45 Alessandro Piccolomini, *In mechanicas quaestiones*, f. LVIII–LVIIIv: *Non enim inconveniens est eandem complexionem sive (ut ita dicam) conclusionem ex pluribus assumptionibus pluribusque veris causis in mathematicis haberi posse, ut scribit Proclus, et ut quilibet cognoscere potest, qui mathematicas disciplinas quam modice degustaverit.*

46 Proclus, *In Euclidem*. For ἀνάλυσις (21 times) see p. 8, l. 7; p. 43, l. 2, and l. 7, and l. 19; p. 57, l. 19; p. 24, l. 67, l. 7, and l. 24; p. 92, l. 12; p. 93, l. 23; p. 211, l. 19–20; p. 242, l. 15; p. 246, l. 9, l. 12, l. 23; p. 255, l. 20; p. 256, l. 1; p. 423, l. 15; p. 523, l. 25. For ἀναλύω (7 times) see p. 73, l. 18; p. 118, l. 15; p. 246, l. 10; p. 282, l. 2, and l. 5; p. 382, l. 2, and l. 5; p. 422, l. 13. For ἀναλυτικός/ή/όν (2 times) see p. 42, l. 20; p. 69, l. 17.

47 Proclus, *In Euclidem*, p. 3, l. 6; p. 5, l. 5; p. 7, l. 23; p. 8, l. 7; p. 9, l. 11; p. 43, l. 3 and l. 18; p. 55, l. 7; p. 57, l. 18; p. 60, l. 25; p. 87, l. 16; p. 245, l. 13 and l. 16; p. 246, l. 8; p. 255, l. 20 and l. 23.

A look at some of these passages can cast light on Proclus' views on the vague and manifold methodology of ἀνάλυσις-σύνθεσις. For instance, in p. 211, l. 18–23 he refers to the fruitful results of the application of this method by ancient mathematicians, amongst whom Plato and Leodamas are quoted by name:

> Nevertheless, there are certain methods that have been handed down, the best being the method of analysis, which traces the desired result back to an acknowledged principle. Plato, it is said, taught this method to Leodamas, who also is reported to have made many discoveries in geometry by means of it.[48]

And in f. 8, l. 4–6 Proclus describes this method as follows:

> And, certainly, beauty and order are common to all branches of mathematics, as is the method of proceeding from things better known to things we seek to know and the reverse path from the latter to the former, the methods called analysis and synthesis.[49]

Lastly, we can raise the comment found in f. 43, l. 1–2:

> Being thus endowed and led towards perfection, mathematics reaches some of its results by analysis, others by synthesis.[50]

Two main traits related to Proclus' understanding of ἀνάλυσις-σύνθεσις can be inferred from these passages, namely (1) that it is closely related to the nature and structure of mathematical sciences, whose order and clarity allow a thorough deductive process from an assumed proposition to its suppositions and/or principles (axioms), and *vice versa* (2) that, because of that, it is a methodology both of research and justification.

---

48 Proclus, *In Euclidem*, p. 211, l. 18–23: μέθοδοι δὲ ὅμως παραδίδονται, καλλίστη μὲν ἡ διὰ τῆς ἀναλύσεως ἐπ'ἀρχὴν ὁμολογουμένην ἀνάγουσα τὸ ζητοὺμενον, ἣν καὶ ὁ Πλάτων ὡς φασὶν Λεωδάμαντι παραδέδωκεν, ἀφ'ἧς καὶ ἐκεῖνος πολλῶν κατὰ γεωμετρίαν εὑρετὴς ἱστόρηται γενέσθαι.

49 Proclus, *In Euclidem*, p. 8, l. 4–6: καὶ μὴν καὶ τὸ κάλλος καὶ ἡ τάξις κοινὰ πάντων ἐστὶ τῶν μαθημάτων καὶ ἡ ἀπὸ τῶν γνωριμωτέρων ὁδὸς ἐπὶ τὰ ζητούμενα καὶ ἡ ἐκ τούτων ἐπ'ἐκεῖνα μετάβασις, ἃς δὴ καλοῦσιν ἀναλύσεις καὶ συνθέσεις.

50 Proclus, *In Euclidem*, p. 43, l. 1–2: ἀφ'ὧν χορεγουμένη καὶ τελεουμένη ἡ μαθηματικὴ τὰ μὲν δι'ἀναλύσεως εὑρίσκει, τὰ δὲ διὰ συνθέσεως [...].

As described by Pappus of Alexandria,[51] one of Proclus' sources,[52] ἀνάλυσις and σύνθεσις in mathematics was the method of, firstly, recasting a proposition-to-be-solved in such a way that it could be reduced, derived from, or constructed upon (in the case of problems) already known propositions or principles. The analytic procedure sometimes even led to a new axiom required for the researched proposition to be resolved. In the second synthetic stage, the proposition can therefore be deduced from the previous propositions to which it has been reduced. Hence, broadly speaking, analysis-synthesis was held both as a heuristic and justificatory methodology in its different steps, although, as pointed out by Hintikka, ancient mathematicians deemed analysis to be mainly a heuristic tool.[53] Nevertheless, the Neoplatonic understanding of analysis as presented in Proclus' *In Euclidem* adds some philosophically relevant aspects which were not so pronounced in his mathematical sources (such as the Stoic Posidonius, Geminus, and Pappus himself). As Hintikka remarks:

> On the other hand, Proclus' opinions about analysis stem partly from Porphyry, who, although a Neoplatonist, was strongly influenced by Aristotelian ideas of the nature of science. No wonder, then, if Proclus' description of analysis is a mixture of elements which have little to do with actual mathematical practice.[54]

---

51 For Pappus' notion of analysis (including the original Greek text) see Hintikka, Remes, *The Method of Analysis*, especially pp. 7–22; Hintikka, "Method of Analysis".

52 O'Meara, "Mathematics and the sciences", pp. 169–171.

53 Hintikka, Remes, *The Method of Analysis*, p. 7.

54 Hintikka points to the following section to signal the more philosophical (logical/metaphysical) than mathematical understanding of analysis by Proclus; *In Euclidem*, p. 255, l. 8–23: ὅλως γὰρ εἰδέναι, χρὴ ὅτι πᾶσαι αἱ μαθηματικαὶ ἢ ἀπὸ τῶν ἀρχῶν εἶσιν ἢ ἐπὶ τὰς ἀρχάς, ὡς που φησὶ καὶ ὁ Πορφύριος. καὶ αἱ μὲν ἀπὸ τῶν ἀρχῶν διτταὶ καὶ αὐταὶ τυγχάνουσιν, ἢ γὰρ ἀπὸ τῶν κοινῶν ἐννοιῶν ὥρμηνται καὶ τῆς ἐναργείας μόνης τῆς αὐτοπίστου, ἢ ἀπὸ τῶν προδεδειγμένων. αἱ δὲ ἐπὶ τὰς ἀρχὰς ἢ θετικαὶ ἢ ἀναιρετικαί, ἀλλὰ θετικαὶ μὲν οὖσαι τῶν ἀρχῶν ἀναλύσεις καλοῦνται, καὶ ταύταις αἱ συνθέσεις ἀντίκεινται, δυνατὸν γὰρ ἀπὸ τῶν ἀρχῶν ἐκείνων προελθεῖν εὐτάκτως ἐπὶ τὸ ζητούμενον, καὶ τοῦτό ἐστιν ἡ σύνθεσις. In Morrow's translation (slightly modified): "In general we must understand that all mathematical arguments proceed either from or to the principles, as Porphyry somewhere says. Those that proceed from the principles are themselves of two kinds, as it happens, for they proceed either from common notions, that is, from self-evident clarity alone, or from things previously demonstrated. Those that proceed to the principles are either affirmative of them or destructive. But those that affirm first principles are called analyses, and their reverse procedures syntheses". More notes on the Neoplatonic understanding of mathematical analysis are found in J. Hintikka, U. Remes, *The Method of Analysis*, p. 21.

Proclus highlights both the heuristic function of ἀνάλυσις and the deductive, justificatory role of σύνθεσις. He also asserts that, as part of the analytic process, mathematical figures can be visually or schematically analysed – decomposed into their geometrical constituent parts – as a strategy for learning or discovery.[55]

Proclus integrates philosophical aspects into ἀνάλυσις-σύνθεσις, grounding them in a Platonic understanding of knowledge as recollection and the metaphysical derivation of realities from higher ontological principles. This Neoplatonic understanding of the structure of mathematical analysis can be illuminated through the following passage:

> By contrast, mathematics, though beginning its recollection from the outside world (τῆς μὲν ἀναμνήσεως ἔξωθεν ἄρχεται), ends with the reasons within; it is awakened to activity by lower realities, but its destination is the higher being of Forms. Its activity is not motionless, like that of the intellect, but because its motion is not a change of place or quality, as is that of the senses, but a life-giving activity, it unfolds and traverses the immaterial order of the reasons, now moving from first principles to conclusions, now proceeding in the opposite direction, now advancing from what it already knows to what it seeks to know, and again referring its results back to the principles that are prior in knowledge (τότε μὲν ἀπὸ τῶν ἀρχῶν ἐπὶ τὰ ζητούμενα, τότε δὲ ἀπὸ τῶν ζητουμένων ἐπὶ τὰ προηγούμενα).[56]

In this section, Proclus articulates the two aforementioned Platonic principles through the language of analysis and synthesis (ἀρχαί, τὰ ζητούμενα, τὰ προηγούμενα). First, (a) he posits that mathematical learning is a form of recollection that ultimately guides us to the comprehension of higher realities. Second, (b) he asserts that the ordered structure facilitating the interplay of mathematical propositions and enabling analysis is not exclusive to mathematics but extends to the study of the Forms.

As Dubouclez points out, one of the most remarkable features of Proclus' views on analysis-synthesis is that it not only stands as a fruitful methodology for the research and justification of true propositions within the scientific field

55 Proclus, *In Euclidem*, p. 118, l. 15; p. 442, l. 13–21. See also J. Hintikka, U. Remes, *The Method of Analysis*, especially pp. 94–95.

56 Proclus, *In Euclidem*, p. 18, l. 17–28: ἡ δ'αὖ μαθηματικὴ τῆς μὲν ἀναμνήσεως ἔξωθεν ἄρχεται, τελευτᾷ δὲ εἰς τοὺς ἔνδον λόγους, καὶ ἀνεγείρεται μὲν ἀπὸ τῶν ὑστέρων, καταντᾷ δὲ εἰς τὴν προηγουμένην οὐσίαν τῶν εἰδῶν, καὶ ἀκίνητος μὲν αὐτῆς οὐκ ἔστιν ἡ ἐνέργεια καθάπερ ἡ νοερά, διὰ δὲ κινήσεως οὐ τοπικῆς οὐδὲ ἀλλοιωτικῆς ὥσπερ αἱ αἰσθήσεις, ἀλλὰ ζωτικῆς ἀνελίσσεται καὶ διέξεισι τὸν ἀσώματον τῶν λόγων διάκοσμον, τότε μὲν ἀπὸ τῶν ἀρχῶν ἐπὶ τὰ ζητούμενα, τότε δὲ ἀπὸ τῶν ζητουμένων ἐπὶ τὰ προηγούμενα κατὰ τὴν γνῶσιν.

of mathematics, but it also reflects a deeper, grounding aspect of reality.[57] In the same manner that theorems can be conjectured, reduced to prior principles, and then demonstrated from those prior principles, so metaphysical propositions can be reduced to always prior principles, producing thereby an analytic-synthetic picture of the hierarchy of principles.

In a revealing passage of the *Theologia Platonica*, Proclus makes explicit that the use of the geometrical arrangement is not an arbitrary choice in the way of exposition, but a requirement of the contemplative sciences (dialectics, in this example). Proclus explains how Plato adopted the geometrical mode of reasoning in dealing with metaphysics and sustains that this mathematical method was proposed as an intellective model (παράδειγμα προτείνει νοερόν).[58] This is justified by invoking Plato's position in the *Timaeus* (29b, l. 2–3) that words/reasonings (λόγοι) are images of things. Proclus concludes that analytically-synthetically arranged discourses structurally resemble the deeper syntax of reality, whose order is captured by demonstrations from simpler principles to increasingly complex propositions:

> If, therefore, the discourses [about things] bear an image of [those] things (λόγοι τῶν πραγμάτων εἰκόνα φέρουσιν) for which they are the explanations, and as the unfolding of demonstrations is, so too must the order demonstrated things be, then it is necessary, I think, that those things which begin from very simple principles be absolutely the most basic and that they are arranged in connection with the One, while those that are more numerous and dependent on various demonstrations have come

---

57 O. Dubouclez, *Descartes et la voie de l'analyse* (Paris: Presses Universitaires de France, 2013), pp. 83–99.

58 Proclus, *Théologie platonicenne*, eds H. D. Saffrey, L. G. Westerink, vol. 1 (Paris: Belles Lettres, 1968), I, pp. 45, l. 19–46, l. 2: ἔτι τοίνυν κἀκ τοῦ τρόπου τῶν ἀποδείξεων τὸ αὐτὸ καταδησώμεθα. τὰ μὲν πρῶτα τῶν συμπερασμάτων δι'ἐλαχίστων ὡς οἷόν τε καὶ ἁπλουστάτων καὶ γνωριμωτάτων καὶ οἷον κοινῶν ἐννοιῶν εὐθὺς κατάδηλα γίνεται, τὰ δὲ τούτων ἐφεξῆς διὰ πλειόνων καὶ ποικιλωτέρων, τὰ δὲ ἔσχατα παντελῶς ἐστὶ συνθετώτατα. χρῆται γὰρ [ὁ Πλάτων] ἀεὶ τοῖς πρώτοις συμπεράσμασιν εἰς τῶν ἐχομένων ἀποδείξεις καὶ τῆς ἐν γεωμετρίᾳ τάξεως ἢ τοῖς ἄλλοις μαθήμασι παράδειγμα προτείνει νοερὸν τὴν τῶν συμπερασμάτων τούτων πρὸς ἄλληλα συνάρτησιν. In my own translation: "Therefore, let us also follow the same method of demonstrations. The initial conclusions are immediately evident through very simple and familiar common notions; the ones that follow are shown through more steps and variations, and the final ones are the most complex. He [Plato] always uses the initial conclusions as demonstrations in his arguments and presents the mutual connection amongst conclusions in the order of geometry or other mathematical disciplines as an intellective model".

> after a considerable interval from the One, if it is permissible to speak in this way.[59]

The persuasion that the relations of production, dependence, and reversion of non-mathematical realities are captured by the logical structure of analysis-synthesis explains Proclus' fondness for producing *elementationes* and his aspirations to a complete corpus of contemplative sciences ordered in geometrical(-like) manner. Indeed, this zeal would lead him to the production of two (still lacking) propositionally arranged systems for the two other Aristotelian contemplative sciences, namely physics and theology, with his *Elementatio physica* and *Elementatio theologica*. As Opsomer has pointed out, each of these *elementationes* possesses its idiosyncratic traits, for the object of study of each of the physical and metaphysical realities respectively requires different argumentative strategies.[60] Proclus' *elementationes* were to be added to the mathematical *elementatio*, already axiomatized by Euclid,

---

59 Proclus, *Theologia Platonica*, 1, p. 46, l. 3–9: εἰ τοίνυν οἱ λόγοι τῶν πραγμάτων εἰκόνα φέρουσιν ὧν εἰσὶν ἐξηγηταὶ καὶ ὡς ἔχουσιν αἱ [ἀπὸ] τῶν ἀποδείξεων ἀνελίξεις, οὕτως ἀνάγκη καὶ τὰ δεικνύμενα τάξεως ἔχειν, ἀναγκαῖον οἶμαι τὰ μὲν δι᾽ἁπλουστάτων ἀρχόμενα πάντως ἀρχοειδέστατα καὶ τῷ ἑνὶ συνενωμένα τετάχθαι, τὰ δὲ ἀεὶ πληθυόμενα καὶ ποικίλων ἀποδείξεων ἠρτημένα πορρώτερον προεληλυθέναι τῆς τοῦ ἑνὸς ἀποστάσεως, εἰ θέμις εἰπεῖν. On this locus and the history of axiomatics as a philosophical method, see J. L. Solère, "L'ordre axiomatique comme modèle d'écriture philosophique dans l'Antiquité et au Moyen Age", in *Revue d'Histoire des Sciences* 56/2 (2003), pp. 323–345. Concerning Proclus and axiomatics see Opsomer, "Organiser la philosophie selon ses éléments: Structures argumentatives dans les *Éléments de Théologie*", in G. Aubry, L. Brisson, P. Hoffmann, L. Lavaud (eds), *Relire les Éléments de théologie de Proclus: Réceptions, interprétations antiques et modernes* (Paris: Hermann, 2021), pp. 133–176; Opsomer, "Proclus' *Elements of Physics* and the axiomatization of kinematics", in F. Baghdassarian, I. Papachristou, S. Toulouse (eds), *Relectures néoplatoniciennes de la théologie d'Aristote* (Sankt Augustin: Academia, 2020), pp. 83–102; Martijn, "Proclus' geometrical method"; S. A. Kiosoglou, "Proclus' Theory of Seemingly "Useless" Theorems", in *Méthexis* 34/1 (2022), pp. 148–165; Nikulin, "Physica more geometrico demonstrata: Natural philosophy in Proclus and Aristotle", in *Proceedings of the Boston Area Colloquium in Ancient Philosophy* 17 (2003), pp. 183–209. For other passages in which the geometric method is proposed as the canonical methodology for contemplative sciences, see *In Platonis Timaeum commentaria*, ed. E. Diehl, Vol. 1–2 (Lepzig: Teubner, 1903–1904), 1, p. 272, l. 10–17. On Proclus' account of geometrical method in his *In Timaeum*, see Martijn, *Proclus on Nature: Philosophy of Nature and Its Methods in Proclus' Commentary on Plato's Timaeus* (Leiden Boston: Brill, 2010), chap. 3.

60 Opsomer, "Proclus' *Elements of Theology* and Platonic Dialectic", in D. Calma (ed.), *Reading Proclus and the Book of Causes*, Vol. 3, pp. 17–36. See especially part 3, where the particular, un-hypothetical nature of Platonic dialectics and the absence of definitions, postulates, and axioms in the *Elementatio theologica* are discussed.

to contribute to the theoretical aspiration of an analytic-synthetic picture of the world through the three contemplative sciences.

2.1.3 *Proclus' Ontology of Mathematics*

As has been highlighted in scholarship, Proclus' ontology of mathematics, as presented in his *In Euclidem*, is one of the theoretical and authoritative pillars of *De certitudine*.[61]

It is widely acknowledged that Proclus' views on the nature of mathematics are an elaboration of Syrianus' philosophy, partly preserved through the latter's commentary *In Metaphysica*.[62] The positions of Proclus regarding the ontological status of mathematical realities relevant to our inquiry are (i) in-betweenness, (ii) innatism, (iii) phantastic projectionism, (iv) insubstantiality. We will explore them in the following sections.

2.1.3.1 In-betweenness

The first trait of mathematical entities according to Proclus' philosophy is that they stand, ontologically and epistemologically, as a bridge between the sensible and the noetic strata of reality. Two classical Neoplatonic sources used to justify this view are the Divided Line in *Respublica* (VI, 509c–511e), where mathematical items are placed between sensible objects that are subject to opinion and the truly knowable Forms, and the creation of the World Soul based on harmonic proportions from the model of the Forms in the *Timaeus* (34c–37c).[63] Proclus argues for the in-betweenness of mathematics by

---

61 De Pace, *Le matematiche e il mondo*, chap. 1; Higashi, *Penser les mathématiques*, pp. 135–150; Claessens, "Clavius, Proclus, and the Limits of Interpretation"; Claessens, "Sixteenth- and Seventeenth-Century Views on Mathematical Imagination"; Claessens, "Proclus in the Renaissance"; Felgner, *Philosophy of Mathematics in Antiquity and in Modern Times*, chap. 7.

62 Mueller, "Aristotle's Doctrine of Abstraction".

63 See, for instance, Syrianus' argumentation of mathematical in-betweeness in Syrianus, *On Aristotle Metaphysics 13–14*, p. 82, l. 20–30: "These discursive forms Plato clearly says in the *Timaeus* that the Demiurge implants in souls, structuring them by means of geometric and arithmetic and harmonic proportions; while in the division of the Line in the *Republic* he declares that they are 'images' of the intelligibles (and for this reason he has not disdained to call them on occasion 'intelligible'), but that they preside over sense-objects as models; and in the *Phaedo* he says that they are causes for us of recollection […]". Original Greek in Syrianus, *In Metaphysica commentaria*, ed. W. Kroll (Berlin: Reimer, 1902): ταῦτα δὴ τὰ διανοητὰ εἶδη ἐν μὲν Τιμαίῳ σαφῶς ὁ Πλάτων ἐντιθέναι φησὶ τὸν δεμιουργὸν ταῖς ψυχαῖς δι' ἀναλογιῶν γεωμτρικῶν τε καὶ ἀριθμητικῶν καὶ ἁρμονικῶν αὐτὰς διακοσμοῦντα. ἐν δὲ τῇ κατατομῇ τῆς ἐν Πολιτεῖᾳ γραμμῆς εἰκόνας εἶναι τῶν νοητῶν (καὶ διὰ τοῦτό ποτε νοητὰ καλεῖν οὐκ ἀπεξίωσε) προστάναι δὲ παραδειγματικῶς τῶν αἰσθητῶν. ἐν δὲ Φαίδωνι τῆς ἀναμνήσεως ἡμῖν αἴτια γίγνεσθαι.

appealing to the hierarchical distribution of the cosmos, concluding that the middle ground between the Intellect and the visible realm should be identified with both soul and mathematics. Therefore, the very substance of the soul and mathematical Forms meet here in one; they occupy the same ontological stratum and are intimately related to one another (see below). The mathematical soul, or ensouled mathematics, fills the gap between the indivisible, unchangeable, and un-extended Forms of the Intellect and the divisible, visible, extended bodily realities:

> Mathematical being necessarily belongs neither amongst the first nor amongst the last and least simple of the kinds of being but occupies the middle ground between the partless realities – simple, incomposite, and indivisible – and divisible things characterised by every variety of composition and differentiation.[64]

64 Proclus, *In Euclidem*, p. 4, l. 1–7: τὴν μαθηματικὴν οὐσίαν οὔτε τῶν πρωτίστων ἐν τοῖς οὖσι γενῶν οὔτε τῶν ἐσχάτων εἶναι καὶ παρὰ τὴν ἁπλῆν διῃρημένων, ἀναγκαῖον, ἀλλὰ τὴν μέσην χώραν ἀπειληφέναι τῶν τε ἀμερίστων καὶ ἁπλῶν καὶ ἀσυνθέτων καὶ ἀδιαιρέτων ὑποστάσεων καὶ τῶν μεριστῶν καὶ ἐν συνθέσεσιν παντοίαις καὶ ποικίλαις διαιρέσεσιν ἀφωρισμένων. See also Proclus, *The Elements of Theology: A Revised Text with Translation, Introduction, and Commentary*, ed. E. R. Dodds (Oxford: Clarendon Press, 1992), prop. 190, p. 167: πᾶσα ψυχὴ μέση τῶν ἀμερίστων ἐστὶ καὶ τῶν περὶ τοῖς σώμασι μεριστῶν. In Dodds' translation: "Every soul is intermediate between the indivisible principles and those which are divided in association with bodies". Proclus also argues mathematical in-betweenness while outlining the "monad of the soul" in *In Platonis Parmenidem commentaria*, ed. C. Steel (Oxford: Oxford University Press, 2007–2009), III, p. 827, l. 11–18: καὶ γὰρ οὐδὲ ἦν δυνατὸν ἐκ τῶν ἡνωμένων εὐθὺς ὑποστῆναι τὰ πάντη διῃρημένα καὶ εκ τῶν κοινοτάτων τὰ μερικώτατα, ἀλλ'ἐν ταῖς μεταξὺ φύσεσι κατὰ τὸν λόγον τῆς ὑφέσεως ἀναγκαῖον ἦν γενέσθαι καὶ διαίρεσιν. καὶ σχήματος οὖν αἰτίαν θετέον πάντων σχημάτων οἰστικήν, καὶ ἀριθμῶν μονάδα μίαν γεννητικὴν πάντων ἀριθμῶν· ὅπου γὰρ καὶ ἡ ἐνταῦθα μονὰς ἀρτία δείκνυται καὶ περιττὴ καὶ πάντα ἔχουσα τὰ εἴδη καθ' ἕνωσιν, τί χρὴ φάναι περὶ τῆς μονάδος ἐκείνης; ἆρ'οὐχ ὡς πάντων ἑνοειδῶς ἐστιν αἰτία, καὶ ὡς τὸ ἀπειροδύναμον αὐτῆς γεννᾷ καὶ ἐν ἡμῖν τὸν ἄπειρον ἀριθμόν; καὶ πῶς οὐκ ἀναγκαῖον, εἴπερ καὶ ἡ ἐνταῦθα μονὰς ἐκείνης προῆλθεν εἰκών; English translation (slightly modified) by Morrow and Dillon in Proclus, *Commentary on Plato's Parmenides*, eds G. R. Morrow, J. M. Dillon (Princeton: Princeton University Press, 1992): "Since it is not possible for beings that are divisible through and through to be constituted directly from the unified realities, nor for the most particular beings to come immediately from the most universal, their differentiation must occur in the intermediary natures as they descend into plurality. We must therefore posit a cause that confers figure upon all figures, and a single generative monad of all the numbers. When it has been proved that the monad here is both odd and even and, in its unity, contains all species of number, can we not say of that intelligible monad that it is the unitary cause of all things and that it is its infinite power that generates infinite number in us also? Are we not compelled to say this, if this monad here has come forth from that monad as its image?" On this section see A. Kobec, "Proclus on the Forms of Attributes: The Case of Figures and Numbers (*in Prm.* 3.826.19–827.18)", *Mnemosyne* 70/5 (2017), pp. 775–807.

Given its specific ontological location, mathematics would also possess an epistemic place of its own as the cognitive power between perception and noetic apprehension, i.e., the dianoetic faculty.

2.1.3.2 Innatism

The postulate of in-betweenness leads to the second trait of Proclus' philosophy of mathematics, namely innatism. As said, Proclus refers to the principles of mathematical knowledge as μαθηματικοὶ λόγοι and locates them as part of the intimate structure of the soul.[65] Putting Plato's doctrine of recollection together with the Neoplatonic conception of reality as a cascade in which every ontological level stands as a blurred reflection of its predecessor, Proclus declares that every soul bears within itself the blueprints of the intellectual structure reflecting the order of superior realities of which it derives its existence as an image. This means that a unitary, condensed mapping of the intellectual realm is to be found within every soul, waiting there to be discovered. Proclus describes this ensouled source of the mathematical Forms as the "generative monad" (μονὰς γεννητική): it grounds the entire reality of the soul and contains all the ensouled Forms or reasons of mathematics as an image of the intellectual monad, which in itself contains all the intellectual Forms.[66] This makes the process of learning mathematics a sort of introspective acknowledgement of what is already present as a disclosure of the soul's substance into the soul's activity.[67]

65 For some places in which Proclus uses the term λόγος (either in singular or plural) in referring to ensouled mathematical reasons see *In Euclidem*, p. 13, l. 8 and l. 23; p. 14, l. 1; p. 17, l. 15, l. 11, l. 17, and l. 23; p. 18, l. 1; p. 22, l. 5; p. 24, l. 2; p. 25, l. 2; p. 32, l. 8; p. 33, l. 13 and l. 17; p. 43, l. 16; p. 45, l. 22–23; p. 55, l. 12 and l. 16–17; p. 56, l. 7; p. 57, l. 12–13; p. 55, l. 27; p. 66, l. 12; p. 69, l. 16; p. 89, l. 11; p. 92, l. 3; p. 121, l. 6; p. 141, l. 6, l. 16–17; p. 145, l. 1; p. 167, l. 19 and l. 26; p. 186, l. 4–5; p. 201, l. 3.

66 Proclus, *In Parmenidem*, III, p. 828, l. 24–30: τῶν τεχνῶν ὅσαι μὲν ἀνάγουσιν τὴν ψυχὴν καὶ ὁμοιοῦσιν πρὸς νοῦν, τούτων εἴδη θησόμεθα, πρὸς ἃ καὶ οἰκειοῦσιν ἡμᾶς. ὁμοίως γάρ ἐστι σχῆμα καὶ ἡ νόησις τοῦ σχήματος, καὶ ἀριθμὸς καὶ ἡ τοῦ ἀριθμοῦ νόησις. καὶ ἀριθμητικῆς οὖν καὶ μουσικῆς καὶ γεωμετρίας καὶ ἀστρονομίας εἴδη θήσομεν, εἰ μὴ τὰς δεδημευμένας λέγοι τις τούτων ἑκάστην, ἀλλὰ τὰς νοερὰς καὶ ποιητικὰς τῶν θείων εἰδῶν. English translation (slightly modified): "Of all the arts that elevate the soul and make it like Intellect we shall posit Forms to which they make us like. For there is likeness between figure and the intellection of figure, and between number and the conception of number. So, we will postulate forms of arithmetic music and geometry and astronomy, if we mean, in each case, not the vulgar kinds but, the intellectual arts that are productive of divine Forms". See also *In Euclidem*, 92, l. 26–93, l. 2; and *In Parmenidem*, III, p. 827, l. 11–18 above.

67 Helmig, *Forms and Concepts*, chap. 6: Syrianus' and Proclus' Attitude towards Aristotle; Steel, "Breathing Thought"; Helmig, "Proclus et les arguments pour et contre l'hypothèse des ideés", in *Revue de Philosophie Ancienne* 2 (1984), pp. 3–27.

According to Proclus, two moments can be isolated in the soul: its essence (οὐσία) and its activity (ἐνέργεια).[68] This tenet of Proclus' psychology and its connexion to his epistemology is clearly expressed in propositions 191, 194, and 195 of the *Elementatio theologica*:

> Prop. 191: Every participated soul has an eternal substance (οὐσίαν αἰώνιον) but a temporal activity (ἐνέργειαν κατὰ χρόνον).
> Prop. 194: Every soul possesses all Forms which Intellect priorly possesses.
> Prop. 195: Every soul is all things, the sensible things after the manner of an exemplar (παραδειγματικῶς) and the intelligible things after the manner of an image (εἰκονικῶς).[69]

From proposition 191 we learn that every participating soul, as a hinge between divisible and indivisible realities, has a double existence: whereas its essence remains in an unchangeable state of eternity, its activity is temporally unfolded.

Proposition 194 declares that the soul appropriately (see below) contains all the noetic Forms. In accounting for this, Proclus explains that the Intellect transfers to the soul all its "substantial reasons" – introducing the syntagm οὐσιώδεις λόγους, which also includes mathematical reasons.

Proposition 195 establishes that the soul contains every noetic reality in a way of existence proper to its ontological stratum, namely "in the manner of an image" (εἰκονικῶς). Because mathematical reasons exist in the substance of the soul as an image or icon of the noetic reality, they lie beyond discursivity, definition, and demonstration: these all are logical operations which belong to the psychic activity, displayed when ensouled images of the Intellect are transplanted from the essence to the temporal activity of the soul.[70] Therefore, mathematical reasons must not be understood as mathematics in the common

68 This distinction comes from Aristotle, *De anima*, ed. I. Bekker, vol. 1 (Berlin: Accademia regia Borusica, 1831), II, 5. On Proclus' theory of the soul's activity and its epistemic implications see section III in Steel, "Breathing Thought".

69 Proclus, *Elements of Theology*, ed. E. R. Dodds (Oxford: Oxford University Press, 1963). English translation slightly altered. Original Greek: Prop. 191, pp. 166–167: πᾶσα ψυχὴ μεθεκτὴ τὴν μὲν οὐσίαν αἰώνιον ἔχει, τὴν δὲ ἐνέργειαν κατὰ χρόνον; Prop. 194, pp. 168–169: πᾶσα ψυχὴ πάντα ἔχει τὰ εἴδη, ἃ ὁ νοῦς πρώτως ἔχει; Prop. 195, pp. 171–172: πᾶσα ψυχὴ πάντα ἐστὶ τὰ πράγματα, παραδειγματικῶς μὲν τὰ αἰσθητά, εἰκονικῶς δὲ τὰ νοητά. See also Proclus, *In Parmenidem*, IV, p. 894, l. 318. For an account of Proclus' general epistemology see also Helmig, "Proclus on Epistemology, language, and Logic", in P. D'Hoine, M. Martijn (eds), *All From One*, pp. 183–296; Chlup, *Proclus, An Introduction*, chapter 4.

70 *In Euclidem*, p. 16, l. 10–16; p. 45, l. 22; *In Alcibiadem*, ed. V. Cousin (Paris: Minerva, 1864), p. 189, l. 7; p. 192, l. 2–24; *In Parmenidem*, p. 897, l. 37–39.

sense of the word (numbers, problems, theorems, figures). Rather, mathematical reasons are the psychic pre-conditions for the science of mathematics to be practised, the patterns whose manifestation as quantitative beings produces what we ordinarily call "mathematics".[71]

This kind of mathematical innatism is presented by Proclus in explicit opposition to Aristotle's abstractionism through a concatenation of arguments in his first prologue.[72] Proclus there intends to show that mathematical objects cannot be *a posteriori* generation of the soul by a process of abstraction (ἀφαίρεσις). The core of his argument consists of a list of four traits of mathematical objects that are unattainable via abstraction. The four non-abstractable mathematical traits are (a) accuracy and certainty (ἀκρίβεια),[73] (b) spatial or relational features impossible to obtain in sensible objects, such as indivisibility (point), non-three-dimensionality (lines and planes) and absolute equality,[74] (c) ontological and epistemic immutability (if theorem τ is true, then τ is always true),[75] and (d) the axiological necessity of mathematics' ontological priority over particulars.[76] Therefore, Proclus concludes that mathematicians do not gain access to their object of study through abstraction or collection of sensible things. Thus, again in confrontation with Aristotle, Proclus states that the soul is not a blank slate,[77] but a self-writing slate which receives its contents by the actuality of the Intellect, containing all conceivable mathematical knowledge or the plenitude[78] of mathematics (πλήρωμα) in a condensed form:

71 The origin of the division between quantitative entities as those handled by mathematicians and a superior eidetic, yet not quantitative mathematical being which is the source of mathematical beings was already present in Plato's distinction between ideal and mathematical numbers (cf. *Metaphysica*, XIII, 6), treated by Plotinus in *Enneades*, V, 5, 4–5.

72 Proclus, *In Euclidem*, pp. 12, l. 2–18, l. 4.

73 Proclus, *In Euclidem*, p. 12, l. 13–14: [...] πόθεν ἡ ἀκρίβεια καὶ τὸ ἀνέλεγκτον ὑπάρχει τοῖς λόγοις.

74 Proclus, *In Euclidem*, p. 12, l. 19–20: ποῦ γὰρ ἐν τοῖς αἰσθητοῖς τὸ ἀμερὲς ἢ τὸ ἀπλατές ἢ τὸ ἀβαθές [...].

75 Proclus, *In Euclidem*, p. 12, l. 21–22: ποῦ δὲ οἱ ἑστῶτες ἀεὶ λόγοι [...].

76 Proclus, *In Euclidem*, p. 12, l. 27–13, l. 1: [...] πῶς οὖν τοῖς ἀκινήτοις λόγοις ἐκ τῶν κινουμένων καὶ ἄλλοτε ἄλλως ἐχόντων αὐτὴν τὴν μόνιμον οὐσίαν δώσομεν.

77 Aristotle, *De anima*, III, 4 in I. Bekker (ed.), *Aristotelis opera*, Vol. 1–5 (Berlin: Accademia regia Borusica, 1831–1870).

78 Proclus consistently refers to the soul by using the adjective πλήρης (filled) and many of its derivative verbal forms, ex. gr., *In Euclidem*, p. 16, l. 6: πλήρωμα τῶν εἰδῶν αὐτήν [...]; p. 16, l. 25: [...] πλήρωμα τῶν πάντων ἡ ψυχή; p. 17, l. 23–24: [...] οἱ λόγοι συμπληροῦντες τὰς ψυχάς [...]; *In Timaeum*, II, p. 200, l. 21: λόγων πλήρωμα.

> The soul therefore was never a writing tablet bare of inscriptions; she is a tablet that has always been inscribed and is always writing itself and being written on by the intellect.[79]

This last statement implies that being part of the eternal and impassible essence of the soul, mathematical reasons are always actualities. Thus, one more time in opposition to Aristotle,[80] Proclus rejects the existence of any potential intellect in which un-actualised knowabilities lie: mathematical reasons are the actual "tissue" out of which the immortal soul consists.

2.1.3.3 Phantastic Projectionism

How, then, do these ensouled mathematical principles, understood discursively, become amenable to definition, systematic exposition, and demonstration? In what way do they transform into the subjects addressed by the mathematical sciences?

The answer to such questions is to be found in perhaps the most famous trait of Proclus' philosophy of mathematics, namely the theory that Mueller labelled as "projectionism".[81] According to Proclus, ensouled mathematical reasons are spatially displayed with the aid of phantasy (φαντασία) and so rendered fit to be worked out by the discursive mind of human beings.[82] Following the preceding Neoplatonic tradition at least since Iamblichus,[83] Proclus adopts

---

79 Proclus, *In Euclidem*, p. 16, l. 8–10: καὶ οὐκ ἄρα ἦν ἡ ψυχὴ γραμματεῖον καὶ τῶν λόγων κενόν, ἀλλὰ γεγραμμένον ἀεὶ καὶ γράφον ἑαυτὸ καὶ ὑπὸ νοῦ γραφόμενον. Proclus also denies that the soul is as a blank-slate in *In Alcibiadem*, pp. 280, l. 19–281, l. 6; Proclus, *In Platonis Cratylum commentaria*, ed. G. Pasquali (Leipzig: Teubner, 1908), p. 61, l. 26–27; *In Parmenidem*, p. 892, l. 19–24.

80 *De anima*, III, 5, 430a, l. 24.

81 Mueller, "Aristotle's Doctrine of Abstraction".

82 A detailed exposition of the process of phantastic projectionism in geometry is provided by Proclus in the second prologue (amply used by Piccolomini in chapter VII of *De certitudine*). See Proclus, *In Euclidem*, pp. 49, l. 7–56, l. 22. I will provide an analysis of the concept of *phantasia* in Proclus, Piccolomini, and its sources in section 2.2. For a general history of the concept of phantasy in Neoplatonic thought see G. Watson, *Phantasia in classical thought* (Galway: Galway University Press, 1988), chap. 5; and A. Sheppard, "Phantasia and Mathematical Projection in Iamblichus", in H. J. Blumenthal, J. F. Finamore (eds), *Iamblichus: The Philosopher* (Iowa City: University of Iowa, 1997), pp. 113–120.

83 In *De communi*, 11, Iamblichus describes the process of recollection by using the verb προβάλλω ([…] ἀφ'ἑαυτοῦ προβάλλει; […] προβάλλει τοὺς οἰκείους αὐτῶν λόγους). In chapter 22, Iamblichus introduces φαντασία as a power assisting mathematical practice ([…] καὶ φαντασίᾳ χρῆται μᾶλλον […]) – nevertheless, he does so to criticise overly empirical mathematics.

the technical term προβολή to name the process of migration of ensouled reasons from the substance of the soul into the soul's temporal activity.

Helmig has produced the most detailed account so far of the terms προβολή and προβάλλω as found in Proclus' works.[84] I succinctly summarise here some of Helmig's findings as they relate to the general epistemology of Proclus and, particularly, to his epistemology of mathematics.

Firstly, (I) Helmig points out that this term possesses three meanings in Proclus' writings, namely (a) ontologically, as the effect produced by a higher reality or faculty in a lower one, as a procession from the superior to the inferior, (b) in a vital sense, as the endorsement of a type of life (Proclus uses it in his interpretation of the myth of Er), and (c), epistemically, as the unfolding of innate reasons. This includes (c.i) the particular case of the phantastic disclosure of mathematical reasons.

Secondly, (II) Helmig also points out that, whereas the term ἐπιβολή is adopted by Proclus for every cognitive power, the term προβολή is only attributed to opinion and discursive thinking.

Lastly, (III) because of this, Helmig proposes that προβολή and προβάλλω are more fittingly translated in their general meaning as "putting forward" (I also find "bringing forth" appropriate). Helmig suggests using the translation of προβολή as "projection" solely for (c.i), i.e., for the process of obtaining phantastic mathematical items. In my wording, I will follow Helmig's suggestion and use the more general "putting forward" for other uses of προβολή/προβάλλω when treating Proclus' theory, reserving "to project", "projection", and "projectionism" only for Proclus' theory of mathematics. It is to be noted that (c) and (c.i) are also cases of (a): mathematical projection is a case of the general "putting forward" of the innate reasons; on its behalf the epistemic "putting forward" of the innate reasons is a case of the ontological "putting forward" (a), i.e., of a higher reality being appropriately enacted through an inferior power (such as δόξα or φαντασία).

Proclus offers the following description of the process of projecting mathematical reasoning through the faculty of phantasy:

> Thus, thinking in geometry occurs with the aid of phantasy (νόησις αὐτῆς μετὰ φαντασίας). Its syntheses and divisions of the figures are phantastic; and its knowing, though on the way to understandable being, still does not reach it, since the understanding is looking at things outside itself. At the same time, the understanding sees them by virtue of what it has

84 Helmig, *Forms and Concepts*, pp. 291–299.

> within; and though employing projections of its reasons (προβολαῖς μὲν χρωμένης λόγων), it is moved by itself to make them external.[85]

The projectionist theory serves to satisfy and more deeply explain three pivotal Platonic views: (i) it is a highly sophisticated version of Platonic recollection, which explains why knowledge (with emphasis on mathematics) is acquired through remembrance and self-contemplation. Secondly, (ii) it explains the long-standing Platonic tendency to conceive mathematics as an intellectual training preparing the student for the exercise of dialectics. Indeed, the activity of inner discovery, projection, and manipulation of phantastic mathematics prepares the mind for the even more abstract labours of noetic speculation.[86] Thirdly, (iii) it preserves the objectivity of mathematics given that every soul pre-contains the same inborn reasons. Because of that, although the phantastic representations and the ways of demonstration may vary in each individual, they all reach the same conclusions. Secondly, since mathematical reasons do not belong to the material realm but to the superior, extra-temporal psychic stratum, mathematics is, accordingly, free from change, contingency, and inaccuracy. Hence, the timeless truth of mathematical conclusions.

2.1.3.4 Insubstantiality

But what are the entities we perceive in our phantasy as geometrical or, more generally, as projections of substantial mathematical reasons? What is the ontological status of these phantastic images we usually call "mathematical

---

85 Proclus, *In Euclidem*, p. 55, l. 6–13: ὅθεν καὶ ἡ νόησις αὐτῆς μετὰ φαντασίας αἵ τε συνθέσεις τῶν σχημάτων καὶ αἱ διαιρέσεις φανтασταὶ καὶ ἡ γνῶσις ὁδὸς μὲν εἰς τὴν διανοητικήν ἐστιν οὐσίαν, οὔπω δὲ εἰς ἐκείνην ἀναδεδράμηκε, τῆς διανοίας εἰς τὰ ἔξω βλεπούσης καὶ ταῦτα κατὰ τὰ ἔνδοθεν θεωρούσης καὶ προβολαῖς μὲν χρωμένης λόγων ἀλλ'ἀφ' ἑαυτῆς εἰς τὸ ἔξω κινουμένης. Yet again in p. 78, l. 22–25: τὴν γὰρ τῆς διανοίας ἡμῶν κίνησιν καὶ τὴν προβολὴν τῶν ἐν αὐτῇ λόγων γένεσιν τῶν ἐν φαντασίᾳ σχημάτων εἶναι φάμεν καὶ τῶν περὶ αὐτὰ παθημάτων. In Morrow's translation (slightly modified): "[…] for the movement of our thought in projecting its own ideas is a production, we have said, of the figures in our phantasy and of their properties". See also pp. 49, l. 27–50, l. 2, where Proclus depicts how indivisible ensouled reasons are made extended, divisible, and mobile by phantasy.

86 Plato, *Respublica*, ed. J. Burnet, vol. 2 (Oxford: Clarendon, 1992), VII, 524d, l. 9–531c, l. 7; 531d, l. 7–535a, l. 1; Plotinus, *Enneades*, P. Henry, H. R. Schwyzer (eds), *Plotini opera*, vol. 1–2 (Oxford: Clarendon, 1989–1990), I, 3, 3, l. 6–7; Iamblichus, *De communi mathematica scientia*, ed. N. Festa (Leipzig, Teubner, 1891), p. 55, l. 8–22; p. 59, l. 13–18. On the use of mathematics as a means of representing superior realities in the Neoplatonic tradition, see S. Rappe, *Reading Neoplatonism: Non-discursive Thinking in the Texts of Plotinus, Proclus, and Damascius* (Cambridge: Cambridge University Press, 2007), chap. 6. See also Iamblichus, *On the General Science of Mathematics*, eds. J. M. Dillon, J. O. Urmson, S. Gertz (London: Bloomsbury Academic, 2020).

objects" and about which we produce demonstrations and theorems? This question leads us to the fourth trait of mathematical items in Proclus' philosophy, namely their alleged insubstantiality.

The acceptance of this trait can be considered part of the Neoplatonic attempt to (at least partially) harmonise the teachings of Aristotle with those of the divine Plato. One of the most relevant results of Aristotle's inquiry concerning the nature of mathematical objects is that they are not substances.[87] Both Syrianus and Proclus would partly accept Aristotle's claims: mathematical objects – as we perceive and deal with them in mathematical practice – are not substances. But this does not deny (nor is it incompatible with) the substantial status of the innate reasons lying in the atemporal activity of the soul. This view is clearly stated by Syrianus' commentary *In Metaphysica* III, 996a12–15:

> For indeed one might see in the perceptible works of nature both figure and number and natural surface and its limits. Furthermore, these are constituted also in our phantasy and in opinion, either by being taken, as Aristotle thinks, by abstraction (ἐξ ἀφαιρέσεως) from perceptibles, or by being produced in us from the substantial forms of the soul (οὐσιωδῶν τῆς ψυχῆς εἰδῶν). Now the objects of phantasy and opinion share in being but are not substances (οὐσίαι δὲ οὔκ εἰσίν).[88]

Here Syrianus establishes a clear ontological distinction between mathematical objects as built upon our phantasy and the mathematical reasons inborn to the soul. Importantly, he affirms that the phantastic entities upon which

---

87 Aristotle, *Metaphysica*, ed. I. Bekker, vol. 2 (Berlin: Accademia regia Borusica, 1831), 1077a, l. 32. For a detailed discussion on Aristotle's views on mathematical objects and abstraction, see Cleary, *Aristotle and Mathematics: Aporetic Method in Cosmology and Metaphysics* (Leiden New York Köln: Brill, 1995), especially chap. 3, 5, and 7. On Syrianus's approach to interpretation and his stance towards Aristotle see P. D'Hoine, "Syrianus and Proclus on Aristotle", in A. Falcon (ed.), *Brill's Companion to the Reception of Aristotle in Antiquity* (Leiden: Brill, 2016), pp. 374–393; Helmig, "The Truth Can Never Be Refuted: Syrianus' View(s) on Aristotle Reconsidered", in A. Longo, L. Corti, N. D'Andrès, D. Del Forno, E. Maffi, A. Schmidhauser (eds), *Syrianus et la métaphysique de l'antiquité tardive. Actes du colloque international, université de Genève, 29 septembre–1er octobre 2006*, Elenchos 51 (Napoli: Bibliopolis, 2009), pp. 347–380.

88 O'Meara's and Dillon's translation (slightly modified) in *Syrianus: On Aristotle Metaphysics 3–4*, p. 33. Original at Syrianus, *In Metaphysica commentaria*, W. Kroll (Berlin: Reimer, 1902), Seventh problem, p. 12, l. 29–35: καὶ γὰρ ἐν τοῖς αἰσθητοῖς τῆς φύσεως ἔργοις ἴδοι ἄν τις καὶ σχῆμα καὶ ἀριθμὸν καὶ ἐπιφάνειαν φυσικὴν καὶ ταύτης πέρατα, ἔτι δὲ καὶ ἐν τῇ φαντασίᾳ ἡμῶν καὶ ἐν τῇ δόξῃ ταῦτα συνίσταται, εἴτε ἐξ ἀφαιρέσεως τῶν αἰσθητῶν, ὡς αὐτῷ ἀρέσκει, ληφθέντα, εἴτε καὶ τελειωθέντα παρὰ ἡμῖν ἐκ τῶν οὐσιωδῶν τῆς ψυχῆς εἰδῶν. ταῦτα οὖν τὰ φανταστὰ καὶ τὰ δοξαστὰ τοῦ μὲν εἶναι μετέχει, οὐσίαι δὲ οὔκ εἰσιν.

the mathematician works in obtaining mathematical knowledge are not substances. Nevertheless, Syrianus has already pointed out that the ensouled reasons whence mathematical objects come as projections are to be held as substantial. Thus, the items generated by mathematical phantasy are depicted as shadows, phantastic manifestations whose reality is dependent on the really substantial ensouled reasons.[89] Although mathematical objects participate in being to some degree, they are not *sensu stricto* substances.

This doctrine was inherited by Proclus. When he describes in the first prologue of the commentary the nature of the inborn mathematical reasons, he adopts Syrianus' terminology:

> The mathematical reasons that make up the complement of souls are therefore substantial (οὐσιώδεις) and self-moving. By emitting and unfolding them the understanding brings into being all the variety of the mathematical sciences.[90]

As a second token, we may refer to the passage where the process of examining phantastic geometrical figures leads back to the source-reasons of which they are expressions. There Proclus calls ensouled Forms the "substantial hypostasis" of mathematical reasons, hence pointing out that the substantiality of mathematical items does not reside in themselves as expressed in phantasy but, rather, in the ensouled root from which they are projected:

> It [understanding] investigates their variety, exhibiting the forms of figures shaped from them in phantasy it comprehends within fixed boundaries and refers back to the essential hypostasis (i.e., being) of the reasons (τὴν οὐσιώδη τῶν λόγων ὑπόστασιν).[91]

Proclus never refers to "phantasisable" mathematical items (φανταστά) as "substantial", but he describes them as εἴδωλα, i.e., "images" or "simulacra".[92]

---

89 As put by Steel, Iamblicus is the first author who uses the expression λόγος οὐσιώδης in this sense. See Steel, "Breathing Thought", p. 295, note 9. The same syntagm, but with a different meaning, appears in Plotinus, *Enneades* VI, 3, 3, l. 2. The expression ἀριθμὸς οὐσιώδης also occurs in Plotinus' *Enneades*, V, 5, 4, l. 17; and VI, 6, 16, l. 26.

90 Proclus, *In Euclidem*, p. 17, l. 22–26: οὐσιώδεις ἄρα καὶ αὐτοκίνητοι τῶν μαθημάτων εἰσὶν οἱ λόγοι συμπληροῦντες τὰς ψυχάς, οὓς δὴ καὶ προβάλλουσα ἡ διάνοια καὶ ἐξελίττουσα πᾶσαν τὴν ποικιλίαν ὑφίστησι τῶν μαθηματικῶν ἐπιστημῶν […].

91 Proclus, *In Euclidem*, p. 62, l. 15–18: […] ἀφ'ὧν δὴ καὶ τὰς φανταστὰς διαμορφώσεις σχημάτων ἐν πέρασιν ὡρισμένοις περιλαμβάνει καὶ ἀνάγει πρὸς τὴν οὐσιώδη τῶν λόγων ὑπόστασιν.

92 Some places in which Proclus calls the products of phantasy εἴδωλα are Proclus, *In Euclidem*, p. 11, l. 1–5; p. 13, l. 17; p. 15, l. 1; and p. 141, l. 6.

The semantics of this Greek term confers a sense of unreality and existential dependency: if something is an εἴδωλον, it must be a simulacrum or a representation of *something else*, upon whose reality the simulacrum is dependent in its derivative, fictitious nature.

The derivative reality of phantastically displayed mathematics is particularly clear in Proclus' comments on definition XIV in his commentary *In Euclidem* (the definition of a figure as that which is contained by any boundaries):

> the soul, exercising her capacity to know, projects on phantasy, as on a mirror (εἰς κάτοπτρον), the reasons of the figures; and phantasy, receiving in pictorial (ἐν εἰδώλοις) form these impressions of the reasons within the soul, by their means affords the soul an opportunity to turn inward from the pictures (τῶν εἰδώλων) and attend to itself. It is as if a man looking at himself in a mirror (ἐν κατόπτρῳ) and marvelling at the power of nature and at his own appearance should wish to look upon himself directly and possess such power as would enable him to become at the same time the seer and the object seen. In the same way, when the soul is looking outside itself at phantasy, seeing the figures depicted (ἐσκιαγραφημένα σχήματα) there and being struck by their beauty and orderliness, it [the soul] is admiring its own reasons from which they are derived; and though it adores their beauty, it dismisses it as something reflected (ὡς ἐν εἰδώλοις) and seeks its own beauty. It wants to penetrate within itself to see the circle and the triangle there.[93]

Firstly, Proclus compares phantasy to a mirror (κάτοπτρον) upon whose surface the ensouled principles of geometry are reflected, enabling the soul to perceive itself as something external. Here, the derivative nature of mathematical objects is highlighted: without something to serve as the basis for reflection, upon which the image depends, there would be no reflection at all.

Secondly, Proclus refers to phantastic mathematical objects as εἴδωλα three times in this section and employs the passive perfect participle of the Greek

93 Proclus, *In Euclidem*, p. 141, l. 4–20: […] ὡς ἡ ψυχὴ κατὰ τὸ γνωστικὸν ἐνεργοῦσα προβάλλει περὶ τὴν φαντασίαν ὥσπερ εἰς κάτοπτρον τοὺς τῶν σχημάτων λόγους, ἣ δ'ἐν εἰδώλοις αὐτὰ δεχομένη καὶ ἐμφάσεις ἔχουσα τῶν ἔνδον ὄντων διὰ τούτων τῇ ψυχῇ παρέχεται τὴν εἰς τὸ εἴσω στροφὴν καὶ πρὸς ἑαυτὴν τὴν ἀπὸ τῶν εἰδώλων ἐνέργειαν· οἷον εἴ τις ἑαυτὸν ὁρῶν ἐν κατόπτρῳ καὶ θαυμάσας τὴν τῆς φύσεως δύναμιν καὶ τὴν ἑαυτοῦ μορφὴν ἑαυτὸν ἰδεῖν θελήσειεν καὶ λάβοι δύναμιν τοιαύτην, ὥστε ὅλως ὁρῶν καὶ ὁρατὸν ἀποτελεσθῆναι. καὶ γὰρ ψυχὴ τοῦτον τὸν τρόπον ἔξω ἑαυτῆς εἰς φαντασίαν βλέπουσα καὶ ἐσκιαγραφημένα σχήματα θεωμένη καὶ τὸ κάλλος αὐτῶν ἐκπλαγεῖσα καὶ τὴν τάξιν ἄγαται τοὺς ἑαυτῆς λόγους, ἀφ'ὧν καὶ ταῦτα, καὶ ἀγασθεῖσα τὸ μὲν τούτων κάλλος ὡς ἐν εἰδώλοις φερόμενον ἀφίησι, ζητεῖ δὲ τὸ ἑαυτῆς, καὶ εἰς τὸ εἴσω παρελθεῖν ἐθέλει καὶ τὸν κύκλον ἐκεῖ καὶ τὸ τρίγωνον ἰδεῖν […].

verb σκιαγραφῶ to characterise geometrical figures. This verb denotes a pictorial technique that creates an illusion of depth and distance through variations in size and color. By likening phantastic projections to an optical illusion, Proclus underscores yet again the surrogate nature of phantastic mathematical objects.

Lastly, in indicating the awakening of the soul to its inner journey, Proclus points out that it eventually dismisses (ἀφίησι) the beauty of phantastic figures "as something reflected" or "as in images" and attempts to grasp the nature of mathematical objects within itself (sc., in its inborn substantial reasons).

Therefore, much like Syrianus, Proclus attributed a lesser degree of reality to projected mathematical items. Much like physical realities, mathematical objects can only equivocally be labelled as "substances".

2.1.4 *The Common Mathematical Science*

The third Proclean theme of which Piccolomini would make use is the notion of a common mathematical science. In the first prologue, Proclus introduces the topic of the so-called "total mathematics" or "common mathematics" (ὅλη/κοινὴ μαθηματική).[94] In his history of this idea, Rabouin has shown that it is at first hinted at in Aristotle's *Metaphysica* (VI, 1, 1026a, l. 30–31 and XI, 7, 1064b, l. 13), where the Stagirite seems to ponder the possibility of a sort of maximally general mathematical science which would include as its species all the sciences dealing with quantity.[95]

The second great episode in the history of common mathematics took place amongst the Neoplatonists. Three factors might have led to more profound speculation on this topic: (1) the increasing attention paid to Aristotle's works by Neoplatonist authors ever since Plotinus and Porphyry, (2) the revival of Pythagorean sacred numerology within Platonic philosophy ever since Numenius of Apamea (2nd century AD), and (3) some scattered suggestions in the Platonic corpus that could be interpreted as hinting to a common mathematical science, particularly in Plato's *Respublica* VII (534e) that there is a "summit of mathematics" (θριγκὸς τοῖς μαθήμασιν – sometimes also understood as Platonic dialectics) and in the pseudo-Platonic *Epinomis* (991e), where the author affirms that there is a "link" (δεσμός) reuniting all

94 It is referred to by this name in Proclus, *In Euclidem*, p. 18, l. 10.

95 Rabouin, *Mathesis universalis: L'idée de « mathématique universelle » d'Aristote à Descartes* (Paris: Presses Universitaires de France, 2009), especially chap. I, II, and III. See also L. M. Napolitano Valditara, *Le idee, i numeri, l'ordine: La dottrina della «Mathesis universalis» dall'Accademia antica al Neoplatonismo* (Padua: Bibliopolis, 2010); G. Crapulli, *Mathesis universalis: Genesi di una idea nel XVI secolo* (Rome: Edizioni dell'Ateneo Roma, 1969).

mathematical disciplines. Plotinus elaborated the notion of a superior kind of dialectical (i.e., metaphysical) speculation on "intelligible number" (νοητὸς ἀριθμός – posited at the level of the first Monad) and defended Plato's proposition of the supra-mathematical, ideal numbers of which mathematical number participates.[96]

The two most significant Neoplatonic figures whose works have survived, and who integrated Plato's "summit of mathematics" with Aristotle's "common mathematics", positioning it as a discipline epistemically preceding Plotinus's speculation on eidetic number, are Iamblichus in his *De communi mathematica scientia* and, following him, Proclus in his prologues to *In Euclidem*.[97]

Proclus' first prologue contains two pertinent passages on this topic. When read together, these sections offer an account of Proclus' position:

> It is clear, then, that the principles primary in mathematics are those that preside over all things. Just as we have noted these common principles and seen that they pervade all classes of mathematical objects, so let us enumerate the simple theorems that are common to them all, that is, the theorems generated by the single science that embraces alike all forms of mathematical knowledge; and let us see how they fit into all these sciences and can be observed alike in numbers, magnitudes, and motions.[98]

And:

> It cannot be that we have sciences of particular areas of being and knowledge of them but have no single science of the immaterial objects that stand much closer to intellectual inspection. Knowledge of those objects is by far the prior science, and from it, the several sciences get their common propositions, our knowledge ascending from the more partial to the

96 For the introduction of the "intelligible number" as an object of speculation see Plotinus, *Enneades*, V, 9, 11, l. 13. Regarding Plotinus' philosophy of mathematics in VI, 6 see S. Slaveva-Griffin, *Plotinus on Number* (Oxford: Oxford University Press, 2009), especially chapter 5.

97 For an account of Iamblichus' common mathematics see L. M. Napolitano Valditara, *Le idee, i numeri, l'ordine*, chap. 6.

98 Proclus, *In Euclidem*, p. 7, l. 13–22: ἀλλ'ὅτι μὲν ἀρχαὶ καὶ τῶν μαθημάτων αὗται προεστήκασιν, αἵ καὶ τῶν ὄντων ἁπάντων, φανερόν. ὥσπερ δὲ τὰς κοινὰς ἀρχὰς αὐτῶν τεθεωρήκαμεν καὶ διὰ πάντων διηκούσας τῶν μαθηματικῶν γενῶν, οὕτω δὴ καὶ τὰ κοινὰ αὐτῶν θεωρήματα καὶ ἁπλᾶ καὶ τῆς μιᾶς ἐπιστήμης ἔγγονα τῆς πάσας ὁμοῦ τὰς μαθηματικὰς γνώσεις ἐν ἑνὶ ἐπεχούσης ἀναλογισώμεθα, καὶ ὅπως ἐφαρμόττει πάσαις καὶ δύναται καὶ ἐν ἀριθμοῖς καὶ ἐν μεγέθεσι καὶ ἐν κινήσεσι θεωρεῖσθαι σκοπήσωμεν.

> more general until at last we reach the science of being as being (τὴν τοῦ ὄντος, ᾗ ὄν ἐστιν, ἐπιστήμην).[99]

Firstly, Proclus asserts that there exists a set of common theorems governing every branch of mathematics, including number, magnitude, and mathematical motion. He then justifies the existence of this common science by arguing that it is inconceivable to have unified sciences and knowledge of sensible entities (i.e., physics) while lacking a unified science of intermediary mathematical entities, which, being closer to Intellect, are more proper objects of scientific inquiry. Lastly, Proclus offers his interpretation of the preparatory role of mathematical knowledge for superior metaphysical understanding. He suggests a link between common mathematics, and metaphysics, allowing the learner to ascend from the contemplation of the highest mathematical theorems to the science of being *qua* being.

Although Proclus' passages are inspired by *De communi mathematica scientia*,[100] his account adds something which was not explicitly stated by Iamblichus. While Iamblichus treats common mathematics as a propaedeutic but distinct discipline from the higher science of metaphysics, Proclus seemingly positions ὅλη μαθηματική as a bridge between mathematical sciences and the science of being *qua* being. Proclus seems to believe that the degree of generalisation achieved over quantity and its properties in the practice of common mathematics would lead to a transition from meta-mathematical speculation to discussions regarding purely immaterial principles and methods, such as the analytic-synthetic structure of reality, division, quantity in itself, or equality in itself. That common mathematics is the link between mathematics and metaphysics might have been suggested to Proclus by Iamblichus' elenchus of the objects of study of common mathematics. This index includes methodological and metaphysical concepts such as analysis, synthesis, and unqualified quantity (ποσὸν ἁπλῶς).[101]

---

99 Proclus, *In Euclidem*, p. 9, l. 11–19: οὐ γὰρ δήπου τῶν μὲν μεριστῶν εἰσὶν ἐπιστῆμαι καὶ γνώσεις, τῶν δὲ ἀΰλων καὶ τῆς νοερᾶς θεωρίας ἐγγυτέρω τεταγμένων οὐδὲ μίαν ἔχομεν ἐπιστήμην, ἀλλὰ πολλῷ πρότερον ἡ ἐκείνων γνῶσίς ἐστιν ἐπιστήμη καὶ ἀπ'ἐκείνης αἱ πολλαὶ τοὺς κοινοὺς ὑποδέχονται λόγους καὶ μέχρι τοσούτου γνώσεων ἄνοδος ἀπὸ τῶν μερικωτέρων ἐπὶ τὰς ὁλικωτέρας, ἕως ἂν ἐπ'αὐτὴν ἀναδράμωμεν τὴν τοῦ ὄντος, ᾗ ὄν ἐστιν, ἐπιστήμην.

100 O'Meara, "Mathematics and the Sciences"; G. Bechtle, "Sur la mathématique générale comme σκοπός (sujet) du *De communi mathematica scientia* de Jamblique et sur la composition de ce traité", in A. Lernould (ed.), *Études sur le commentaire de Proclus*, pp. 31–44.

101 For a list of the themes of common mathematics, some of them absent from Proclus' account, according to Iamblichus, see *De communi*, p. 19, l. 2 and forth.

2.1.5 *Mathematical Insubstantiality and Causal Idleness: Priscian's* In De anima

After examining the themes from Proclus' *In Euclidem* present in *De certitudine*, it will be beneficial to explore another significant work within the Neoplatonic tradition that influenced Piccolomini's philosophical views on mathematics. This is the rejection of mathematical substantiality, causality, and scientificity as expressed by the 5th-century philosopher Priscian of Lydia in his commentary on Aristotle's *De anima* – traditionally attributed to his contemporary Simplicius. Priscian's scattered remarks on the philosophy of mathematics are highly relevant in accounting for Piccolomini's philosophy in *De certitudine* given that (a) this work was undoubtedly familiar to the Sienese, as it was a leading philosophical authority within the 16th-century Paduan intellectual milieu; (b) Piccolomini references it as one of his sources on the philosophy of mathematics in *De certitudine* (see 2.3.3); (c) furthermore, I will argue that Piccolomini interpreted the philosophy of mathematics in Proclus' *In Euclidem* through the lens of this extensively discussed and well-known Greek source (see section 3.4.3).

Turning to Priscian's commentary, it is not surprising that it addresses certain topics related to the philosophy of mathematics. Traditional discussions on concepts such as certainty (ἀκρίβεια) and the epistemology and ontology of phantasy often engaged with Aristotle's remarks, notably in *De Anima* (1, 402a, l. 2; 1, 419a, l. 16–21). Although Priscian's text touches on the insubstantiality of phantastic mathematical objects and their epistemic implications in various sections, I will propose four key *loci* relevant for better accounting for Piccolomini's (perhaps initially perplexing) Proclean hermeneutics.

The first passage appears as a passing note when Priscian comments on 429a, l. 15, where Aristotle treats the reception of the intelligibles by analogy to sensible forms. There Priscian imparts:

> Since the soul cognizes even then, but cognizes the accidents and properties of the substance, not individual objects in the way that sense-perception and phantasy (φαντασία) do, but common ones (τὰ κοινά), being active in respect of them in a rational way and considering their mutual consequentiality or contrariety, looking at the antecedent and the consequence, as do all crafts which are concerned with things in matter: the so-called mathematical sciences (αἱ μαθηματικαὶ λεγόμεναι ἐπιστῆμαι) already do this. For these do not investigate substances (οὐδὲ γὰρ αὗται οὐσίας), but shapes and sizes, states, movements, and mutual similarities: they do not look at their substantial reason (because they have no extension and are

> unified), but rather at the extensions and divisions that come from as a departure from themselves and as a result of otherness.[102]

This passage restates the tenet of Proclus' philosophy of mathematics (found in his *In Euclidem*) as just explored above that it is by a process of dianoetic unfolding that ensouled reasons are known. Then, when specifying that this is the case also in mathematics, Priscian points out that the phantastic shapes upon which mathematicians operate are not really substances.

Priscian has already synthesised his view that φαντασία falls short of substantiality and causation in his commentary to 428, l. 16 when he bluntly established that:

> Every cause is a substance, and the cause in the fullest sense is also the first substance. Phantasy is a long way from cause and from all substance.[103]

Next, in Priscian's commentary on 431b, l. 12 three passages are worth mentioning. Here the commentator resumes the discussion on the epistemological status of mathematics by comparing it to the study of natural objects:

> In this respect, the student of nature also differs from the mathematician, as has been said in the second book of the *Physica*. Since the mathematician considers accidental properties, such as shapes, sizes, motions, numbers and harmonious ratios, in themselves, he needs sense-perception for his cognition. The student of nature, however, contemplates both the

102 Translation by Blumenthal (slightly modified), Simplicius, *On Aristotle's On the Soul 3.1–3.5* (London: Bloomsbury Academic, 2014), p. 87. See Simplicius, *In libros Aristotelis De anima commentaria*, ed. M. Hayduck (Reimer: Königlich Preussische Akademie der Wissenschaften zu Berlin, 1882), com in III, 429b, l. 18, p. 233, l. 22–30: ἐπεὶ γινώσκει γε καὶ τότε ἡ ψυχή, ἀλλὰ τὰ συμβεβηκότα καὶ ἐπακολουθοῦντα τῇ οὐσίᾳ, οὐχ ὡς ἡ αἴσθησις καὶ ἡ φαντασία τὰ ἄτομα, ἀλλὰ τὰ κοινά, λογικῶς περὶ αὐτὰ ἐνεργοῦσα καὶ τὸ πρὸς ἄλληλα αὐτῶν ἀκόλουθον ἢ ἀντίθετον, τὸ προηγούμενόν τε καὶ ἑπόμενον θεωροῦσα, ὡς πᾶσαι αἱ περὶ τὰ ἔνυλα τέχναι, ἤδη δὲ καὶ αἱ μαθηματικαὶ λεγόμεναι ἐπιστῆμαι. οὐδὲ γὰρ αὗται οὐσίας, ἀλλὰ σχήματα καὶ μεγέθη καὶ πάθη καὶ κινήσεις καὶ πρὸς ἄλληλα συμφωνίας ἐπισκοποῦσιν, οὐ τοὺς οὐσιώδεις αὐτῶν λόγους (ἀδιάστατοι γὰρ καὶ ἡνωμένοι οὗτοι), ἀλλὰ τὰς ἐκβεβηκότως ἀπ'αὐτῶν καὶ δι'ἑτερότητος προελθούσας διαστάσεις καὶ μερισμούς. For a full account of Priscian's psychology and noetics, see Steel, *The Changing Self*, pp. 7–21 and pp. 132–142. See also F. A. J. de Haas, "Priscian of Lydia and Pseudo-Simplicius on the soul" in L. P. Gerson (ed.), *The Cambridge History of Philosophy in Late Antiquity*, Vol. 2 (Cambridge: Cambridge University Press, 2010), pp. 756–764.

103 Translation by Blumenthal (slightly modified), Simplicius, *On Aristotle's On the Soul 3.1–3.5*, pp. 68–69. See Simplicius, *In De anima*, comm. in III, 428a, l. 16, p. 210, l. 5–9: πᾶσα δὲ αἰτία οὐσία, καὶ ἥ γε κυρίως αἰτία καὶ πρώτη οὐσία. πόρρω δὲ αἰτίας καὶ πάσης οὐσίας ἡ φαντασία.

> shapes and the other properties that are proper attributes of the substances from the perspective of the causes (ταῖς οὐσίαις θεωρεῖ) of both the forms and the composite substances as substances. Therefore, the study of nature too is a philosophy, as it is involved with substance and knows the things caused on the basis of their causes; this is not, however, the case with mathematics (οὐχὶ δὲ ἡ μαθηματική), which, as Plato says, is "like a bridge". For, as he explains, "its starting point is what it does not know, and its middle term and end come from what it does not know", as it falls short of the causes because it does not consider the substances (ἀπολείπεται τῶν αἰτιῶν τὸ μὴ τὰς οὐσίας ἐπισκοπεῖσθαι). For the cause in the proper sense is always a substance or even something higher than substance.[104]

The section introduces the puzzle (and an interpretation of the rather unclear answers to it) posed by Aristotle's *Physica* II, 2 193b, l. 23: What is the difference between the mathematician and the natural philosopher?

Priscian's answer is that much like the natural philosopher, the mathematician requires the perception of natural quantities to obtain his proper object, namely quantities in themselves (although how that is achieved is not explained here). Priscian immediately diminishes the philosophical status of mathematics, sustaining that this discipline does not deal with substances, and, therefore, that it does not investigate or explain through real causes. Further authoritative reasons are adduced to sustain this claim: Plato's portrait of the in-betweenness of mathematics in *Respublica* (VI, 510c–511d). Priscian uses the image of the bridge (γέφυρα) to describe the hinging state between the perceptible and the intelligible attributed to mathematics by Plato (an image absent from Plato's texts but present in the interpretative tradition).[105] Then

---

104 Translation by Steel in 'Simplicius', *On Aristotle On the Soul 3.6–13*, transl. C. Steel (London: Bloomsbury Academic, 2013), p. 77. See Simplicius, *In De anima*, comm. in III, 431b, l. 12, p. 276–277, l. 27–35: ᾗ καὶ διαφέρειν ὁ φυσικὸς τοῦ μαθηματικοῦ ἐν τῷ Β τῆς Φυσικῆς ἀκροάσεως ἐλέγετο. ὁ μὲν γὰρ μαθηματικός, ἅτε καθ'αὑτὰ τὰ συμβεβηκότα, οἷον σχήματα καὶ μεγέθη καὶ κινήσεις καὶ ἀριθμοὺς καὶ λόγους ἐναρμονίους ἐπισκοπούμενος, δείται πρὸς τὴν γνῶσιν αὐτῆς τῆς αἰσθήσεως. ὁ δὲ φυσικὸς ὡς ἐξ αἰτιῶν τῶν τε εἰδῶν καὶ αὐτῶν τῶν συνθέτων οὐσιῶν ὡς οὐσιῶν καὶ τὰ σχήματα καὶ τὰ λοιπὰ τὰ οἰκείως ἐπισυμβαίνοντα ταῖς οὐσίαις θεωρεῖ. διὸ φιλοσοφία τις καὶ ἡ φυσικὴ πραγματεία ὡς περὶ οὐσίαν στρεφομένη καὶ ὡς ἐξ αἰτιῶν τὰ αἰτιατά γινώσκουσα, οὐχὶ δὲ ἡ μαθηματική, ἀλλ' ὡς φησὶν Πλάτων, γεφύρᾳ ἔοικεν. ἀρχὴ γὰρ αὕτῃ, ὥς καὶ αὐτὸς ἀποφαίνεται φιλόσοφος, ἃ μὴ οἶδε, μέσα δὲ καὶ τελευτὴ ἐξ ὧν μὴ οἶδε, διότι ἀπολείπεται τῶν αἰτιῶν τὸ μὴ τὰς οὐσίας ἐπισκοπεῖσθαι· τὸ γὰρ κυρίως αἴτιον πάντως οὐσία ἢ καὶ ταύτης κρεῖττον.

105 As Steel has pointed out, the image of the bridge to describe mathematical in-betweenness appears in Ammonius of Alexandria's *In Isagogen* 13, 5 and in Pseudo-Galenus, *De partibus philosophiae*, 9, 3. See Steel, *'Simplicius': On Aristotle On the Soul 3.6–13*, p. 171, note 230.

Priscian stresses that, as Plato asserts (VII, 533b–c), whereas dialectics is an un-hypothetical discourse, mathematics proceeds by the hypothesis or supposition of its objects, for mathematics does not contain in itself and, thus, it is unable to account for, its own principles. In this passage, Plato also suggests that vis-à-vis dialectics, mathematical disciplines are a sort of illusory discipline, unable to grasp real being and rather performing "as dreaming about being, for its waking vision is impossible to them" ([…] ὡς ὀνειρώττουσι περὶ τὸ ὄν, ὕπαρ δὲ ἀδύνατον αὐταῖς ἰδεῖν […]). Priscian, therefore, reads the hypothetical and dream-like status of mathematics as described by Plato as indicating the insubstantiality and causal idleness of its objects.

In the immediately following lines in 431b, l. 12, Priscian makes another highly relevant comment:

> I now speak of mathematics not in the Pythagorean sense, namely, a science that investigates also the living reason-principles, which are capable of projecting distances and shapes, which either are produced in the natural bodies or in phantasy – for such a form of mathematics is a kind of philosophy – I mean that mathematics which is wholly concerned with the projected things themselves (περὶ αὐτὰ τὰ προβαλλόμενα), such as shapes, sizes and motions, which Aristotle is accustomed to call "from abstraction" (ἐξ ἀφαιρέσεως).[106]

Priscian specifies that by "mathematics" he is referring here to the mathematical objects present in phantasy, i.e., to the represented items on which the mathematician works. To the superior (dialectic-like) type of mathematics labelled as "Pythagorean" and whose objects are the substantial reasons themselves, Priscian opposes the speculation on the quantities projected in phantasy.

The most important piece of information for Piccolomini's hermeneutics of the highly problematic term *abstractio* as it appears in *De certitudine* is the last subordinate sentence: namely, that τὰ προβαλλόμενα or projected mathematical items and the items which Aristotle affirms to be obtained via ἀφαίρεσις are the same thing or, to put it more clearly, that the Neoplatonic name προβολή and Aristotle's ἀφαίρεσις actually refer to the same process.

106 Translation by Steel in 'Simplicius', *On Aristotle On the Soul 3.6–13*, p. 77. See Simplicius, *In De anima*, comm. in III, 431b, l. 12, p. 277, l. 1–6: λέγω δὲ νῦν μαθηματικήν, οὐ τὴν κατὰ τοὺς Πυθαγορείους, τὴν καὶ τοὺς ζωτικοὺς λόγους τοὺς τῶν διαστάσεων καὶ σχημάτων προβλητικούς, εἴτε ἐν τοῖς φυσικοῖς σώμασιν εἴτε ἐν φαντασίαις ἐγγίνονται, ἐπισκοποῦσαν (φιλοσοφία γὰρ ἐκείνη), ἀλλὰ τὴν ὅλην περὶ αὐτὰ τὰ προβαλλόμενα, οἷον σχήματα καὶ μεγέθη καὶ κινήσεις, στρεφομένην, ἅπερ ἐξ ἀφαιρέσεως εἴωθεν ὁ Ἀριστοτέλης καλεῖν.

This statement puts some distance between Priscian's and Proclus' views. As mentioned, Proclus (and Syrianus before him) regarded with great hostility Aristotle's ontology of mathematics.[107] Priscian, however, appears to moderate both positions, seeking reconciliation by integrating Aristotle's mathematical abstraction into the framework of phantastic projectionism. The resulting harmonised theory can be described as follows:[108]

(1) An epistemic agent is suggested by a natural quantity, say, a crystal ball, a quantitative object which is not perfectly instantiated in the item, say the sphere.

(2) The epistemic agent tries to conceive what a perfect sphere would be, i.e., he attempts to think of it as stripped from all of its material traits and flaws except the relevant quantity, i.e., he attempts "to separate" or "to abstract" it.

(3) This works as a stimulus which triggers the unfoldment (i.e., recollection) of the ensouled substantial reason of the sphere, whose visual expression is the phantastic sphere as an independent quantitative entity.

(4) The perfect sphere is instantiated (i.e., projected) onto phantasy as a quantity separated from natural substance. This grants to the mathematical object the objectual precision or ἀκρίβεια that sensible items lack (see below), producing the image of a solid figure in which infinite punctual sides are actually equidistant from a central point. Thus, the sensibly stimulated inborn substantial reason is now dianoetically knowable through its extended phantastic "icon" – which only exists with the highest degree of precision in phantasy.

Put in this way, all the traits of phantastic projectionism are preserved. The process of projecting a mathematical object onto phantasy would be a functional description of the powers of the soul involved in what Aristotle calls "abstraction". On the other hand, the process of mathematical abstraction as variously proposed (and not always *prima facie* coherently so; see 3.5.1) in Aristotle's works is reinterpreted as a case of projection – including the assumption that Aristotle meant that the inborn reasons are the origin of mathematical items and that phantasy is the place where they are displayed. Thus, Aristotle's ἀφαίρεσις is a depiction of projection emphasising the resemblance between the triggering sensible input and the final phantastic outcome. This kinship arises because both the geometrical and the sensible sphere are expressions of the

107 *Metaphysica*, VIII, 1, 1042a, l. 29; *De anima*, III, 7, 431a, l. 8, l. 15–19; 432a, l. 20; 10, 433b, l. 24–25.

108 Simplicius, *In De anima*, pp. 213, l. 31–214, l. 12; p. 224, l. 24–30; p. 233, l. 5–25; p. 277, l. 24–33; p. 278, l. 29–39; pp. 283, l. 37–284. l. 5.

same substantial reason as it is instantiated in two different substrates. This is a reading of Aristotle's expressions ἀφαίρεσις/λόγῳ χωρίζειν to which, as we shall see, Proclus' account of projection in the second prologue (particularly p. 54, l. 14–26) can also be assimilated.

But in Priscian's interpretation, is mathematics anterior or posterior to natural quantity? This is, in fact, the problem which moved both Syrianus and Proclus to oppose Aristotle's mathematical epistemology as a subversion of the procession of beings. What is Priscian's view?

I find that Priscian manages with great subtlety to sustain that mathematics is, in some respect, prior and, in another respect, posterior to perceptible quantity, so that he can affirm that both his Neoplatonic sources and Aristotle speak truly. For, as we just saw, Priscian distinguishes between two kinds of mathematical discourse: one labelled as "Pythagorean" dealing with substantial reality, the second one devoid of substantiality and causality in its objects – the mathematics of the mathematicians. Priscian is wary even to concede the name of "mathematics" to the discourse on the projected quantitative items which are the objects of common arithmetic and geometry. As he puts it, "the reason for calling these objects ['mathematical'] is that they are thus considered by those who are usually called 'mathematicians' by ordinary people (τοῖς πολλοῖς)".[109] The implication is that the properly mathematical science is the one about the causes of phantastic quantities, i.e., about the ontologically prior, ensouled οὐσιώδεις λόγοι. On the other hand, ordinary mathematical practice, concerned with the representations of such λόγοι as expressed in τὰ φανταστά, has as its object the dream-like items Plato spoke of. Thus, ontologically speaking, mathematical substances are prior to natural realities, whereas the mathematical projections of phantasy, objects of ordinary mathematical practice, are ὑστερογενεῖς. Therefore, whereas the objects of "Pythagorean" mathematics are ontologically prior and truly substantial, their phantastic expressions, improperly known as "mathematical", are posterior generations of the soul and somewhat dependent on the perception of sensible quantity.

But to what extent are these phantastic items dependent on sensible quantity as a cause? This question leads us to the last passage relevant to our inquiry:

> For the [mathematical] objects of phantasy, too, are related to living reason-principles, which are substances, as is demonstrated by their precision (τὸ ἀκριβὲς αὐτῶν), which is greater than that of perceptible objects,

109 Translation by Steel (slightly modified) in 'Simplicius', *On Aristotle On the Soul 3.6–13*, p. 78. See Simplicius, *In De anima*, comm. in III, 431b, l. 12, p. 278, l. 2–3: […] ὃν δὴ καὶ μαθηματικῶν ἄν τις εἴποι νοῦν κατὰ τὸν τοῖς πολλοῖς συνήθως ὀνομαζόμενον μαθηματικόν.

> and by the fact that they add what is lacking [in perceptible objects] and remove what is superfluous. These [mathematical] objects could not be produced from outside (οὐκ ἂν ἔξωθεν ἐγγινόμενα) nor have what is better from the worse, but they are clearly the immediate result of the projection (ὑπ'αὐτῶν προβαλλόμενα) by the living reason-principles themselves, as is shown also by their critical activity; for a critical judgement is never an effect, but pure activity, and comes from within rather than from without (καὶ ἔνδοθεν ἀλλ'οὐκ ἔξωθεν).[110]

Priscian declares that the kinship between phantastic mathematicals and the ensouled reasons (adjectivised here in the Neoplatonic manner as ζωτικοί, i.e., "living" or "vital") is shown by their possession of τὸ ἀκριβές, translated into Latin as *certitudo* (see section 2.3). The term ἀκριβές indicates in this context what we may call "objectual precision" or the degree of exactness with which a mathematical object such as a line or a right angle is instantiated. Thus, when this "precision" is attributed to phantastic items, it is meant to stress the fitting instantiation of geometrical objects as represented by phantasy vis-à-vis their sensible correlates. The sensible right angle is never a perfect right angle; a right angle can only be flawlessly instantiated – with the highest degree of precision – in phantasy. As we will see, Priscian, much like Proclus and Syrianus,[111] sustains that the objectual accuracy proper to phantastic mathematicals is a symptom of their superior lineage in the ensouled reasons: objectual accuracy is a superior epistemic trait handed down to the phantastic expressions of the λόγοι as the sign of their more epistemically transparent origin.

Secondly, although the necessary intervention of sensible quantity in obtaining phantastic mathematicals has been acknowledged, Priscian now observes that this does not make sensible stimuli an effective cause of phantasised mathematics. He indicates this by pointing out that phantastic quantities cannot derive from something external (ἔξωθεν). Rather, mathematical items are produced from within, from the cooperation of phantasy with the ensouled reasons. The production of mathematical objects is conceived as an internal operation occasionally triggered by sensible stimuli. The perception of the

110 Translation by Steel (slightly modified) in 'Simplicius', *On Aristotle On the Soul 3.6–13*, p. 78. See Simplicius, *In De anima*, comm. in III, 431b, l. 12, p. 277, l. 24–33: ὅτι γὰρ καὶ τὰ φανταστὰ περὶ τοὺς ζωτικοὺς λόγους, οὐσίας ὄντας, δηλοῖ τὸ ἀκριβὲς αὐτῶν, ὑπὲρ τὰ αἰσθητὰ ὄν, καὶ ἡ τοῦ ἐλλείποντος αὐτοὺς προσθήκη καὶ ἡ τοῦ περιττοῦ ἀφαίρεσις, οὐκ ἂν ἔξωθεν ἐγγινόμενα καὶ ἀπὸ τῶν χειρόνων τὸ κρεῖττον ἔχοντα, ἀλλὰ δῆλον ὡς προσεχῶς ὑπ'αὐτῶν προβαλλόμενα τῶν ζωτικῶν λόγων, ὡς καὶ αὐτὴ ἡ κριτικὴ αὐτῶν δηλοῖ ἐνέργεια, οὐδέποτε πάθος οὖσα ἡ κρίσις, ἀλλὰ καθαρὰ ἐνέργεια, καὶ ἔνδοθεν ἀλλ'οὐκ ἔξωθεν.

111 Proclus, *In Euclidem*, p. 12, l. 13–14; Syrianus, *In Metaphysica*, pp. 95, l. 29–96, l. 14.

crystal sphere is the condition, but not the efficient or paradigmatic cause, of the phantastic projection of the geometrical sphere. Thus, rather than replicating a perceived quantity as it exists in perceptible matter (for that would mean also replicating all its inaccuracies), the inner sensible-like power of phantasy, set off by perceptible stimuli resembling the material expression of a reason-principle, generates a phantastic item which expresses the very same reason-principle with more precision than the sensible token.

A further proof adduced by Priscian of the independence, detachment, and priority of mathematics over sensible items is its use as a rule of judgment (κρίσις).[112] This means that to produce a judgement about a quantitative trait (is this straight? is this spherical?), we use the traits of unfolded phantastic objects as the criterion for discerning which is (more-or-less) straight or (more-or-less) spherical amongst natural objects, and not the other way around. This, according to Priscian, is further proof of the priority of mathematical reasons and their phantastic expressions over natural objects.

Lastly, it is also worth noting that, although Priscian denies substantiality and causation to projected mathematical items, he considers them to be more than arbitrary fictions. We may ask: what is the difference between the phantasised satyr and the phantasised sphere? Which features (if any) set phantastic mathematicals apart from poietic fiction and dreams? Or are they just the same in their fictional, unsubstantial way of being?

Priscian's view targets some unnamed Peripatetics who not only deprived phantastic mathematicals of substantial status but also denied to them every substantially-related trait ([…] τῶν Περιπατητικῶν ἔνιοι φαίνονται τὰ φανταστὰ ὡς ἀνούσια πάντῃ […]).[113] Priscian's position against the Peripatetic extreme reading of mathematical insubstantiality-tenet could be expounded as follows: unlike the satyr depicted in phantasy, the phantastic sphere possesses an intimate connection to an ensouled reason, which bestows upon its image its reflected nature and provides it with accuracy. In other words, whereas there is not such a thing as an ensouled reason whose expression is the phantastic satyr, the phantastic sphere can "iconically" reveal something about the substantial reality which it expresses. Hence, although phantastic mathematical entities share the same substratum as satyrs, sirens, and dreams, and are neither substances nor bear any form of causality, they nevertheless occupy a position closer to real beings than mere fictions, as they are reflections of substantial realities. Epistemically speaking, phantastic mathematics conveys, in a

112 See Steel's clarifications in 'Simplicius', *On Aristotle On the Soul 3.6–13*, p. 172, note 241.

113 Simplicius, *In De anima*, comm. in III, 431b, l. 12, p. 277, l. 22–23.

diluted way, some insight about real beings – something which the phantastic satyr does not.

Indexing Priscian's declared positions on the philosophy of mathematics, we have seen that:

(1) Phantastic mathematical items are not substances.
(2) They, hence, are deprived of causation.
(3) Therefore, mathematics conceived as a discourse about phantastic items is not a proper science (i.e., it does not demonstrate through real causes).
(4) Mathematical certainty stems from the nature of its object of study as its objects are closer to the inborn substantial reasons.
(5) Sensible input is not the efficient cause of, but rather a condition for, phantastic quantity to be projected. The efficient cause of phantastic items is phantasy itself, whereas the paradigmatic cause is the inborn reason.

Additionally, regarding Priscian's Aristotelian hermeneutics, we have established that:

(6) "Projected mathematical objects" is another name for Aristotle's "abstracted" or "separated mathematical objects". Hence the Neoplatonic προβολή and the Peripatetic ἀφαίρεσις meet in one.

Tenets (1), (2), and (3) would be endorsed as such by Piccolomini as the core of his argumentation in *De certitudine*, whereas (4) and (5) would be adapted in his philosophy of mathematics. Piccolomini defends both that the certainty of mathematics has its origin in the ontological status of its object of study and that sensible input is a necessary, yet occasional, condition for obtaining phantastic mathematical items. Nonetheless, ensouled substantial reasons have no philosophical role in *De certitudine*.

Crucially, tenets (1) to (5) will be justified by appealing to or straightforwardly quoting from Proclus' *In Euclidem* rather than by invoking any section from Pseudo-Simplicius' *In De anima*.

Tenet (6) will be paramount in tackling Piccolomini's use of Proclus in chapter VII of *De certitudine* and in addressing the problems presented by Piccolomini's notion of *abstractio*. I will show that, in all likelihood, Piccolomini read Proclus' mathematical προβολή following Priscian's assimilation of it to ἀφαίρεσις. As said, the Sienese was instructed during his years at Padua by Genua De' Passeri, and, thus, he was certainly acquainted with Priscian's commentary, its language, and its teachings on the philosophy of mathematics. Therefore, as I will show in sections 3.4.3 and 3.5.2, to answer the question "How did Piccolomini interpret Proclus' philosophy of mathematics?" we need to ask the further question "What is in Pseudo-Simplicius' *De anima* concerning the metaphysics, causality, akribology, and scientificity of mathematics through which Piccolomini could have read Proclus?"

2.2 *Phantasia*: Ontology and Epistemology

In this section, I will provide a brief history of *phantasia* leading to Piccolomini's understanding of the concept as it is deployed in *De certitudine*. I will focus on both its ontological status and epistemological roles. This sketch will provide the required insight to eventually conclude that the theory of *phantasia* contained in *De certitudine* is the result of a "Simplician" Averroistic reading of the second prologue of Proclus' *In Euclidem*.

2.2.1 *Aristotle and Alexander*

Although the Greek term "φαντασία" appears to have been coined by Plato,[114] it is unquestionable that the most influential Greek elaboration on this theme is that of Aristotle's – and indeed this theory is highly influential in Proclus' and Piccolomini's accounts. Aristotle's most thorough account of this concept is contained in *De anima* (mostly in book III), but the Stagirite deals with it in other texts of the *Parva naturalia*, namely *De insomniis* and *De memoria*.

Aristotle's concept of φαντασία has given rise to several debates. Amongst the most prominent is whether φαντασία is a *sui iuris* capacity of the soul or, instead, a modification of the faculty of perception. This is so because, according to Aristotle's standards in *De anima* II, 4, what makes an activity able to be considered as a capacity in its own right is that such a power has an object of its own which it does not share with other capacities: hearing captures audible things, sight visible things, and intellect intelligible things.

But what is the object of φαντασία? Depending on the metaphysics of phantastic items one assumes, the answer will be different. If the object of φαντασία is a perceptible thing, it will not be a power of its own, but a mode or alteration of another power (namely, perception) that has perceptible items as its own object. If, contrarily, we postulate the independent existence of phantastic items as having their origin in, but ontologically detached from, perceptible items, φαντασία will be a power of its own (Proclus' and Piccolomini's view, at least regarding mathematical φαντασία).

In this section, I will focus on presenting the elements of Aristotle's views on φαντασία, leaving aside the question of what Aristotle's actual opinion might have been. In the following exposition, we will find how other philosophers adopted, interpreted, and transformed Aristotle's account in various (and mutually exclusive) ways.

---

114 *Theaetetus*, 152c, l. 1; 161e, l. 8; *Sophista*, 260c, l. 9; 260e, l. 4; 263d, l. 6; 264a, l. 6; *Respublica*, 382e, l. 10.

To begin with, Aristotle presented his theory of φαντασία in contrast to that of Plato.[115] Plato's epistemological analysis of φαντασία is of special relevance in the *Theaetetus* (161d–162a), where, in the process of refuting Protagoras' views, it is stated that φαντασία is the result of perception plus opinion (αἴσθησις + δόξα). This arguably means that it is a judgement of an epistemic agent attributing truth value (x is the case, x is not the case) to one sensible perception x. Plato's view is targeted by Aristotle who claims that φαντασία cannot be the combination of perception and opinion because some animals possess φαντασία without the faculty of judgement allowing them to produce δόξα.[116] In other words, φαντασία, as characterised by Aristotle, is a condition rather than the result of the opinative capacity.

Aristotle discusses his doctrine of φαντασία, providing some historically relevant details of its ontology and epistemic role, in *De anima*. Aristotle's account in this work can be further nuanced and completed by his scattered mentions of φαντασία in other treatises. The overall resulting picture can be succinctly expressed in the following seven points:[117]

Firstly, (1) regarding the entities possessing φαντασία, in *De anima* I, 3, 406b this notion is presented as linked to perception and, thus, as present in living beings able to perceive. Nonetheless, Aristotle does not attribute φαντασία to all animals, for he considers that extremely simple forms of animal life (that he calls imperfect) do not possess it.[118] From this Aristotle would conclude that, although dependent upon it, φαντασία is neither the capacity of perception nor perception itself, but a non-necessary consequence of it.[119]

Secondly, (2) from *De insomniis* (461a, l. 20–30) we learn that, although etymologically Aristotle related φαντασία to visual perception by associating it to φάος/φῶς (light), there is φαντασία of other senses. For instance, there is φαντασία of sounds.

---

115 G. Watson, *Phantasia in classical thought* (Galway: Galway University Press, 1988), chap. 2.

116 Aristotle, *De anima*, III, 428a, l. 24–29.

117 For a thorough analysis of Aristotle's concept of phantasy and a clear elucidation of the puzzles raised by it, see T. K. Johansen, *The Powers of Aristotle's Soul* (Oxford: Oxford University Press, 2012), especially chapter 10. See also D. Charles, *The Undivided Self. Aristotle and the Mind-Body Problem* (Oxford: Oxford University Press, 2021), chap. 6; R. Lefebvre, "La phantasia chez aristote: Subliminalité, indistinction et pathologie de la perception", in *Les Etudes Philosophiques. Aristote sur la imagination* (1997), pp. 41–58; D. Frede, "The Cognitive Role of Phantasia in Aristotle", in M. C. Nussbaum, A. O. Rorty (eds), *Essays on Aristotle's* De Anima (Oxford: Oxford University Press, 1995), pp. 279–296; K. White, "The Meaning of *Phantasia* in Aristotle's *De Anima*, III, 3–8", in *Dialogue: Canadian Philosophical Review* 24/3 (1985), pp. 483–505.

118 Aristotle, *De anima*, 428a, l. 10–11; 433b, l. 31 ff.

119 Aristotle, *De anima*, 428a, l. 6.

Thirdly, (3) again in *De insomniis* (450b, l. 11–451a, l. 2) we gain some insight regarding the ontological status and causes of φαντασία. Aristotle states that dreams are produced as φαντασία by the remnants of perceptive affections in the body. He explicitly mentions that blood carries the movement responsible for generating these images (461a, l. 14 ff.). Regarding dreams, Aristotle explains that these movements continue to produce lingering representations of recently perceived (now absent) items after the perceptive activity of the sense organs ceases during sleep. Thus, we can conclude that, for Aristotle, φαντασία generally refers to the representation of an absent, perceptible item and is, in this way, dependent on sense perception. This is further confirmed by Aristotle's statement in *De anima* III, 3 that "phantasy seems to be some sort of movement not without perception".[120] This train of thought would lead Aristotle to conclude that the definition of φαντασία is "a movement caused by actual perception" (κίνησις ὑπὸ τῆς αἰσθήσεως τῆς κατ' ἐνέργειαν).

Fourthly, (4) as stressed by Johansen,[121] φαντασία has an important role in the survival and motivations of animals and, overall, in Aristotle's theory of action. In *De anima* 429a, l. 4–7 Aristotle explains the behaviour of animals by their use of φαντασία. This indicates that some animals equipped with φαντασία can use it to represent desirable or unpleasant items or events that they remember. In *De memoria*, Aristotle explains that memory functions by reproducing stored φαντάσματα, a term he sometimes uses to refer to the products of φαντασία as distinct from φαντασία itself. A pleasant or unpleasant φάντασμα or representation will motivate the course of action that an animal will follow. For instance, a bear seeing a honeycomb may trigger his φαντασία for representing himself with a desirable good that he might not be directly perceiving at the moment: honey. This φαντασία of a desirable thing moves the bear towards it as a goal, motivating his following actions.

Fifthly, (5) Aristotle points out that φαντασία, unlike perception and opinion, is "up to us" (ἐφ'ἡμῖν ἐστιν).[122] This suggests that, while a perceiving agent cannot choose to perceive colours differently or decide whether an opinion appears true or false to him, φαντασία allows for a certain degree of free, creative exercise. It can be used to represent possible situations or items, regardless of their actual existence. This interpretation is confirmed by Aristotle's description

120 Aristotle, *De anima* III, 3, 428b, l. 10–11: ἡ δὲ φαντασία κίνησίς τις δοκεῖ εἶναι καὶ οὐκ ἄνευ αἰσθήσεως. See also *De insomniis*, 459a, l. 15.

121 Johansen, *The Powers of Aristotle's Soul*, pp. 207–209.

122 Aristotle, *De anima*, III, 427b, l. 17–21: τοῦτο μὲν γὰρ τὸ πάθος ἐφ'ἡμῖν ἐστιν, ὅταν βουλώμεθα (πρὸ ὀμμάτων γὰρ ἔστι τι ποιήσασθαι, ὥσπερ οἱ ἐν τοῖς μνημονικοῖς τιθέμενοι καὶ εἰδωλοποιοῦντες), δοξάζειν δ' οὐκ ἐφ'ἡμῖν· ἀνάγκη γὰρ ἢ ψεύδεσθαι ἢ ἀληθεύειν.

of at-will phantasising agents as "εἰδωλοποιοῦντες", that is to say, as "producers of images". *Nota bene* that εἴδωλον is the term used by Proclus to describe the phantastically projected objects of geometry (see 2.1.3 above).

It remains, though, unexplained how this poietic use of φαντασία coheres with the former definition. Conceded that φαντασία is "a movement caused by actual perception" which lingers in the body, how is it that a phantasising agent can use it at will? Would the movement underpinning the at-will generated φάντασμα be an alteration caused by the agent itself (or its memory, or its reason) in its own body? Would not this force it to change the definition? Arguably, Aristotle held that the possession of reason somehow qualified φαντασία in different kinds or sub-species. For instance, in accounting for his theory of action in human and non-rational animals, he introduces two sorts of qualified φαντασία: αἰσθητική and λογιστική.[123] Thus, Aristotle considered that there is such a thing as "rational phantasy", clearly indicating the existence of qualified uses of φαντασία in the presence of other powers.

Sixthly, (6) φαντασία will be revealed as a momentous piece in Aristotle's theory of knowledge. In *De anima* III, 431a, 9–15, Aristotle posits that the intellective capacity (νοῦς, νοεῖν) requires φαντασία and, thus, φαντάσματα for being led to act. The exact meaning of this phrase would be subject to various interpretations throughout the text's history. It turns even more problematic when the precise role of φαντασία needs to be put together with the obscure passages on νοῦς ποιητικός, νοῦς παθητικός and the δύναμις for intellection as they are depicted in *De anima* III, 4 and 5. I will return to the long-standing exegesis of these sections later, for some of its details will be crucial for understanding Piccolomini's notion of *phantasia*.

Seventhly, (7) given that intellection is dependent upon it, Aristotle attributes an ambiguous epistemological role to φαντασία. He claims that φαντασία is neither opinion, nor opinion about a perception (against Plato), nor is it just perception, nor intellection or knowledge. Probably moved by his view that the works of φαντασία have a paramount role in gaining knowledge, that it is susceptible to rational qualifications (see 5 above), and that it is rooted in perception (Aristotle refers to φαντάσματα also as αἰσθήματα, see 3 above), he allows that one could postulate that φαντασία is a sort of intellection (εἴ τις τὴν φαντασίαν τιθείη ὡς νόησίν τινα).[124]

Broadly speaking, these are the essential elements that Aristotle imparted to the tradition that built upon his writings.

123 Aristotle, *De anima*, III, 433b, l. 27–30.
124 Aristotle, *De anima*, III, 433a, l. 10–11.

One aspect of our research still requires attention: Aristotle does not openly relate φαντασία with the production of mathematical items. The only section in the *corpus Aristotelicum* which could be interpreted as hinting that mathematical objects are found in phantasy is *De memoria* where the author uses the case of a triangle which we "draw determined according to quantity" (γράφομεν ὡρισμένον κατὰ τὸ ποσόν) in explaining the intervention of the φάντασμα in intellection.[125] Nor does Aristotle relate ὕλη νοητή – the underpinning stuff of geometrical objects invoked in *Metaphysica*[126] – to φαντασία. Such associations would be later elaborations.

The most influential author in the reception of Aristotle's notion of φαντασία in Late antiquity is, of course, Alexander of Aphrodisias. In his *De anima* he attempted to cast light on some of the obscurities of Aristotle's exposition.[127]

Firstly (I), regarding the questions of whether φαντασία is a capacity or not and how poietic φαντασία can be coherently squared with the rest of Aristotle's account, Alexander introduces the distinction between φανταστόν and φανταστικόν. His exegesis runs as follows (my own translation and punctuation):

> In relation to phantasy, something (a) is the phantasisable (φανταστόν) (this is what would be a product of the residue left by actual perception, analogous to both the perceptible and intelligible); something else is (b) the phantastic (φανταστικόν), being itself analogous to both the sensory and the intellectual; and yet something else (c) phantasy (φαντασίαν), being also itself analogous to both the sensory and the intellectual, being

125 Aristotle, *De memoria*, 449b, l. 30–450a, l. 7: ἐπεὶ δὲ περὶ φαντασίας εἴρηται πρότερον ἐν τοῖς περὶ ψυχῆς, καὶ νοεῖν οὐκ ἔστιν ἄνευ φαντάσματος συμβαίνει γὰρ τὸ αὐτὸ πάθος ἐν τῷ νοεῖν ὅπερ καὶ ἐν τῷ διαγράφειν· ἐκεῖ τε γὰρ οὐθὲν προσχρώμενοι τῷ τὸ ποσὸν ὡρισμένον εἶναι τοῦ τριγώνου, ὅμως γράφομεν ὡρισμένον κατὰ τὸ ποσόν, καὶ ὁ νοῶν ὡσαύτως, κἂν μὴ ποσὸν νοῇ, τίθεται πρὸ ὀμμάτων ποσόν, νοεῖ δ' οὐχ ᾗ ποσόν. English translation by Sorabji in Aristotle, R. Sorabji (ed.), *Aristotle on Memory* (Chicago: University of Chicago Press, 2006): "An account has already been given of imagination in the discussion of the soul, and it is not possible to think without an image. For the same effect occurs in thinking as in drawing a diagram. For in the latter case, though we do not make any use of the fact that the size of the triangle is determinate, we none the less draw it with a determinate size. And similarly someone who is thinking, even if he is not thinking of something with a size, places something with a size before his eyes, but thinks of it not as having a size".

126 Aristotle, *Metaphysica*, VII, 10, l. 1036a, 10–12; 11, 1037a, l. 2–4.

127 For an account of Alexander's theory of phantasy see A. Hangai, *Alexander of Aphrodisias on Phantasia: An Aristotelian Account of Mental Representation in 2nd–3rd Centuries CE*, Ph.D. (Budapest, Central European University, 2016); Hangai, "Alexander of Aphrodisias' Criticism of the Stoic Theory of Perception: typos and typōsis", in *Elenchos* 43/2 (2022), pp. 339–362.

> the actuality of the phantastic capacity (τῆς φανταστικῆς δυνάμεως) concerning phantasisable things (τὰ φανταστά).[128]

Alexander distinguishes in this passage three aspects of φαντασία for which he tries to establish both the commonalities and particularities. He claims that:

(a) One kind of object of φαντασία is defined as φανταστόν: those are the remnants of perceptive activity in the body which can potentially be the objects of φαντασία.

(b) A second kind of object of φαντασία is named φανταστικόν which is not defined as the lingering effect of actual perception.

(c) There is a capacity called φανταστική whose activity is φαντασία.

Concerning (c), there is no doubt that Alexander considered that φαντασία is the activity of a discrete power of the soul and not just an affection of perception; he even refers to it as the ψυχή φανταστική.[129] For solving the hindrance of the proper object as the criterion for being a *sui iuris* faculty, Alexander proposes that (i) φαντασία is analogous to both perception and intellection, but (ii), nevertheless, that it possesses an object of its own, namely the φανταστόν – a term absent from Aristotle's texts.[130] This abstract noun is introduced by Alexander in indicating that, as intellect has the intelligible as its object and perception the perceptible, so must also φαντασία have the phantasisable as its proper object.[131] In this way, Alexander manages to detach the objects of perception from the objects of phantasy, rendering it a *sui iuris* faculty: whereas the perceptible is external, the phantasisable is internal to the subject, as it is an alteration in the subject's own body.

Alexander introduces (b), sc., the φανταστικόν. The specific trait of this kind of φαντασία seems to be that it is not a direct effect of the activity of perception, although it necessarily depends on it, as all φαντασία does. Thus, the most likely interpretation is that the φανταστικόν refers to objects of φαντασία other than the spontaneous representations of post-perceptive motions of the blood, to wit, the poietic and the mnemonic activities. Whereas the refrain of a catchy melody is constantly represented in the φαντασία of an (unwilling) subject who has recently perceived it or a dream appears without being intended, recalling

---

128 Alexander of Aphrodisias, *De anima*, p. 68, l. 30–35: […] ἐπὶ τῆς φαντασίας τὸ μέν τι εἶναι φανταστόν (τοῦτο δ' ἂν εἴη τὸ ἀπὸ τῆς κατ' ἐνέργειαν αἰσθήσεως γινόμενον ἐγκατάλειμμα, ἀνάλογον ὂν τῷ αἰσθητῷ τε καὶ νοητῷ), τὸ δὲ φανταστικόν, ἀνάλογον ὂν καὶ αὐτὸ τῷ αἰσθητικῷ τε καὶ τῷ νῷ, τὸ δὲ φαντασίαν, οὖσαν καὶ αὐτὴν ἀνάλογον τῇ αἰσθήσει καὶ τῇ νοήσει, ἐνέργειαν οὖσαν τῆς φανταστικῆς δυνάμεως περὶ τὰ φανταστά.

129 Alexander of Aphrodisias, *De anima*, ed. I. Bruns (Berlin: G. Raimer, 1887), p. 97, l. 11–12.

130 Hangai, *Alexander of Aphrodisias on Phantasia*, pp. 17–18.

131 Alexander of Aphrodisias, *De anima*, p. 68, l. 21–25.

a melody and mentally depicting the Pegasus are cases of φαντασία in which the movements in the body underpinning the images would have their origin not in perception, but in the subject itself. This would be the φανταστικόν.

Lastly, Alexander fully endorses Aristotle's suggestion that other powers of the soul qualify phantasy in producing φαντασία αἰσθητική or λογιστική. He even proposes that all the faculties of perception, opinion, science and intellect can be predicated on φαντασία.[132] Alexander's view seems to be that φαντάσματα can be epistemically qualified depending on the faculty which they help to enact. A clear example of this would be φαντασία νοητική or the name corresponding to the φάντασμα allowing the actualization of a universal concept in the material intellect.

2.2.2 *Plotinus and Porphyry*

Plotinus' reception and understanding of φαντασία would accentuate the twofold nature, noetic and aesthetic, suggested by Aristotle and, analogously, accepted by Alexander. Plotinus follows Alexander's interpretation in linking φαντασία to different cognitive powers: he connects it to perception,[133] literally identifying it with opinion,[134] and relates it to the logical and dianoetic processes of representation (εἰκονική) of noetic realities.[135] Nonetheless, one of the most remarkable additions made by Plotinus is his introduction of the concept of a twofold φαντασία (or two fused phantasies).[136]

Plotinus discusses two types of φαντασίαι based on the ontological level that each of our two souls occupies – superior and inferior. Each soul has its own distinct phantasy, but the two phantasies are melded into the one we constantly experience. Plotinus' discussion focuses on the clash between the memories stemming from both souls, i.e., empirical memory and recollection of the contents of the superior soul:

> But if memory belongs to the phantastic (τοῦ φανταστικοῦ), and we have said that each soul remembers, there will be two [types of] phantastic

---

132 Alexander of Aphrodisias, *De anima*, p. 66, l. 24: [...] κατὰ αἰσθήσεως καὶ κατὰ δόξης καὶ κατὰ ἐπιστήμης καὶ νοῦ τὴν φαντασίαν κατηγοροῦμεν.

133 Plotinus, *Enneades*, IV, 3, 29.

134 Plotinus, *Enneades*, III, 6, 4, l. 19–20.

135 Plotinus, *Enneades*, III, 6, 18, l. 33; IV, 3, 13, l. 11–16 ; IV, 3, 23, l. 33.

136 On this question, see the excellent chapter in M. Di Pasquale Barbanti, *Ochema-Pneuma e phantasia nel neoplatonismo. Aspetti psicologici e prospettive religiose* (Catania: Symbolon, 1998), pp. 76–96; see also Watson, *Phantasia in classical thought*, pp. 97–105; H. J. Blumenthal, "Neoplatonic Interpretations of Aristotle on "Phantasia"", in *The Review of Metaphysics* 31/ 2 (1977), pp. 242–257.

> items (δύο τὰ φανταστικά). Let us grant that when the souls are apart they each have one, but when they are in the identical place, in us, how are there two, and in which of them do the memories happen? If in both, there will always be duplicate phantasies, for it would certainly not be the case that [the phantasy] of one soul would be of intelligibles, while that of the other was of sensibles. In this way, there would be two living beings having nothing in common with each other. If, then, both souls have [phantasy], what will be the difference between them? And then, if so, how do we not recognize such a difference? In fact, when one soul agrees with the other, and the two faculties of phantasy are not separate, and the higher one is dominant, the resemblance becomes one, as if a shadow were following the other, or as if a weaker light were being subsumed into a stronger one. But whenever there is conflict and discord, then the other one manifests itself on its own account, while we do not realise that it is in a different faculty. And, in general, the duality of the souls escapes our notice. For the two of them have come to be one thing (εἰς ἕν), and one of them rides dominant over the other.[137]

Plotinus portrays φαντασία as a sort of common ground bridging the noetic and the corporeal natures of the human being. It has its origin in two different, not always harmonious, streams or two "memories". Moreover, although both the noetic and the perceptive phantasies meet in one, the perfect balance between those two powers is not allowed: one of them must always prevail over the other. According to Plotinus, this explains our misleading sense of the unity of φαντασία (and of our soul in general). The phantastic elements that originate from perceptual experiences are often conflated with those that reflect our superior self and are revealed through recollection.

Beyond the ethical implications of this doctrine – which invite us to free ourselves of the deleterious rule of perceptive phantasy through philosophy –, it is worth noting that, because of its amphibious status, the fused φαντασία

137 Plotinus, *Enneades*, IV, 3, 31, l. 1–13. Trans (altered) by Ll. Gerson. Original Greek: ἀλλ'εἰ τοῦ φανταστικοῦ ἡ μνήμη, ἑκατέρα δὲ ἡ ψυχὴ μνημονεύειν εἴρηται, δύο τὰ φανταστικά. Χωρὶς μὲν οὖν οὖσαι ἐχέτωσαν ἑκάτερα, ἐν δὲ τῷ αὐτῷ παρ' ἡμῖν πῶς τὰ δύο καὶ τίνι αὐτῶν ἐγγίνεται; Εἰ μὲν γὰρ ἀμφοτέροις, διτταὶ ἀεὶ αἱ φαντασίαι· οὐ γὰρ δὴ τὸ μὲν τῆς ἑτέρας τῶν νοητῶν, τὸ δὲ τῶν αἰσθητῶν· οὕτω γὰρ ἂν παντάπασι δύο ζῷα οὐδὲν ἔχοντα κοινὸν πρὸς ἄλληλα ἔσται. εἰ οὖν ἀμφοτέραις, τίς ἡ διαφορά; εἶτα πῶς οὐ γινώσκομεν; ἢ ὅταν μὲν συμφωνῇ ἡ ἑτέρα τῇ ἑτέρᾳ, οὐκ ὄντων οὐδὲ χωρὶς τῶν φανταστικῶν, κρατοῦντός τε τοῦ τῆς κρείττονος, ἓν τὸ φάντασμα γίνεται, οἷον παρακολουθούσης σκιᾶς τῷ ἑτέρῳ, καὶ ὑποτρέχοντος οἷον σμικροῦ φωτὸς μείζον. ὅταν δὲ μάχη ᾖ καὶ διαφωνία, ἐκφανὴς ἐφ' αὑτῆς καὶ ἡ ἑτέρα γίνεται, λανθάνει δὲ ⟨ὅ τι⟩ ἐν ἑτέρῳ. [ὅτι] καὶ ὅλως τὸ διττὸν τῶν ψυχῶν λανθάνει. Εἰς ἓν γὰρ ἦλθον ἄμφω καὶ ἐποχεῖται ἡ ἑτέρα.

of Plotinus is defined as a middle-ground (μεταξύ) between natural impressions and noetic activity.[138] Moreover, what was an epistemic qualification in Aristotle's account was read by Plotinus also as ontological: φαντασία is not just an alteration of the body with noetic aspects or applications, but the frontier filling the ontological gap between the perceptive, embodied soul and the noetic, undescended soul. In other words, φαντασία as the highest (most noetic-like) power of the embodied soul ontologically coincides – and overlaps – with the lowest power (the most perception-like) of the noetic soul.

Porphyry would also greatly contribute to the interpretation of φαντασία in a way that would influence Proclus' views. As Watson points out, this development regards a more elaborated ontology of φαντασία and its connection with πνεῦμα.[139] Allegedly following Aristotle's usage of this notion in *De generatione animalium*,[140] Porphyry holds both that (a) πνεῦμα shares a kinship with celestial bodies (something Aristotle stated by analogy) and, (b) that it works in the living being as a sort of link between the soul and the movements and activities of the living body.[141] Porphyry adds two further points, namely (c) that the πνεῦμα present in the living being adheres to the soul during its celestial descent to the corporeal world,[142] and (d) that it is the ontological substratum where φαντασία is located.[143] This theory allowed Porphyry to elaborate on the nature of prophetic dreams by appealing to the so-called ὄχημα τῆς ψυχῆς or the "vehicle of the soul".[144] As studied by Barbanti, the link between πνεῦμα, the vehicle of the soul, and φαντασία greatly influenced the later tradition, being accepted and elaborated on by authors such as Iamblichus, Sinesius (374–414), and Hierocles of Alexandria (fl. *c.*430).[145]

---

138 Plotinus, *Enneades*, IV, 4, 13, l. 12–13: ἡ δὲ νόησις φαντασίας κρεῖττον· φαντασία δὲ μεταξὺ φύσεως τύπου καὶ νοήσεως.

139 Watson, *Phantasia in classical thought*, p. 107.

140 Aristotle, *De generatione animalium*, II, 3, 736b, l. 29–737a, l. 1.

141 Aristotle, *De motu animalium*, 10, *passim*.

142 Porphyry, *Sententiae*, 29, l. 8–9: ἐξελθούσῃ γὰρ αὐτῇ τοῦ στερεοῦ σώματος τὸ πνεῦμα συνομαρτεῖ, ὃ ἐκ τῶν σφαιρῶν συνελέξατο.

143 See ex. gr. *De antro nympharum*, 11, l. 17: καὶ ἐκ τῶν τοιούτων αἳ συναντῶσί τισι κατὰ φαντασίαν χρώζουσαι τὸ πνεῦμα εἰδώλων ἐμφάσεις, αἱ μέντοι καθαραὶ γενέσεως ἀπότροποι. See also *Sententiae*, 29, l. 13.

144 Porphyry, *De abstinentia*, II, 38 and 39. An account of Porphyry's theory of the vehicle of the soul is preserved by Proclus in *In Rempublicam*, II, 169, l. 28 ff. and in *In Timaeum*, III, 234, l. 20 ff. See Di Pasquale Barbanti, *Ochema-Pneuma*, pp. 109–130; Watson, *Phantasia in classical thought*, pp. 108–109; E. R. Dodds "Appendix II: The Astral Body in Neoplatonism", in *The Elements of Theology*, pp. 313–321.

145 Di Pasquale Barbanti, *Ochema-Pneuma*, pp. 131–204.

2.2.3 *Proclus*

The notions of the twofold nature of phantasy of Plotinus and its pneumatic substratum of Porphyry were inherited by the later tradition. Both tenets are amongst the sources of Proclus' views, which could be justly considered as delivering a many-sided synthesis of the Neoplatonic theories of φαντασία. Although this inquiry focuses on Proclus' doctrine of mathematical phantasy, it will be necessary to summarily address also the non-mathematical aspects of φαντασία in the Diadochus' writings other than *In Euclidem* to attain a more precise account of its role and metaphysics.

Firstly, what is the exact ontological status of φαντασία according to Proclus? Where is it exactly placed? Proclus gives meagre details about it in his *In Euclidem* (pp. 51, l. 20–52, l. 4.) – a section well known by Piccolomini. Proclus states both that:

(a) The power of φαντασία exists "with body and in the body" (μετὰ σώματος καὶ ἐν σώματι).
(b) And that φαντασία is to be identified with that which Aristotle calls νοῦς παθητικός.

This characterisation of φαντασία will be pivotal in grasping how Piccolomini read Proclus' theory. I will come back to it when analysing Piccolomini's use of the second prologue of *In Euclidem* (see section 3.5.2). Nevertheless, a more nuanced understanding of both points (a) and (b) from *In Euclidem* can be achieved by delving into Proclus' insights on πνεῦμα as the vehicle of the soul.[146]

Proclus' exposition of his theory occurs in book III of *In Timaeum*.[147] The discussion concerns an eschatological question: if the soul is impassible without its temporal activities, how is it possible, as Plato teaches in the myth of Er (*Respublica*, 614a–621d), that it would be subject to punishment, retribution, purification, and new-life election after its separation from the body? Thus, the discussion focused on the non-somatic part of the individual human being (with all his experiences and memories) which is, nonetheless, passible after its separation from the body, namely the vehicle of the soul. Proclus proposes (amongst others) the two conflicting views of Porphyry and Iamblichus: whereas the former had considered that the vehicle of the soul is mortal and that it dissipates into the heavens soon after the body perishes, the latter had sustained that the vehicle of the soul, directly generated by the Demiurge, is immortal – as everything generated by the Demiurge is eternal. Proclus'

146 B. Bohle, "Proclus on the Pneumatic Ochema", in S. Coughlin, D. Leith, O. Lewis (eds), *The Concept of Pneuma after Aristotle* (Berlin: Edition Topoi, 2020), pp. 343–364; Opsomer, "Was sind irrationale Seelen?", in *Perkams/Piccione* (2006), pp. 136–166.

147 Proclus, *In Timaeum*, III, 310e, l. 6 ff.

harmonising elaboration, which he seems to attribute to Syrianus (ὁ ἡμέτερος διδάσκαλος), contains three points worth mentioning:[148]

Firstly, (1) Proclus considers that the πνεῦμα/ὄχημα is an intermediary entity between the corporeal and incorporeal realities, sharing in both natures without being identical to either of them: it is a kind of being which fills the ontological gap between soul and body and, thus, more body-like than the soul but much soul-like than the body.

Also, (2) Proclus reconciles the perspectives of Porphyry and Iamblichus that the vehicle can undergo change and that it is a sempiternal, demiurgic creation. To resolve this, Proclus postulates that each human soul in reality possesses two vehicles. That is to say, two intermediary degrees of being are required to close the ontological hiatus between the corporeal and the intelligible without any sort of discontinuity: one superior and god-like produced by demiurgic causation, another inferior and mortal-like (θνητοειδές), which is identified with the πνεῦμα and made by the lesser gods. Proclus also elaborates on the superior vehicle, congenitally (συμφυῶς) joined to the soul, in the last propositions of his *Elementatio theologica* (207 to 210).

Lastly, (3) the inferior, pneumatic ὄχημα is described in *In Timaeum* as susceptible to passions and change and, thus, as the one in which φαντασία occurs (κατὰ φαντασίαν).[149] Proclus also attributes the function of somatic-like perception to the pneumatic vehicle of the soul and establishes that

> Perception and phantasy have the same nature: the "being" of both is common [i.e., both are a state of the πνεῦμα]. But when it is directed to the external world, then it is called "αἴσθησις"; when it remains within contemplating forms and figures in the pneuma, [then it is called] "φαντασία" (my own translation).[150]

This pneumatic vehicle fulfils the requirements postulated by Proclus for being the locus of mathematical phantasy: the πνεῦμα stands as the middle entity between the body and the noetic nature of the human being, i.e., between the living body and the impassible vehicle naturally conjoined to the actual essence of the soul. When the perception-related powers of πνεῦμα

---

148 Proclus, *In Timaeum*, III, 236, l. 32; see Syrianus, *In Metaphysica*, p. 86, l. 3 ff.

149 Proclus, *In Timaeum*, III, 311e, l. 28.

150 Proclus, *In Timaeum*, III, p. 286, l. 24–28: καὶ ἐκείνη μὲν ἡ αἴσθησις τῇ φαντασίᾳ τὴν αὐτὴν ἔχει φύσιν. τὸ γὰρ εἶναι κοινὸν ἀμφοῖν. ἀλλ᾽ἔξω μὲν προϊοῦσα καλεῖται αἴσθησις, ἔνδον δὲ μένουσα καὶ ἐν τῷ πνεύματι θεωροῦσα τὰς μορφὰς καὶ σχήματα, φαντασία [...]. That Proclus uses here the verb θεωρέω may suggest that he is referring specifically to the epistemic, contemplative function of φαντασία (see below).

are inwardly used – as happens in mathematical projection – then it is called "φαντασία".

Therefore, πνεῦμα/φαντασία, as put in (a) above, exists in the body, for πνεῦμα is part of the living organism, and acts with the body, since it depends upon the impressions of perception for its phantastic, sensible-like activity. This intermediary position also accounts for the mention of (b) "νοῦς παθητικός" as another name given to φαντασία: it might be considered that φαντασία is a sort of intellect which is susceptible to passions and change, although this is just a metaphor, for intellect is always activity and it is impassible.

The ontological in-betweenness of φαντασία makes it the place of a wide collection of cognitive powers situated between the corporeal and the incorporeal. Because of this range of powers, Proclus refers to φαντασία through different names, depending on the aspect that he wants to stress. He uses "φαντασία" (the most general and prevalent),[151] "νοῦς μορφωτικός" (when describing the capacity for spatial and, thus, also geometrical, representation),[152] "νοῦς παθητικός" (as said),[153] "νοῦς φανταστικός",[154] "ὕλη νοητή" (as the subject-matter of phantasised items),[155] and, when describing its ontological basis, "ὄχημα πνευματικόν".[156]

To put it concisely, Proclus attributes to φαντασία a general power for the generation of images or "icons". As a power for representing images, it possesses some cognitive aspects:

(i) It is the superior power of the irrational soul, able both to receive and shape unified sense-data from the κοινὴ αἴσθησις as well as ensouled reasons from διάνοια.[157]

---

151 Étienne Évard has pointed out the irregular presence of the very term φαντασία. To note some of the philosophical works, the term frequently appears in Proclus' *In Alcibiadem*, *In Rempublicam*, *In Timaeum*, *In Cratylum*, and *In Euclidem*; it is mentioned in *In Parmenidem*, *Theologia Platonica*, in *Tria opuscula* whereas it is completely absent from the *Elementatio theologica* and the *Elementatio physica*. Cf. E. Évard, "Φαντασία chez Proclus", in M. Fattori, M. Bianchi (eds), *Phantasia-Imaginatio. Saggi su Marsilio Ficino, Pomponazzi, Platone, Aristotele. V Colloquio Internaz. Del Lessico Intellettuale Europeo* (Roma: Ediz.dell'Ateneo, 1988), pp. 57–68.

152 Proclus, *In Euclidem*, p. 51, l. 20 (φαντασία μορφωτική); Proclus, *In Platonis Rempublicam commentarii*, ed. W. Kroll (Leipzig: Teubner, 1899), II, p. 235, l. 7.

153 Besides *In Euclidem*, φαντασία is put as νοῦς παθητικός in *In Rempublicam*, II, p. 52, l. 6; *In Timaeum*, I, p. 244, l. 21; p. 255, l. 18; III, p. 158, l. 9.

154 Proclus, *In Rempublicam*, II, 107, l. 17.

155 Proclus, *In Euclidem*, p. 18, l. 19; p. 86, l. 12.

156 Proclus, *In Timaeum*, III, p. 237, l. 25.

157 For an illuminating exposition of Proclus' epistemology and the role of φαντασία in it (including Proclus' innovative theory of δόξα as a form of recollection) see Helmig, *Forms and Concepts*, chap. IV.

(ii) It intervenes in the creation of words (φαντασία λεκτική) as a representation of non-verbal realities (*In Cratylum*, 51, 18–20).
(iii) It allows the creation and understanding of myths as fabulous, discursive representations of superior truths adapted to the conditions of incarnated souls, i.e., as reflections of a superior truth graspable by people mostly led by the perceptible and affective parts of the soul rather than by their rational soul (*In Rempublicam*, I, p. 80, l. 23; p. 86, l. 10; and II, p. 107, l. 16–23 where "νοῦς φανταστικός" occurs).
(iv) Contrarily, when the gods desire to manifest themselves to human beings, it also occurs through a theophanic function of φαντασία. In this case, φαντασία provides form and shape (μετὰ φαντασίας καὶ σχῆμα καὶ μορφήν) to the shapeless divine beings, which are thereby manifested in the form of inspired visions (*In Rempublicam*, II, p. 241, l. 23–27).
(v) Thus, as a power able to give shape and symbolically represent as a spatial-temporal process shapeless entities (μορφωτικὴ κίνησις), it includes not only the gods themselves but also the ensouled mathematical reasons (*In Euclidem*, p. 54 ff.).
(vi) Lastly, there is also a deceptive aspect of φαντασία mentioned by Proclus (*In Alcibiadem*, p. 39, l. 9–16). He indicates that, as far as φαντασία is linked with exteriority, it might work as a hindrance to our philosophical progression: indeed, too much attention to the φαντάσματα as a source of bodily pleasures, pains, or (real or fictitious) mundane worries may work as an anti-epistemic burden, hiding the pathway to the noetic realm.

Taking (vi) into consideration and the passage from *In Alcibiadem* vis-à-vis the positive epistemic capacities of φαντασία, it follows that φαντασία has two ethical and epistemic sides in Proclus' philosophy, one beneficial and active, another deleterious and passive.[158] I explain this position in the following terms: phantasy can be led to occupy itself with corporeal items and the moving affections therefrom induced (ἐπιθυμία, φόβος, λύπη, ἡδονή). These passions will besiege the human being by occupying his φαντασία, passively drawing his attention downwards and enslaving his decision-making and even his way of life. On the other hand, the same φαντασία can be actively pulled upwards when focused on discovering the ensouled reasons in collaboration with διάνοια by generating the appropriate φαντάσματα to acquire science and godly communion. In this sense, the theoretical sciences, whose practice requires the use of φαντασία for triggering recollection, would liberate the human being from passivity. Hence, speculative science (including mathematics) stands as the most active, superior use of φαντασία. In this way, I find that Proclus

158 Helmig, *Forms and Concepts*, pp. 230–231; Di Pasquale Barbanti, *Ochema-Pneuma*, p. 242.

preserved Plotinus' intuition of a twofold, conflicting φαντασία, one captured by the passiveness of the sensible world, the other stretching to the actuality of the undescended soul.

The most relevant aspect of Proclus' φαντασία is its role as a hinge between the sensible and the intelligible: it is something akin to perception (showing traits such as shape or colour) but free from both the ontological and the epistemic limitations that put sensible items apart from intelligible ones. This allows an entire ontological stratum of phantastic items to have their specific hypostasis in phantasy without being faithful representations either of sensible items or of intelligible Forms. Those natural inhabitants of the realm of phantasy are mathematical objects: the one-dimensional line can neither be enacted in sensible matter, nor (therefore) can it be sensibly known; nonetheless, a geometrical line is not a pure, shapeless Form, as it needs one dimension to be instantiated and, so, understood. The line, much like all mathematical objects *qua* mathematical objects, properly belongs to the middle region of φαντασία, i.e., to the pneumatic vehicle of the soul.

2.2.4 *From φαντασία to* imaginatio

It is now convenient to introduce some terminological clarifications regarding the reception in Latin of the term "φαντασία".

This abstract noun was introduced into Latin as a Greek importation taken from the philosophical context, as it appears in the works of authors such as Cicero, Quintilianus, and Gellius.[159] In his *Academica*, while discussing the Stoic and Sceptic epistemological positions, Cicero states that φαντασία is that which Latin speakers call *visum*, defining it as "things which are and that receive the assent of souls as they are accepted by the senses" ([…] *quae visa sunt et quasi accepta sensibus assensionem adiungit animorum*).[160] Quintilianus explicitly refers to φαντασίαι as *visiones*, defining them as the "images of absent things so represented by the soul as to appear to us that we see them and that they are present to us".[161]

159 Cf. Cicero, *Academica*, 1, 40, l. 4, in *M. Tulli Ciceronis Scripta Quae Manserunt Omnia*, 42, ed. O. Plasberg (Leipzig: Teubner, 1922); Cicero, *Lucullus*, 18, l. 12; Cicero, *Epistulae ad Familiares*, ed. D. R. Shackleton Bailey (Leipzig: Teubner, 1977), XV, 16, 1, l. 6; Cicero, *Epistulae ad Atticum*, ed. D. R. Shackleton Bailey (Leipzig: Teubner, 1965–1968), IX, 6, 5, l. 5; M. F. Quintilianus, *Institutio Oratoria*, ed. M. Winterbottom (Leipzig: Teubner, 1970), VI, 2, 29, l. 3; VIII, 3, 88, l. 4; X, 7, 15, l. 2; XII, 10, 6, l. 5; A. Gellius, *Noctes Atticae*, ed. K. Marshall (Leipzig: Teubner, 1968), III, 16, 8, l. 5; XI, 5, 6, l. 6; XIX, 1, 15, l. 2; 1, 18, l. 2.

160 Cf. Cicero, *Academica*, 1, 40, l. 7–8.

161 M. F. Quintilianus, *Institutio Oratoria*, II, 29, l. 3–6: *Quas φαντασίας Graeci vocant* (*nos sane visiones appellemus*), *per quas imagines rerum absentium ita repraesentantur animo ut eas*

It is noteworthy that, whereas the Latin verb *imaginari* exists in classical Latin, the abstract noun *imaginatio* seems a later, infrequent addition. It appears for the first time in Tacitus' *Annales* (1 time)[162] and in the *Naturalis historia* of Pliny (3 times).[163] This term was destined to be adopted as the standard translation of the Greek φαντασία.

Regarding the late ancient Latin philosophical tradition, Bundy has indicated that Augustine of Hippo (354–430) showed a clear proclivity for the use of *imaginatio* together with *sensus, impressio* or *visio*, seemingly trying to avoid the ambiguity of the Greek φαντασία as found in Aristotelian and Stoic texts: there it indicates both the power itself and the activity of the various uses of the power.[164] Nevertheless, the transliteration *phantasia*, as well as *phantasma,* occurs in the Augustinian corpus.[165] The term *imaginatio* was also inherited by the posterior Latin tradition through the works of Boethius.[166]

As we consider the lexical and philosophical history of both terms in Latin, we find that the Latin translation movement introduced a variety of spellings for the pair *phantasia-imaginatio*. For instance, in the corpus of the *Aristoteles Latinus*, we find that φαντασία is rendered as *fantasia, phantasia*, or *ymaginatio/imaginatio*.[167] As Spinoza has shown in her detailed study of the question,[168] James of Venice (*ob.* 1147) adopts the term *imaginatio* in 5 cases out of the 34 occurrences in his long-lived translation of *De anima*; in the other 29 occurrences, φαντασία is just transliterated as *phantasia*. *Phantasia* (over *imaginatio*) and *phantasma* are included in Isidore of Seville's *Libri differentiarum*.[169] Other

---

cernere oculis ac praesentes habere videamur, has quisquis bene ceperit is erit in adfectibus potentissimus.

162 Tacitus, *Annales*, ed. C. D. Fisher (Leipzig: Teubner, 1906), XV, 36, l. 3: *Nec multo post omissa in praesens Achaia (causae in incerto fuere) urbem revisit, provincias Orientis, maxime Aegyptum,* ***secretis imaginationibus*** *agitans.* For a history of the terms in classical Latin see P. Flury, "*Phantasia* und *imaginatio* im antiken Latin", in M. Fattori, M. Bianchi (eds), *Phantasia-Imaginatio*, pp. 68–79.

163 Plinius, *Naturalis Historia*, ed. C. Mayhoff (Leipzig: Teubner, 1892–1909), X, 166, l. 2; XX, 68, l. 4; XXIV, 161, l. 4.

164 M. W. Bundy, *The theory of imagination in classical and mediaeval thought* (Illinois: Folcroft library, 1927), pp. 158–163.

165 Augustine, *De Musica*, VI, 11, ff.

166 Boethius, *Consolatio Philosophiae*, V, 4 ff.; Boethius, *In Isagogen Porphyrii commenta*, II.1, pp. 136, l. 13–137, l. 11. On the early Latin theories of imagination see R. Palmén, *Richard of St. Victor's Theory of Imagination* (Leiden: Brill, 2012), chap. 2.2 and 2.3.

167 G. Spinoza, "*Phantasia* e imaginatio nell'*Aristotele latino*", in M. Fattori, M. Bianchi (eds), *Phantasia-Imaginatio*, pp. 117–133.

168 Spinoza, "*Phantasia* e imaginatio nell'*Aristotele latino*", pp. 127–128.

169 Isidore of Seville, *Liber differentiarum*, ed. M. A. Andrés Sanz (Turnhout: Brepols, 2006), 1, F, 216: *Inter phantasiam et phantasma. Phantasia est imago alicuius corporis visa, et*

examples are Eriugena's *Periphyseon* (III, 657d; V, 962c) and Moerbeke's Latin rendition of Proclus' *De providentia et fato*, where terms related to φαντασία are transliterated even in the verbal form (*fantasiamus*, III, 8, l. 5). Following this trend, most philosophers – such as Aquinas – would also adopt both terms as totally synonymous.[170]

Hamesse indicates that there are some (rare) exceptions to the general tendency of taking *phantasia* and *imaginatio* as equivalents.[171] For instance, probably influenced by the Arabic medical theories, Dominicus Gundissalinus distinguishes in his *De anima* between *phantasia* and *imaginatio* as different powers located in different parts of the brain: whereas *phantasia* is a second name for designating the unified perception of the animal or *sensus communis*, the poietic capacity of images is rendered as *imaginatio*.[172] Another case is that of Albert the Great: in his commentary on Aristotle's *De anima* he posits that *imaginatio* and *phantasia* belong to different bodily organs and that their activities possess different degrees of abstraction and, thus, different epistemic values (being *phantasia* more abstract than *imaginatio*).[173]

---

cogitando postea in animo figurata, ut puta, avi vel patris species, quem aliquando vidimus, ac dum cogitando memoramus, phantasiam dicimus. Phantasma vero est ex imagine cognita, aliqua, quam vidimus, imago formata, ut puta, species avi quem nunquam vidisse meminimus; sed tamen eius species non memoria, sed motu animi figuratur. De cognitis ergo speciebus memoria collecta, phantasia est; de incognitis species animo figurata, phantasma. Nam figurata phantasmata nihil aliud sunt quam de specie corporis, corporeo sensu abstracta; figmentoque memoriae, ut accepta sunt, vel partiri, vel multiplicare, vel contrahere, vel distendere, vel ordinare, vel turbare, vel quidlibet figurare cogitando facillimum est, sed, cum verum quaeritur, cavere et vitare difficile. Item phantasia est incognitarum rerum ex cognitis coniectura, phantasma vero rerum incognitio cognitarum.

170 R. Busa SI, "De phantasia et imaginatione iuxta S. Thomam", in M. Fattori, M. Bianchi (eds), *Phantasia-Imaginatio*, pp. 136–152.

171 J. Hamesse, "Imaginatio et phantasia chez les auteurs philosophiques du 12e et du 13e siècle", in M. Fattori, M. Bianchi (eds), *Phantasia-Imaginatio*, pp. 153–184.

172 Dominicus Gundissalinus, *De anima*, ed. J. T. Murckle, in *Mediaeval Studies* 2 /1 (1940), pp. 23–103, see p. 71. See also Hamesse, "Imaginatio et phantasia", pp. 170–171.

173 Albert the Great, *De anima*, ed. A. Borgnet (Paris: Ludovicus Vives, 1890), III, 1, chap. 3, p. 318a: *Oportet igitur phantasiam secundum aliquid esse differentem ab imaginatione et aestimativa, licet enim omnes istae potentiae passivae animae sint et patiantur a formis individui, quae sunt principia cognitionis sensibilis, quae formae eaedem esse videntur, tamen motus passionis non est idem, sed est secundum differentem gradum abstractionis, propter quod et ipsae potentiae non penitus sunt eaedem, sed, sicut superius diximus de sensu communi, videtur tota formalitas sensibilis virtutis esse phantasia, et hoc materialiter differentiam habet secundum quod inchoat ab organo unius motus eius, vel ab organo alterius, et hoc modo videntur omnes istae vires animae sensibilis esse interiores in una essentialitate communi et substantia, differentes autem secundum esse materiale in diversis partibus cerebri, in quo organizantur istae potentiae, quae omnes sunt organicae.*

Those cases notwithstanding, the prevalent usage from the 13th century on would be to consider both *phantasia* and *imaginatio* as synonyms. In the later translation of Aristotle's *De anima* by John Argyropoulos (*c.*1393–*c.*1487), the hendiadys *imaginatio et phantasia* (so emphasising the synonymy) is adopted in some passages when the Greek φαντασία occurs (in singular), although the translator shows a clear preference for the single Latin *imaginatio*.[174] Significantly, in his 1501 *Liber de imaginatione* (perhaps the most thorough account on the question produced in the 16th century) Giovanni Francesco Pico della Mirandola (1470–1533) establishes from the outset the total synonymy of both terms:

> So, the power of the soul, which the Greeks call φαντασία, is called *imaginatio* in Latin, and this name is received from its function of creating and shaping images within itself.[175]

Thus, the synonymy between *phantasia* and *imaginatio*, initiated in the ancient world and present in the Latin tradition of *De anima*, was preserved and well-established in the 16th century, undoubtingly reaching the northern-Italian milieu of Piccolomini.[176] As a token, the occurrence in Nifo's commentary on *De anima* (1544) not only of *phantasia*, but also of the passive participle *phantasiata* (together with *phantasiabilis* and *phantasiari*) reveals that the exact verbal form adopted by Piccolomini to describe mathematical objects – *quantum phantasiatum* – was already in circulation in the debates surrounding *phantasia* in Aristotle's *De anima*.[177] Thus, as an inheritor of the predominant Latin lexical tradition, Piccolomini does not draw any semantic difference between *phantasia* and *imaginatio*. Noteworthy is, though, that the latter only occurs once in the *Quaestiones mechanicae*;[178] for the rest, in his 1547 publication, Piccolomini always uses the Greek transliteration *phantasia*.

174 Spinoza, "*Phantasia* e imaginatio nell'*Aristotele latino*", p. 129.

175 Giovanni Francesco Pico della Mirandola, *Liber de imaginatione* (Rome: Aldus, 1501), p. 8: *Ea igitur animae vis, quam Graeci φαντασίαν nuncupant, Latine imaginatio dicitur idque nomen de officio sortita est ex imaginibus scilicet quas conficit et effingit in se*.

176 It should be noted that, although the assimilation of both terms was prevalent, some Renaissance authors distinguished between *phantasia* and *imaginatio* as different faculties. A notable example of this distinction is Marsilio Ficino. See Marsilio Ficino, *Théologie Platonicienne de l'immortalité des âmes*, ed. R. Marcel (Paris: Les Belles Lettres, 2007), VIII, I, p. 286 ff.

177 See ex. gr. Agostino Nifo, *Super libros De anima* (Venice: Giunti, 1544), III, p. 23a: [...] *nisi singularia sint phantasiata* [...].

178 Alessandro Piccolomini, *In mechanicas quaestiones*, f. XX–XXV.

2.2.5 *Averroes*

The so-called Paduan Averroism or the tendency in the intellectual milieu surrounding the University of Padua from the 14th to 16th century to adopt Averroes' interpretation as a default starting point in dealing with Aristotle's philosophy is part of Piccolomini's philosophical background. I will argue that it is highly likely that this was also the case in Piccolomini's understanding of the doctrine of *phantasia*.

In Averroes' widely read and discussed *Magnum commentarium in Aristotelis De Anima*, the question of *phantasia* (put as *ymaginatio* in the Latin text, التوهم *at-tawahhum* in the original Arabic) is addressed as a piece of the Cordovan's interpretation of *De anima* III. 4 and III. 5. The relevant passage for our purpose is Averroes' elaboration on the concept of *intellectus passibilis* (العقل المنفعل, *al-ʿaql al-munfaʿil* – or νοῦς παθητικός) as mentioned in III, 5, 430a, l. 24. He interprets the section in this manner:

> He meant here by passible intellect the forms of the imagination insofar as the cogitative power proper to human beings acts upon them. For that power is a kind of reason and its activity is nothing but the placing of the intention of the form imagined in its individuality in memory or the discerning of it from [the individual] in conception and imagination. And it is evident that the intellect which is called material receives the imagined intentions after this discernment. That passible intellect, therefore, is necessary for conceptualization. He, therefore, rightly said: "And we do not remember, because that is not passible, while the passible intellect is corruptible, and without this nothing understands". That is, without the imaginative power and the cogitative [power] the intellect, which is called material, understands nothing. For these powers are, as it were, things which prepare artistry's matter for receiving the activity of artistry.[179]

179 Translation by Taylor in Averroes, *Long Commentary on the De Anima of Aristotle*, ed. R.C. Taylor (London: Yale Library of Medieval Philosophy Series, 2009), III, p. 361. Latin in Averroes, *Commentarium magnum in Aristotelis De anima*, ed. F. Stuart Crawford (Cambridge, Massachusetts: The Medieval Academy of America, 1953), III, 5, l. 173–178: *Et intendebat hic per intellectum passibilem formas ymaginationis secundum quod in eas agit virtus cogitativa propria homini. Ista enim virtus est aliqua ratio, et actio eius nichil est aliud quam ponere intentionem forme ymaginationis cum suo individuo apud rememorationem, aut distinguere eam ab eo apud formationem et ymaginationem. Et manifestum est quod intellectus qui dicitur materialis recipit intentiones ymaginatas post hanc distinctionem. Iste igitur intellectus passibilis necessarius est in formatione. Recte igitur dixit: Et non rememoramur, quia iste est non passibilis et intellectus passibilis est corruptibilis, et sine hoc nichil*

Averroes' interpretation of the "passive intellect" can be synthesised in these four points:

(a) It is the part of the intellective human capacity belonging to the individual.
(b) It includes a range of powers dependent upon the body and, thus, perishable.
(c) Those powers are imagination, memory, and the cogitative power (*virtus cogitativa*).
(d) The activity of those powers enables the acquisition of knowledge with the cooperation of two external intellects, sc. the material intellect (*intellectus materialis*, العقل الهيولاني, *al-ʿaql al-hiyūlānī*, νοῦς ὑλικός) and the agent intellect (*intellectus agens*, العقل الفعال, *al-ʿaql al-faʿʿāl*, νοῦς ποιητικός).

Famously, Averroes defended both the separation and unicity of the material intellect as a substance making possible the acquisition of noetic forms which are shared by all human beings.[180]

Regarding *phantasia*, it must be emphasised that, much like Proclus, although because of different reasons, Averroes puts *ymaginatio* (or, most exactly, its intellection-related powers) under the label of *intellectus passibilis* (or *passivus*) in his commentary.

This conceptual alignment with Averroes, one of Piccolomini's major authorities, would significantly ease Piccolomini's reading of Proclus' theory of geometric φαντασία. In other words, there are only three traits of φαντασία mentioned in Proclus' prologues: (a) that it happens in the body and with the body, (b) that it is also called "passive intellect", and (c) that phantastic mathematical items possess a middle status between the perception and intellect. Proclus' more idiosyncratic theories, such as the two vehicles of the soul, the two uses of phantasy, or phantastic mythopoesis and theophany are never explicitly stated in his *In Euclidem*. All three features (a), (b) and (c) can be read as in agreement with Averroes, so facilitating Piccolomini's assimilation and reinterpretation of Proclus' theory of phantasy within the theoretical framework of *De certitudine*.

---

intelligit. Id est, et sine virtute ymaginativa et cogitativa nichil intelligit intellectus qui dicitur materialis; hec enim virtutes sunt quasi res que preparant materiam artificii ad recipiendum actionem artificii.

180 See, for instance, S. O. Ogden, *Averroes on Intellect. From Aristotelian Origins to Aquinas' Critique* (Oxford: Oxford University Press, 2022); Taylor, "Introduction" in Averroes, *Long Commentary on the De Anima of Aristotle*, pp. xv–cix; Taylor, 'Averroes on psychology and the principles of metaphysics' in *Journal of the History of Philosophy* 36 /4 (1998), pp. 507–523.

2.3 *Certitudo mathematica*: a Historical Inquiry

The akribology of mathematics is one of the main topics of Piccolomini's opusculum. Piccolomini holds that the *certitudo* of mathematical demonstrations and conclusions stems from the special status of its object of study which he calls *quantum phantasiatum*. In this respect, Piccolomini's view echoes those of a variety of Greek commentators, including Proclus and Priscian of Lydia. De Pace provides in her seminal study a revealing scholium attributed to Agostino Nifo in the 1527 edition of Themistius' paraphrase of *De anima*. In it, some notes on the meaning of the term together with a summary of the diversity of opinions amongst commentators regarding mathematical *certitudo* are expounded when clarifying the meaning of the term "ἀκρίβεια":

> The word "ἀκρίβεια", which Latin puts as "evidence" (*evidentiam*) or "certainty" (*certitudinem*) and "constancy" (*constantiam*), has been variously explained by authors. For the Latins and Averroes report it in the demonstrations of this doctrine [geometry], as also seems Themistius in this place [i.e., chap. 1, *On the usefulness and presence of the science of the soul*]. However, Simplicius and Philoponus do not put it in demonstrations but in its subject.[181]

Nifo refers to the newly roused debate through the opinions of the recently rediscovered commentaries that do not always agree with Averroes' view on the topic. As explained by Nifo, Themistius, much like Averroes, located the cause of mathematical *certitudo* in its demonstrative procedures or methods. We are also told that other Geek authorities such as Simplicius (probably he refers to Priscian's *In De anima*) and Philoponus grounded mathematical certainty in the ontological status of its subject-matter. Piccolomini's *De certitudine*, taking into consideration this late-ancient debate, would add Proclus as the pivotal and ultimate authority for sustaining the latter opinion.

As we just saw, Nifo puts *certitudo* together with "evidence" and "constancy" in defining ἀκρίβεια. But what did Piccolomini mean by the word *certitudo*? How was it linked to mathematics? This Latin term is invoked 14 times in the

---

181 The scholium is written to c.55r. col. a–b: ἀκρίβειαν *verbum, quod evidentiam vel certitudinem constantiamque interpretatur Latinus* [*sermo*], *varie exposuerunt auctores. Latini namque et Averroes referunt illud in demonstrationes huius doctrinae* [*geometriae*], *sicut etiam videtur Themistius hoc loco* [*cap. 1, De utilitate et praesentia scientiae de anima*]. *Simplicius autem Philoponusque non in demonstrationes, sed in eius subiectum.* See De Pace, *Le matematiche e il mondo*, pp. 49–50. See Themistius, *In Libros Aristotelis De anima paraphrasis*, ed. R. Heinze (Berlin: Reimer, 1899).

body of Piccolomini's opusculum and 2 times in the *Praefatio* to *In mechanicas quaestiones*. In this section, I will explore how the syntagm *certitudo mathematica* came to be part of the philosophical lexicon and what its precise meanings are in Piccolomini's text.

2.3.1 *The Coining of a Latin Concept*

Certitudo is a non-existent word in classical Latin, coined from the Latin adjective *certum*, meaning "firm", "unshakable" or "stable".[182] For its part, the adjective seems to derive from the verb *cerno*, and, thus, its primary sense would be something "seen", "examined", or "decided". It can be also related to the Proto-Indo-European root **krey*- and so related to Greek terms such as κρίνω (to discern), κρίτον (distinguished), or the Latin *cribrum* (filter).[183]

Certitudo is commonly rendered in Latin as corresponding to the Greek adjective ἀκριβές (see below), sometimes substantivised as τὸ ἀκριβές for meaning "acute", "precise", or "exact". Although its etymology is unclear, it has been related to ἄκρος, primarily meaning the topmost part of a mountain and, from this, deriving as "in the most acute part of something".[184] There is also a proper abstract term, ἀκρίβεια, and verbal forms such as ἀκριβόω (to investigate accurately), ἀκριβεύω (to speak precisely), ἀκριβάζω (to investigate accurately). From this come nominal forms such as ἀκρίβωσις (the act of making a precision), ἀκρίβωμα (precise account), ἀκριβασμός (thorough and precise research), and composed terms such as ἀκριβολογία (saying with precision). In its adverbial use (δι' ἀκριβείας; ἀκριβῶς) it also stands for "with minuteness" or "with great detail" or "with precision".

Although the abstract noun *certitudo* was non-existent in classical Latin, philosophical disputations of the nature of *certum*, either attributed to things (*certae res*) or epistemic agents (*incertus*/*certus sum*) are already present in Cicero's works[185] and the anti-Sceptic discussions of Augustine.[186] Related

182 According to Forcellini, *certitudo* appears for the first time and only once in the writings of Ammianus Marcellinus (*c.*330) but it might be a later misreading. See "Certitudo", in E. Forcellini, *Lexicon Totius Latinitatis* (Padua: J. Manfrè, 1771): *Certitudo-inis, f. 3. Apud Ammian. 30. 1. "quidam leg. In eadem fidei certitudine mansit immobilis". Sed abesse eam vocem in* MSS *atque aliter legendum esse Valesii fratres adnotarunt.*

183 M. de Vaan, *Etymological Dictionary of Latin* (Leiden: Brill, 2008), Vol. 7, p. 111, *certus*.

184 R. Beekes, *Etymological Dictionary of Greek* (Leiden: Brill, 2010), Vol. 1, p. 56, see ἀκριβής and p. 57 ἄκρος. On the use of ἀκρίβεια in Greek in various contexts see also A. Harris, L. Taub, "Quantification and precision: a brief look at some ancient accounts", in *Annals of Science* 81 (2023), pp. 10–29.

185 See, for instance, Cicero, *Academica*, I, 12; Cicero, *De natura deorum*, I, 1–5.

186 See ex. gr. Augustine, *Contra Academicos*, eds W. Green, K. Daur (Turnhout: Brepols, 1970), II, 5, 11: *Ex quo confici, ut nulli etiam rei sapiens assentiatur: erret enim necesse est, quod*

terms also occur in Boethius' *De consolatione philosophiae* meaning unchangeable future events.[187] As indicated by Biard, some early Latin pieces of philosophical literature in which the very abstract term *certitudo* was adopted and analysed are found in Peter Abelard's (1079–1142) *Tractatus de intellectibus*, where the subjective certainty characteristic of the subjective *habitus* of sciences is analysed.[188] Another of the term's first appearances in the Latin philosophical context is found in the works of Alan of Lille (*c.*1128–1202).[189]

2.3.2 *The Latin Aristotelian Tradition*

Certitudo is also present in the body of the *Aristoteles Latinus*: it appears 22 times in the anonymous 12th-century translation of *Topica*[190] and in some highly relevant sections of the translations of the *Metaphysica* (see below).

Noteworthy also regarding the wide semantics of the term within the Latin philosophical tradition is that *certitudo* also rendered the Arabic term يقين (yaqīn), which involves the semantics of "un-doubtfulness" and "faithfulness".[191] As Black has shown, the discussion surrounding certainty and how it is attained in scientific discourse became a philosophical question of the uttermost importance amongst Islamic thinkers after al-Fārābī's كتاب البرهان وكتاب شرائط اليقين (*Kitāb al-Burhān wa-Kitāb Šarā'iṭ al-yaqīn, Book of Demonstration and Book on the Conditions of Certitude*). Early appearances of *certitudo* in the Latin philosophical context coming from Arabic sources are found in Gerard of

---

sapienti nefas est, si assentiatur rebus incertis. Et omnia incerta esse non dicebant solum, verum etiam copiosissimis rationibus affirmabant. Sed verum non posse comprehendi, ex illa stoici Zenonis definitione arripuisse videbantur, qui ait id verum percipi posse, quod ita esset animo impressum ex eo unde esset, ut esse non posset ex eo unde non esset. See also Augustine, *Contra Academicos*, III, 10, 23: *Certum enim habeo, aut unum esse mundum, aut non unum; et si non unum, aut finiti numeri, aut infiniti.*

187 Boethius, *De consolatione philosophiae*, V, 4.

188 Biard, "Certitudo", in I. Atucha, D. Calma, C. Koenig-Pralong, I. Zavattero (eds), *Mots médiévaux offerts à Ruedi Imbach* (Porto: Fédération Internationale des Études Médiévales, 2011), pp. 153–162.

189 See ex. gr., Alanus of Lille, *The Plaint of nature*, ed. J. J. Sheridan (Toronto: Pontifical Institute of mediaeval studies, 1980), cap. 2, l. 280: *Quid vero in caligis camisia que, in superioribus vestibus consepultis, picture somniaret industria, nulla* ***certitudinis*** *auctoritate probavi.*

190 Aristotle, *Topica*, transl. Anonymus saec. XII, eds L. Minio-Paluello, B. G. Dod (Turnhout: Brepols, 1969), 4, cap. 5, p. 228 (9 times).

191 D. L. Black, "Knowledge ('ilm) and certitude (yaqin) in al-Farabi's epistemology", in *Arabic Sciences and Philosophy* 16/1 (2006), pp. 11–45.

Cremona's translation of al-Fārābī's *De scientiis*[192] and Avicenna's Latin text of *Philosophia prima*.[193]

Are there any previous cases in which *certitudo* is associated with mathematical disciplines in the Latin philosophical tradition?

As it shall appear, the notion of *certitudo* as linked to *mathematicae* was firmly established in the Latin context through the translations and commentary of particular passages of the Aristotelian corpus. A well-known statement relating Greek terms from ἀκριβ- to mathematical argumentation is found in the *Ethica Nichomachea* I, 3 (1094b). There Aristotle claims that the educated person cannot ask for the same degree of ἀκριβές in every subject:

> for it is the mark of an educated man to look for **precision** in each class of things just so far as the nature of the subject admits (πεπαιδευμένου γάρ ἐστιν ἐπὶ τοσοῦτον **τἀκριβὲς** ἐπιζητεῖν καθ' ἕκαστον γένος, ἐφ' ὅσον ἡ τοῦ πράγματος φύσις ἐπιδέχεται).[194]

Aristotle then proceeds to affirm that τὸ ἀκριβές cannot be expected to the same degree in mathematical demonstrations as in rhetoric and that it is as absurd to expect rhetorical arguments in mathematics (μαθηματικοῦ τε πιθανολογοῦντος) as it is to require demonstrations in rhetoric. Thus, in this section, by proposing extreme cases of certainty in different subjects, Aristotle places

192 Al-Fārābī, *De scientiis*, transl. Gerard of Cremona, eds A. Galnnier, M. Lejbowicz, J. Jolivet (Turnhout: Brebols, 2016), cap. 2, p. 190, l. 133: *Ille autem qui estimat quod usus in orationibus et disputationibus topicis, aut usus in disciplinis, sicut scientia geometrie et arithmetice excusat a scientia canonum dialectice, aut stat loco eius, aut facit operationem ipsius, aut dat homini virtutem ad experiendam omnem rationem et omnem disputationem et omnem sententiam, aut dirigit hominem ad veritatem et* ***certitudinem*** *ita ut non erret in aliqua reliquarum scientiarum penitus, est sicut ille qui estimat quod usus et studium in memoria versuum et epistolarum et multiplicatio reddendi eos sicut ab alio audivit, excuset in rectificatione lingue, et in hoc ne erret homo a regulis grammatice, et stat loco eius, et facit eius operationem, et quod dat homini virtutem qua experiatur declinationem omnis sermonis an recte invenit in eo aut erravit.*

193 See ex. gr., Avicenna, *Liber de philosophia prima sive scientia divina*, transl. Simone van Riet, ed. G. Verbeke, vol. 1–3 (Leiden: Brill, 1977–1983), vol. 1, tract. 1, cap. 1, p. 1, l. 13: *Et diximus quod speculativae sunt illae in quibus quaerit perfici virtus animae speculativa per acquisitionem intelligentiae in effectu scilicet per adeptionem scientiae imaginativae et creditivae de rebus quae non sunt nostra opera nec nostrae dispositiones. In his ergo finis est* ***certitudo*** *sententiae et opinionis.*

194 Translation by Barnes in Aristotle, *Complete Works of Aristotle, the revised Oxford translation*, ed. J. Barnes. Vol. 1–2 (Chichester: Princeton University Press, 1984), Vol. 2 (slightly modified).

mathematical arguments in a greater degree of certainty than the less certain, rhetorical arguments.

The first two completed translations of the *Ethica Nicomachea* known to the Latin West rendered τἀκριβές in this section as *certitudo*. In the early translation of Robert Grosseteste (1175–1253), we find the following:

> Disciplinati enim est in tantum certitudinem inquirere secundum unumquodque genus, in quantum rei natura recipit.[195]

The later translation of Leonardo Bruni (1370–1444) renders the same sentence in the following terms:

> Nam eius qui eruditur est eatenus in unoquoque genere certitudinem exigere, quatenus eius natura rei recipit.[196]

Later in the same book, Aristotle restates his point and invokes geometry as a paradigm of ἀκριβές, reminding the reader that it is not to be expected in the same way in every discourse (τὸ γὰρ ἀκριβὲς οὐχ ὁμοίως ἐν ἅπασι τοῖς λόγοις ἐπιζητητέον [...]).[197] Once again Grosseteste translates this as *certitudo* ([...] *certitudinem non similiter in omnibus exquirere*),[198] whereas, less consistently, Bruni chooses "exact discussion" ([...] *neque exactam discussionem in omnibus eodem modo*).[199] This line from the *Ethica* is invoked by Piccolomini at the very beginning of *De certitudine*.[200]

Another three pivotal *loci* in which Aristotle mentions the question of ἀκριβές and relates it to mathematical disciplines are to be found in the *Metaphysica*. By examining these passages and checking how terms related to ἀκριβές were rendered into Latin in James of Venice's *translatio vetustissima*, the anonymous *translatio media*, and Moerbeke's revision will provide some insight into the association of the term *certitudo* with *mathematicae*, and, more generally, concerning the inclusion of the term in the Latin epistemological lexicon. The first passage is in chapter 2 of the book I:

195 Aristotle, *Ethica Nicomachea*, transl. Robert Grosseteste, ed. R. A. Gauthier (Turnhout: Brepols, 1973), I, cap. 3, p. 143, l. 8.

196 Aristotle, *Ethicorum libri X cum Averrois Cordubensis exactis commentariis*, transl. Leonardo Bruni (Venice: Giunti, 1542), f. 5v.

197 Aristotle, *Ethica Nicomachea*, ed. I. Bekker, vol. 2, I, 7, 1098a.

198 Aristotle, *Ethica Nicomachea*, transl. Robert Grosseteste, libri I, cap. 7, p. 151, l. 23.

199 Aristotle, *Ethicorum libri*, transl. Leonardo Bruni, f. 13r.

200 Alessandro Piccolomini, *De certitudine*, f. LXXIV: *Certitudo mathematica non in omnibus expetenda.*

> And the **most exact** of the sciences are those which deal the most with first principles; for those which involve fewer principles are **more exact** than those which involve additional principles, e.g., arithmetic than geometry ([…] **ἀκριβέσταται** δὲ τῶν ἐπιστημῶν μάλιστα τῶν πρώτων εἰσίν αἱ γὰρ ἐξ ἐλαττόνων **ἀκριβέστεραι** τῶν ἐκ προσθέσεως λεγομένων, οἷον ἀριθμητικὴ γεωμετρίας).[201]

Here, Aristotle adopts terms containing the root ἀκριβ- for expressing exactness or rigour. We are told that the sciences about principles and those using less conjectural statements are more exact and rigorous. He gives arithmetic as an example of a science more exact than geometry. The same passage is translated into Latin by James of Venice, the anonymous 12th- or 13th-century Greek-to-Latin *translatio media*, and Moerbeke in this way:

TABLE 1 *Certitudo* 1

James of Venice's *vetustissima*	***Translatio media***	**Moerbeke's *revisio***
[…] **certissime** autem scientiarum maxime primarum sunt (ex minoribus enim **certiores** sunt his que ex appositione dicuntur, ut ex arismetica geometria).[a]	[…] scientiarum vero **certissime** sunt que maxime prime sunt (nam que sunt ex paucioribus **certiores** sunt ex additione dictis, ut arismetica geometria).[b]	Scientiarum vero **certissime** sunt que maxime primorum sunt; nam que sunt ex paucioribus **certiores** sunt ex additione dictis, ut arismetica geometria.[c]

a Aristotle, *Metaphysica: translatio vetustissima*, transl. James of Venice, ed. G. Vuillemin-Diem (Turnhout: Brepols, 1970), liber I, cap. 2, p. 8.
b Aristotle, *Metaphysica: translatio media*, transl. Anonymus saec. XII vel XIII, ed. G. Vuillemin-Diem (Turnhout: Brepols, 1976), liber I, cap. 2, p. 9.
c Aristotle, *Metaphysica*, recens. William of Moerbeke, ed. G. Vuillemin-Diem (Turnhout: Brepols, 1995), liber I, cap. 2, p. 15.

Thus, in the three cases the superlative ἀκριβέσταται is rendered as *certissime*, and the comparative ἀκριβέστεραι as *certiores* (Moerbeke preserved the *media*'s term).

201 Aristotle, *Metaphysica*, I, 2, 982a, l. 19–28. Translation in Aristotle, *Complete Works of Aristotle*, Vol. 1.

Yet again, in dealing with the degree of minuteness appropriate to each discourse in chapter 3 of book II, Aristotle adopts terms derived from ἀκριβές to mathematics:

> And some want to have everything done accurately (ἀκριβῶς), while others are annoyed by accuracy (certainty), either because they cannot follow the connexion of thought or because they regard it as pettifoggery. For accuracy (τὸ ἀκριβὲς) has something of this character, so that, as it happens in trade, so also in argument some people find it to be mean. Therefore, one must be already trained to know how to take each sort of argument, since it is absurd to seek at the same time knowledge and the way of attaining knowledge; and neither is easy to get. The minute accuracy (τὴν δ'ἀκριβολογίαν) of mathematics is not to be demanded in all cases, but only in the case of things which have no matter.[202]

In this case, Aristotle expounds on the conflicting opinions of those who look for dissimilar degrees of rigour in different sciences and everyday affairs, recalling that the rigorousness and precision in the speech (ἀκριβολογία) required might be annoying for some untrained people or in some contexts. The Stagirite remarks (much like in the section from *Ethica Nicomachea* above) that the educated person should know which degree of detail and precision is to be expected in every discipline and affair, for the rigour/accuracy/precision of mathematics can only be expected in sciences dealing with immaterial things ([…] τοῖς μὴ ἔχουσιν ὕλην). Terms related to ἀκριβ- occur three times in this section, first as an adverb, then in its substantiation from the adjective, and finally in the compound ἀκριβολογία.

James translated the adverb ἀκριβῶς as *certe*, τὸ ἀκριβές consistently as *certum*, and ἀκριβολογία as *certitudo*: this is, thus, the first ever produced translation corresponding *ad litteram* to the Latin syntagm *certitudo mathematica*. The *media* and Moerbeke render the adverbial form ἀκριβῶς as *certe* and τὸ ἀκριβές firstly as *certitudo* and, in its second occurrence, as *certum*. Probably compelled by the challenge of accurately representing a Greek compound in one single Latin word, the translators preserved ἀκριβολογία in its original form, directly transcribing *acribologia*.

202 Translation in Aristotle, *Complete Works of Aristotle*, vol. 1 (slightly modified). Aristotle, *Metaphysica*, II, 3, 995a, l. 8–16: καὶ οἱ μὲν πάντα ἀκριβῶς, τοὺς δὲ λυπεῖ τὸ ἀκριβὲς ἢ διὰ τὸ μὴ δύνασθαι συνείρειν ἢ διὰ τὴν μικρολογίαν: ἔχει γάρ τι τὸ ἀκριβὲς τὸ τοιοῦτον, ὥστε, καθάπερ ἐπὶ τῶν συμβολαίων, καὶ ἐπὶ τῶν λόγων ἀνελεύθερον εἶναί τισι δοκεῖ, διὸ δεῖ πεπαιδεῦσθαι πῶς ἕκαστα ἀποδεκτέον, ὡς ἄτοπον ἅμα ζητεῖν ἐπιστήμην καὶ τρόπον ἐπιστήμης. ἔστι δ'οὐδὲ θάτερον ῥάδιον λαβεῖν. τὴν δ'ἀκριβολογίαν τὴν μαθηματικὴν οὐκ ἐν ἅπασιν ἀπαιτητέον, ἀλλ'ἐν τοῖς μὴ ἔχουσιν ὕλην.

TABLE 2 *Certitudo* 2

James of Venice's *vetustissima*	*Translatio media*	Moerbeke's *revisio*
Et hii quidem omnia **certe**, quidam autem dolent **certum** aut propter id quod non possunt applicari aut propter longam locutionem; habet enim aliquid **certum** huiusmodi, quare, sicut in commerciis, et in rationibus liberum {vel non} aliquid quibusdam esse videtur. [...] **Certitudini** autem mathematice non in omnibus obviandum est, sed in non habentibus materiam.[a]	Et illi quidem omnia **certe**, hiis vero flebilis est **certitudo** aut propter impotentiam complectendi aut propter micrologiam; habet enim aliquid quod **certum** est tale, unde quemadmodum in symbolis et in rationibus non liberum esse quibusdam videtur. Propter quod oportet erudiri quomodo singula sunt recipienda et absurdum est simul quaerere scientiam et modum scientie. Est autem neutrum facile accipere. **Acribologia** vero mathematica non in omnibus est expetenda sed in non habentibus materiam.[b]	Et illi quidem omnia **certe**; hiis vero flebilis est **certitudo** aut propter impotentiam complectendi aut propter micrologiam. Habet enim aliquid quod **certum** est tale, ut quemadmodum in symbolis et rationibus non liberum esse quibusdam videtur. Propter quod oportet erudiri quomodo singula sunt recipienda, et absurdum est simul querere scientiam et modum scientie; est autem neutrum facile accipere. **Acribologia** vero mathematica non in omnibus est expetenda, sed in non habentibus materiam.[c]

a Aristotle, *Metaphysica: translatio vetustissima*, transl. James of Venice, liber II, cap. 3, p. 40.
b Aristotle, *Metaphysica: translatio media*, transl. Anonymus saec. XII vel XIII, liber II, cap. 3, p. 39.
c Aristotle, *Metaphysica*, recens. William of Moerbeke, liber II, cap. 3, p. 47.

This passage from *Metaphysica*, II, 3 is perhaps one the most influential in the history of Latin akribology. This is even indicated by Piccolomini himself who, in the preface to *De certitudine*, refers to Averroes' commentary on this section as the source of the widespread opinion which he will discuss through the treatise, namely that mathematical demonstration possesses the highest degree of *certitudo*.[203] Averroes' comment reads as follows:

203 Alessandro Piccolomini, *De certitudine*, f. LXXII: *Primum et rationibus et authoritatibus demonstrare intendimus mathematicas demonstrationes non esse illas potissimas quas in*

> And it is not fitting for a human being to seek the mode of trust in natural demonstrations to be as the mode of trust in mathematics. For mathematical demonstrations are in the first order of certainty (*ea sunt in primo ordine certitudinis*, هي في المرتبة الأولى من اليقين / *hiya fī l-martaba l-ūlā min al-yaqīn*), and natural demonstrations follow them in this. Certainty varies within the same science, for example, according to unqualified demonstration (*demonstratio simpliciter*) and according to a demonstration "why" and, when it varies within the same science, it varies more within different sciences according to the kind.[204]

In this passage, اليقين (*al-yaqīn*), corresponding to the Greek ἀκριβές, is rendered as *certitudo*. Also paramount, as we shall analyse below, is that Averroes mentions demonstration *simpliciter* or *potissima* in this very same passage in which he concedes the highest degree of certainty to mathematics. This passage would become a pivotal text in the Latin debates regarding mathematical demonstration vis-à-vis demonstration in natural philosophy (see section 2.6).

Coming back to Aristotle's *Metaphysica*, the incipit of book VI contains a third section mentioning both ἀκριβές and mathematics:

> the objects of mathematics have principles and elements and causes, and, in general, every dianoetic science or at all involving reasoning (διανοίας)

Posterioribus analyticis construit Aristoteles. Secundo vero loco, ne propter hoc labefactare videamur Averrois sententiam in II Metaphysicorum iam allegatam causam assignare est animus, qua mathematicae disciplinae in primo esse gradu certitudinis dici posse. The same theme is referred in the title of chapter XII of *De certitudine*: *Quae tandem sit vera causa cur mathematicae disciplinae in primo gradu certitudinis ab Averroe positae fuerint*.

204 Averroes, *In Metaphysica Aristotelis*, lib. II, f. 35v, l. 16–27: *Et non oportet hominem quaerere ut modus fidei in demonstrationibus naturalibus sit sicut modus fidei in mathematicis. Demonstrationes enim mathematicae sunt in primo ordine certitudinis, et demonstrationes naturales consequuntur eas in hoc. Certitudo enim diversatur in eadem scientia verbi gratia secundum demonstrationem simpliciter, et secundum demonstrationem quia et, cum diversatur in eadem scientia, magis diversatur in scientiis diversis secundum genus*. Original Arabic at Averroes Arabicus, *In Aristotelis Metaphysicam* 995a14–16 (α, 3), C. 16: ليس ينبغي ان يطلب الانسان كون نوع التصديق الذي يقع له في البراهين التعاليمية مثل التصديق الذي يقع له في البراهين الطبيعية يريد ان البراهين التعاليمية هي في المرتبة الاولى من اليقين وان البراهين الطبيعية تتلوها في ذلك لا أن براهينها ليست داخلة في العلم اليقين فان اليقين عنده يتفاضل في العلم الواحد بعينه مثل ان نبرهن على الشيء ببرهان مطلق او ببرهان وجود فقط وهو الذي يسمى الدليل واذا كان يتفاضل في العلم الواحد فهو احرى ان يتفاضل في العلوم المختلفة الاجناس. See also P. Adamson, "Yahyá ibn ʿAdi and Averroes on «Metaphysics» Alpha Elatton", in *Documenti E Studi Sulla Tradizione Filosofica Medievale* 21 (2010), pp. 343–374.

> deals with causes and principles which are more exact (ἀκριβεστέρας) or simpler.[205]

Here Aristotle states that different sciences involve different principles, elements, and causes, putting *en passant* the emphasis on mathematics. Following a line of thought like the one followed in book II (see the quotation above), he states that the principles and causes of different sciences entail different degrees of precision (ἀκριβεστέρας) and simplicity. Arguably, mathematics is mentioned in this paragraph as an example of a science endowed with both exactness and simplicity (in its objects). The *media*'s and Moerbeke's versions (James' perhaps was never made) are:

TABLE 3 *Certitudo* 3

Translatio media	Moerbeke's *revisio*
[...] sunt etiam mathematicorum principia et elementa et cause, et totaliter omnis scientia sententialis, alicuius participans scientie circa causas et principia est aut **certiora** aut simpliciora.[a]	[...] sunt etiam mathematicorum principia et elementa et cause, et totaliter omnis scientia intellectualis participans aliquid intellectus circa causas et principia est aut **certiora** aut simpliciora.[b]

a Aristotle, *Metaphysica: translatio media*, transl. Anonymus saec. XII vel XIII, liber VI, cap. 1, p. 116.
b Aristotle, *Metaphysica*, recens. William of Moerbeke, liber VI, cap. 1, p. 125.

In both cases the adjectival form ἀκριβεστέρας is rendered as *certiora.*

These four *loci* and their respective Latin translations provide evidence that, from James of Venice to Moerbeke, and from Grosseteste to Bruni, in the sections of the Aristotelian corpus where terms deriving from ἀκριβές are related to *mathematica*, the translations adopted were almost always that of the Latin root *cert-* (*certissimae*, *certior*, *certitudo*, *certum*, *certe*).

The pair *certitudo* and *mathematica* would be further confirmed and spread in the Latin tradition of commentary using these translations. Moreover, much

205 Translation in Aristotle, *Complete Works of Aristotle*, vol. 1 (slightly modified). Aristotle, *Metaphysica*, VI, 1, 1025b, l. 3–7: [...] μαθηματικῶν εἰσὶν ἀρχαὶ καὶ στοιχεῖα καὶ αἴτια, καὶ ὅλως δὲ πᾶσα ἐπιστήμη διανοητικὴ ἢ μετέχουσά τι διανοίας περὶ αἰτίας καὶ ἀρχάς ἐστιν ἢ ἀκριβεστέρας ἢ ἁπλουστέρας.

like it happened amongst northern Italian scholars, Averroes' commentaries regarding both the highest sort of demonstration and the certainty of mathematics would ignite these philosophical questions amongst 13th- and 14th- century Latin philosophers. As we saw above, in commenting on the sections from Aristotle's book II of *Metaphysica*, Averroes granted supreme certainty to mathematics. The Latin translation of Averroes' comment contained the term *certitudo*.[206] One case of a 13th-century discussion of *certitudo* following both Aristotle and Averroes (undoubtedly influential in the 16th-century philosophical milieu) is that of Thomas Aquinas: he uses the term *certitudo* eight times (or variations from *cert-* such as *certitudinalis* and *certa ratio*) in commenting on the two passages quoted above from book II of the *Metaphysica*. Aquinas preserves the Graecism *acribologia*, however, rendering it otherwise as "certain reason".[207] Another highly relevant case of a 14th-century Latin author discussing the topic is that of Buridan. In his *In Metaphysicen*, book II, q. 1, he addresses the question of certainty in a wide sense and in his *In libros Posteriorum analyticorum*, book I, q. 25, he treats the very issue of mathematical certainty (*Utrum mathematicae scientiae sint aliarum scientiarum certissimae*).[208] In this question, the certainty of mathematics and metaphysics are contrasted, with Buridan proposing and exploring four distinct types of certitude. Nevertheless,

206 Averroes, *In Metaphysica Aristotelis*, II, p. 35v, l. 16–27.

207 Thomas Aquinas, *Sententia libri Metaphysicae*, eds M. R. Cathala, R. M. Spazzi (Turin: Marietti, 1950), lec. 5: *Et hoc etiam est vel propter consuetudinem, vel propter defectum iudicii, quia non possunt diiudicare utrum ratio per* ***certitudinem*** *concludat; et ideo quasi non credentes suo iudicio requirunt iudicium alicuius noti. Sunt etiam aliqui qui omnia volunt sibi dici per* ***certitudinem****, idest per diligentem inquisitionem rationis. Et hoc contingit propter bonitatem intellectus iudicantis, et rationes inquirentis; dummodo non quaeratur* ***certitudo*** *in his, in quibus* ***certitudo*** *esse non potest. Quidam vero sunt qui tristantur, si quid per* ***certitudinem*** *cum diligenti discussione inquiratur. Quod quidem potest contingere dupliciter. Uno modo propter impotentiam complectendi: habent enim debilem rationem, unde non sufficiunt ad considerandum ordinem complexionis priorum et posteriorum. Alio modo propter micrologiam, idest parvorum ratiocinationem. Cuius similitudo quaedam est in* ***certitudinali*** *inquisitione, quae nihil indiscussum relinquit usque ad minima.* [...] *Deinde cum dicit acribologia vero ostendit quod ille modus, qui est simpliciter optimus, non debet in omnibus quaeri; dicens quod* ***acribologia*** *idest diligens et* ***certa ratio****, sicut est in mathematicis, non debet requiri in omnibus rebus, de quibus sunt scientiae; sed debet solum requiri in his, quae non habent materiam. Ea enim quae habent materiam, subiecta sunt motui et variationi: et ideo non potest in eis omnibus omnimoda* ***certitudo*** *haberi. Quaeritur enim in eis non quid semper sit et ex necessitate; sed quid sit ut in pluribus.*

208 For some notes on certainty as used by Aquinas and Buridan see Ch. Grellard, "Certainty", in H. Lagerlund (ed.), *Encyclopedia of Medieval Philosophy* (Dordrecht: Springer, 2011), pp. 206–208; John Buridan, *In Metaphysicen Aristotelis questiones argutissimae* (Paris: Iodicus Badius Ascensius, 1518), repr. in *Kommentar zur Aristotelischen Metaphysik* (Frankfurt am Maine: Minerva, 1964).

Buridan is not mentioned by Piccolomini as one of his sources and there is no evidence pointing that Piccolomini knew Buridan's commentary.

Therefore, it can be reasonably argued that in his understanding of the Greek terms related to ἀκριβ- (including his reading of Proclus' *In Euclidem*), Piccolomini followed the established Latin lexico-philosophical tradition. Direct witnesses, which confirm that Piccolomini understood the terms related to the Greek root ἀκριβ- as *certum* or *certitudo,* are found in the very body of *De certitudine.* Ex. gr., in chapter XII, invoking Aristotle's authority in the passage of the *Metaphysica* quoted above (II, 3, l. 995a, l. 14–16) where he tries to explain the origin of mathematical *certitudo* and the diminution of epistemic transparency of sensible matter, it is affirmed that:

> Aristotle himself more openly assigns the cause of mathematical certainty (*certitudinis mathematicae*) to this, namely, because in mathematics, there is no limited matter, which is the cause of certainty (*causa certitudinis*) and destructive of science. For in book II of the *Metaphysica,* he has these words: "Mathematical accuracy (τὴν δ' ἀκριβολογίαν) is not to be demanded in all things, but in those that do not have matter".[209]

From this, it is clear that Piccolomini took ἀκριβολογία in Aristotle's text as *certitudo.*

2.3.3 *Akribology in Piccolomini's Greek Sources: Priscian and Proclus*

Two other Greek sources (besides Aristotle) are invoked by Piccolomini in expounding on his akribology of mathematics. These are Priscian's commentary on Aristotle's *De anima* (known to Piccolomini as "Simplicius") and Proclus' *In Euclidem.*

In chapter XII of *De certitudine,* Piccolomini invokes (Pseudo-)Simplicius' commentary (on 402b, l. 11) to confirm his view that mathematical *certitudo* has its origin in the idiosyncratic traits of quantity and that mathematical items are known with the greatest certainty.[210] The passage alluded runs as follows:

---

209 Alessandro Piccolomini, *De certitudine,* f. CIXr: *Apertius tamen Aristoteles ipse assignat causam certitudinis mathematicae ex hoc, scilicet quia non est ibi limitata materia, quae est causa certitudinis et scientiae destructiva. II enim Metaphysicorum habet haec verba:* τὴν δ' ἀκριβολογίαν τὴν μαθηματικὴν οὐκ ἐν ἅπασιν ἀπαιτητέον, ἀλλ' ἐν τοῖς μὴ ἔχουσιν ὕλην. See also Biard, "La certitude des mathématiques et ses fondements selon Piccolomini".

210 Alessandro Piccolomini, *De certitudine,* f. CVIII–CVIIIv: *Hoc idem sentit Simplicius, qui I De Anima, XI. dicit causam certitudinis mathematicarum esse, quia versantur circa quantum. Quantitates, enim, ut dicit ipse, sunt res sensatae, et causas sensatas habent, et ideo nobis notas.* Pioccolomini seeminly alludes to the passage following Faseolo's edition

Scientific knowledge is knowledge of activities from essence, but the reverse is clearer to us (πρὸς ἡμᾶς δὲ σαφεστέρα). We could not know the relative activities from their correspondence to the knowable, or to the desirable, or to what nourishes, without knowing the things to which they are related. These are correlates through the relational linkage, in the way that counteraction and resistance are linked with other substances. Knowledge about their related objects [of the soul activities] is clearer than about the activities, since the objects are more sensible (αἰσθητικώτερα), such as the sensible itself, and sensible objects of desire and what is intelligible in sensible things.[211]

---

and division of the text. See Simplicius, *In De anima Aristotelis*, f. IIII–IIIIv: *Scientiae ea sane propria ratio est, quae ex ipsa substantia actionum noscendarum vim suam adipiscitur. Respectu vero nostro ea manifestior existit quae contrario modo nobis manifestatur. At neque actiones, cum ad aliquid referantur propter respectum, aut tanquam ad id cognoscitur, aut quod experitur, aut quod nutrit, cognoscere possemus sine eorum cognitione ad quae referuntur. Quae opposita dicuntur, secundum oppositionem eorum quae ad aliquid dicuntur, ceu recursus ac reflexio quaedam fit, eorumque mutua inter se permanentia. Magisque perspicua haec quae de oppositis habetur cognitio cognitione illa est, quae de actionibus confestim tradi posset. Quandoquidem et sensui magis exposita sunt ipsa opposita, quemadmodum ipsum sensibile, et quod inter haec appetibile; et quatenus ipsum quoque intelligibile inter ea, quae sentiri possunt, connumerandum est.*

211 English translation (slightly modified) by Urmson. See Simplicius, *On Aristotle's On the Soul 1.1–2.4*, transl. J. O. Urmson (London: Bloomsbury Academic, 1994), p. 29. Original Greek in Simplicius, *In libros Aristotelis De anima commentaria*, ed. M. Hayduck (Reimer: Königlich Preussische Akademie der Wissenschaften zu Berlin, 1882), p. 14, l. 15–24: ἐπιστημονικὴ μὲν ἡ ἐκ τῆς οὐσίας τῶν ἐνεργειῶν γνωστική, πρὸς ἡμᾶς δὲ σαφεστέρα ἡ ἀνάπαλιν. ἀλλὰ καὶ τὰς ἐνεργείας πρός τι οὔσας διὰ τὴν ἢ ὡς πρὸς γνωστὸν ἢ ὡς πρὸς ὀρεκτὸν ἢ ὡς πρὸς τὸ τρέφον ἀπότασιν οὐκ ἂν γνοίημεν ἄνευ τῶν πρὸς ἃ λέγεται· ἅπερ ἀντικείμενα κατὰ τὴν τῶν πρός τι ἀντίθεσιν, οἷον ἀντίβασίν τε καὶ ἀντέρεισιν οὖσαν καὶ σὺν ἀλλήλοις ὑπόστασιν. καὶ ἐναργεστέρα τῆς περὶ τῶν ἐνεργειῶν γνώσεως ἡ περὶ τῶν ἀντικειμένων, ὅταν αἰσθητικώτερα ᾖ τὰ ἀντικείμενα, ὡς αὐτὸ τὸ αἰσθητὸν καὶ τὸ ἐν τούτοις ὀρεκτὸν καὶ ὅσον ἐν τοῖς αἰσθητοῖς νοητόν. τὸ δέ γε ἐξῃρημένον τοῦ ἡμετέρου νοῦ ἐκ τοῦ νοῦ ληψόμεθα μᾶλλον. See also the following commentary on 402b, l. 16, where Priscian uses mathematical objects as an example: pp. 14, l. 31–15, l. 5: αἰτία γὰρ τῶν συμβεβηκότων ἡ οὐσία. οὐσίαν δὲ ἀκουστέον νῦν κοινότερον πᾶν εἶδος, κἂν ὑπὸ ἡντιναοῦν τελῇ κατηγορίαν. τὸ γοῦν εὐθὺ καὶ τὸ τρίγωνον εἴληπται, οἷς τὰ συμβαίνοντα ἐκ τοῦ ὁρισμοῦ αὐτῶν γνωρίζομεν. ὁ γὰρ ὁρισμὸς οἰκείως τὸ εἶναι αὐτῶν ἐκφαίνων παρασκευάζει ἡμᾶς ῥᾳδίως καὶ τὰ καθ'αὑτὰ ὑπάρχοντα τῷ τοιῷδε ὄντι γνωρίζειν. ἐπιστημονικὸν οὖν τοῦτο, ἐπειδὴ τὸ ἑκάστου ὂν τῶν ὑπαρχόντων αὐτῷ αἴτιον. ἀλλὰ τὸ ὂν ἐνίοτε οὐκ ἐξ ἑαυτοῦ ἂν γνώριμον ἀνάπαλιν ἐκ τῶν καθ'αὑτὰ συμβεβηκότων αὐτῷ, ὧν καὶ ἡ αἴσθησις καὶ ἡ φαντασία ἀντιληπτική, γινώσκεται. English translation by Urmson in Simplicius, *On Aristotle's On the Soul 1.1–2.4*, pp. 29–30: "For essence is the cause of attributes. 'Essence' here has to be taken in a wide sense as being form, under whichsoever category it falls. Thus the straight line and the triangle are taken as examples, whose consequential attributes we know from their definition. For their definition, which exhibits exactly their being, enables us easily to know what belongs as such to a thing of that sort. So that is scientific knowledge, since the being

Piccolomini would use this passage and the following section in Pseudo-Simplicius' text to reinforce his view that the certainty or "transparency" of mathematical objects stems from the fact that quantity, although insubstantial and inactive, is the most common accident among sensible things, making it "more perceptible" (αἰσθητικώτερα) to us, and, therefore, a more abstractable and better-known trait (see below). As we shall see, Boccadiferro, Piccolomini's teacher, also shared this view (see 2.4.3). This passage confirms that one of the meanings of Piccolomini's *certitudo* is "easy to grasp" and, therefore, "epistemically transparent".

Nevertheless, there are additional passages in Pseudo-Simplicius' commentary that offer reflections on mathematical certainty aligning with Piccolomini's expressed perspectives. I will present three of these.

A locus containing a detailed statement of mathematical ontology and its relation to τὸ ἀκριβές in Pseudo-Simplicius' *In De anima* comes in chapter 4 of book III. As we saw above, the sensible perception of quantity is emphasised by Pseudo-Simplicius as an occasional cause without which mathematical items cannot be generated. From this tenet, he explains the origin and some traits of τὸ ἀκριβές in mathematics. He writes:

> Aristotle denies the substantiality of the impressions [in phantasy] and says they arise from abstraction, as the soul receives them in accordance with an imitation in respect of sensible objects. It is not because the soul does not project these too by its own activity, that is shown by their turning one to truth and precision (ἀκριβὲς), for the perceived shape is not truly cubic, nor does the physical straight line truly have no breadth: only the mathematical one does. For the impressions come from inside, but the soul does not project them without cognition of sensible objects, just as it does not do so for superior or intelligible objects of knowledge. But, because of the very great similarity of the things that are phantasised to the sense-objects they have the place of abstraction (τῆς ἀφαιρέσεως) specifically amongst the phantasised things, as these are as nearly as possible the same as the sense objects, without the physical substance, since the impression in phantasy is not like the form but like what is informed (τὸ εἰδοπεποιημένον).[212]

of each thing is the cause of what belongs to it. But sometimes the being of a thing is not recognisable from itself and is, the other way round, recognised from its consequential attributes, which sense and imagination can seize on. So, in mathematics, the essence of each number is recognised from the qualities and quantities of the definienda, which can be seized on by sense and phantasy".

212 Translation by Steel (slightly modified) in 'Simplicius', *On Aristotle's On the Soul 3.6–13* (London: Bloomsbury Academic, 2014), p. 98. Simplicius, *In De anima*, com in III, 429b,

According to Pseudo-Simplicius (when qualifying Aristotle's views as to reconcile them with phantastic projectionism), corporeal instantiations of mathematical entities fail to embody the real mathematical concepts: a corporeal cube is not cubic, the drawn line is not one-dimensional (without width). He, though, claims that the "true and exact" (τὸ ἀληθὲς καὶ ἀκριβές) mathematical objects arise from the activity of the soul. This means that it is the soul that projects the actual (exact and true) mathematical forms after having encountered a relatable figure, as a wooden cube, in sensible matter. Thus, one of the meanings of ἀκριβές is the representational accuracy of phantastic object vis-à-vis the inadequacy of natural matter to enact perfect geometrical figures. Noteworthy is also that προβολή/προβάλλω are present in Pseudo-Simplicius' passage (I will elaborate on this point below).

The opinion that a greater degree of accuracy (understood as objectual precision) is given in phantasy is touched upon in other places of Pseudo-Simplicius' commentary that might be behind Piccolomini's quotation. When discussing the truth- and falsity-related aspects of phantasy in III, 3, 428a, l. 9 ff., he introduces a brief excursus on mathematics as it is represented in phantasy, highlighting its greater precision when compared to perceptible items.

> Things that can be imagined are more precise (ἀκριβέστερα) than sense-objects, for example, a straight line, a surface, shapes.[213]

A third testimony from Pseudo-Simplicius introduces an implied, though not previously stated, aspect of mathematical accuracy: it is not only found within phantasy but is also provided by phantasy itself as an inherent characteristic, distinct from the sensible items that merely prompt (or awaken, in Priscian's metaphor) phantasy's geometrical activity.

> and the phantasy itself is stimulated to the projection (προβολὴν) of phantastic impressions in accordance with the appropriate reason-principles

---

l. 18, p. 233, l. 13–25: ὁ δὲ Ἀριστοτέλης τῶν τύπων ἀποφάσκει τὴν οὐσίαν καὶ ἐξ ἀφαιρέσεως ἐγγίνεσθαι λέγει τοὺς τύπους, ὡς κατὰ τὴν πρὸς τὰ αἰσθητὰ μίμησιν προσδεχομένης αὐτοὺς τῆς ψυχῆς. οὐ διότι μὴ κατ᾽οἰκείαν ἐνέργειαν καὶ τούτους ἡ ψυχὴ προβάλλεται (δηλοῖ δὲ ἡ εἰς τὸ ἀληθὲς καὶ ἀκριβὲς περιαγωγή. οὔτε γὰρ ἀληθῶς κυβικὸν τὸ αἰσθητὸν σχῆμα οὔτε ἀπλατὴς ἀκριβῶς ἡ εὐθεῖα ἡ φυσική, ἀλλ᾽ἡ μαθηματική. ἔνδοθεν μὲν οὖν καὶ οἱ τύποι, ἀλλ᾽οὐκ ἄνευ τῆς τῶν αἰσθητῶν γνώσεως τὴν προβολὴν αὐτῶν τῆς ψυχῆς ποιουμένης, ὡς οὐδὲ τῶν κρειττόνων ἐπιστητῶν ἢ νοητῶν). ἀλλὰ διὰ τὴν τῶν φανταστῶν πρὸς τὰ αἰσθητὰ πλείστην ὁμοιότητα τὸ τῆς ἀφαιρέσεως χώραν ἔσχεν ἰδίως ἐπὶ τῶν φανταστῶν, ὡς οἷον τῶν αὐτῶν ὄντων τοῖς αἰσθητοῖς χωρὶς τῆς φυσικῆς οὐσίας, ἐπεὶ καὶ ὁ ἐν φαντασίᾳ τύπος οὐχ ὡς τὸ εἶδος ἀλλ᾽ὡς τὸ εἰδοπεποιημένον.

213 Translation by Steel, p. 68. Original at Simplicius, *In De anima*, com in III, 428a, pp. 209–210: ἀκριβέστερα μὲν γὰρ φανταστὰ τῶν αἰσθητῶν, οἷον εὐθεῖα, ἐπίπεδον, σχήματα [...].

> and in a way appropriate to the perceptual forms, sometimes needing the perceptual impressions for the projection (προβολήν) of similar ones not just once or twice but more often. But when it has been stimulated [by sensible input] it [phantasy] projects (προβάλλει) the images from itself and impresses and shapes the figure from itself, either making it like the forms of the sense-objects or adding accuracy (τὸ ἀκριβὲς προστιθεῖσα), as in [projecting] a straight line with no breadth and figures which are maximally accurate (ἀκριβέστατα).[214]

Three aspects must be remarked upon in the philosophy of mathematics conveyed in this short, yet momentous passage of Priscian.

Firstly, (i) the accuracy of mathematical objects, understood as representational accuracy, has its root in phantasy. In other words, the mathematical object (a line in this example) is obtained with the highest degree of accuracy in the use of the poietic capacity of phantasy – and only there.

Secondly, (ii) mathematical phantasy is again portrayed as triggered by sensory input (see 2.1.5). Nevertheless, Priscian never mentions that sensible items are in any fashion a cause of phantastic projection (what kind of cause they might be?), but rather, a condition or stimulus which triggers phantasy as an efficient cause itself. Then, the process of "ἀφαίρεσις" (comm. in 429b, l. 18, l. 13) is not depicted as a sort of quantitative copy-paste operation through which the epistemic agent strips a perceived quantity from its accidents. Rather, "abstraction" grants access to a completely new entity which does not correspond to the stimulus-quantity by which it has been suggested. In other words, there is not such a thing as a perfectly accurate line outside the accuracy-bringing activity of phantasy. Therefore, there is no congruence between the mathematical object and the natural quantities triggering its obtention: the line of the natural philosopher and the line of the mathematician are totally discrete, only vaguely related entities. This detail will be crucial for understanding the link between *certitudo* and *abstractio* in Piccolomini's text.

Thirdly, (iii) regarding Piccolomini's own hermeneutics of Proclus, it must be underscored that the technical terms προβολή and προβάλλω recur three times just in the last section of Pseudo-Simplicius and that, as seen before,

214 Translation by Steel (slightly modified), p. 73. Original at Simplicius, *In De anima*, com in III, 428b, l. 10, p. 214, l. 5–11: [...] καὶ αυτὴ ἡ φαντασία ἐγείρεται πρὸς τὴν κατ' οἰκείους λόγους τῶν φανταστικῶν τύπων προβολὴν οἰκείως τοὺς αἰσθητικοὺς εἴδεσιν, οὐχ ἅπαξ ἢ δὶς ἐνίοτε μόνον τῶν αἰσθητικῶν δεηθεῖσα τύπων ἀλλὰ καὶ πλεονάκις πρὸς τὴν τῶν ὁμοίων προβολήν. ἐγερθεῖσα δ'οὖν καὶ ἀφ'ἑαυτῆς προβάλλει τὰ φαντάσματα καὶ τυποῖ καὶ διαμορφοῖ τὸ σχῆμα ἀφ'ἑαυτῆς ἢ ὁμοίως τοῖς εἴδεσι τῶν αἰσθητῶν ἢ καὶ τὸ ἀκριβὲς προστιθεῖσα, ὡς καὶ τὴν ἀπλατῆ προτείνουσα εὐθεῖαν καὶ σχήματα τὰ ἀκριβέστατα [...].

they are frequently mentioned in his exposition of phantasy and the epistemology and ontology of mathematics. Yet again, this will be highly relevant in accounting for Piccolomini's understanding of Proclus' *In Euclidem* (and a core point of my interpretation of *De certitudine*).

Turning to Proclus' *In Euclidem*, terms related to ἀκριβ- (including verbal forms) appear 43 times in the body of the commentary. Some of these occurrences are quoted (and translated) by Piccolomini in his argumentation. Ex. gr., he introduces in chapter XI Proclus' passage delivering the opinion that not all mathematical demonstrations have the same certainty, translating ἀκρίβειαν as *certitudinem*.[215] This confirms yet again that the terms related to ἀκριβ- in the works of Aristotle, Pseudo-Simplicius, and Proclus were interpreted by Piccolomini as *certitudo*.

Bearing this in mind, and considering his sources, we are able to elucidate what *certitudo* actually means in Piccolomini's text.

2.3.4 *The Three Meanings of* certitudo *in Piccolomini's Text*

Because of both the diverse uses of the Greek terms from ἀκριβ- and the wide tradition conforming the Latin semantics of terms from *cert-*, Piccolomini's opusculum shows many ambiguities regarding its crucial abstract noun: *certitudo*. This notion is never defined by the author, and its interpretation is left to the reader. Some hints of what Piccolomini meant by this word are provided by the context and by an analysis of his main sources, particularly of Proclus' prologues to *In Euclidem*, and Pseudo-Simplicius. Thus, by scrutinising the nuanced meanings of ἀκριβ- in those two sources, three deeply interconnected, but non-equivalent, senses of the term *certitudo* can be inferred.

2.3.4.1 Irrefutability and Necessity

This sense follows the Latin semantics of *cert-*: something is *certum* if it is firm and stable. In epistemic terms, *certum* is that whose truth is not likely to be subverted: if it is *certum* that p is true, it is therefore highly unlikely that p turns out to be false. Thus, a mathematical theorem τ, once demonstrated, is *certum* in the highest degree (*Concedimus igitur mathematicis disciplinis, primum ordinem certitudinis* […]).[216] Thus, all the axioms and precedent theorems being the same, theorem τ will not only be the case, but it will necessarily be the case. In modal terms, this means that non-τ is impossible. Mathematical

---

215 Proclus, *In Euclidem*, p. 43, l. 10–13: οὔτ' οὖν πᾶσαν τὴν μαθηματικὴν τὴν αὐτὴν ἀκρίβειαν ἀπαιτήσομεν / […] *non omnes demonstrationes mathematicas habere eandem certitudinem*. This theme shall be addressed in section 3.4.2.

216 Alessandro Piccolomini, *De certitudine*, f. CVIIv.

propositions are therefore universal and irrefutable propositions ([…] *rationes ipsae universales et inconvincibiles* […]).[217]

This epistemic trait is also expressed by Proclus in his *In Euclidem* when he discusses the origin of mathematical ἀκρίβεια, stressing that it cannot proceed from sensible items. He asks: "Whence come the exactness and irrefutability (certainty and irrefutability) that belong to our [mathematical] ideas?"[218]And on the next page, he asks again: "And how can we get the precision of our precise and irrefutable concepts from things that are not precise?"[219]As will be shown, Piccolomini raised similar questions while laying the foundations of his philosophy of mathematics in chapter VII of *De certitudine*:[220] he introduced the Latin term *inconvincibiles* to translate the Greek ἀνέλεγκτον.

Proclus adopts terms from ἀκριβ- together with ἀνέλεγκτον on both occasions, rhetorically asking how such precision can be obtained from imprecise things, i.e., from sensible entities. When Proclus reiterates this question on the next page, he expresses himself in a manner that closely aligns the term ἀνέλεγκτος with that of ἀκρίβεια. He initially questions both concepts but concludes the same sentence by focusing solely on ἀκρίβεια, or precision. The link between ἀκρίβεια and ἀνέλεγκτον is restated a third time by Proclus in the first prologue when he says: "It brings to perfection all the intellectual insight they contain, making what is certain in them more irrefutable […]".[221] Because of the repeated pairing of ἀκρίβεια and ἀνέλεγκτον, these passages can easily be interpreted as semantically linking the two terms.

Finally, when addressing epistemic necessity, Piccolomini draws upon a significant claim made by Proclus regarding mathematical necessity and the nature of mathematical objects. I refer to the assertion quoted in Greek by Piccolomini in chapter XII, i.e., that geometrical arguments get their necessity (τὸ ἀναγκαῖον) from their subject-matter.[222] This passage of Proclus is embedded in the core akribological argumentation of *De certitudine*. There, Piccolomini translates the term ἀναγκαῖον as *certitudo*.

Therefore, irrefutability and necessity are the pivotal meanings contained in the semantics of Piccolomini's *certitudo*. Arguably, this is the epistemic trait that Nifo calls *consistentia* in his scholium above.

217 Alessandro Piccolomini, *De certitudine*, f. LXXXXVII.

218 Proclus, *In Euclidem*, p. 12, l. 13–14: πόθεν ἡ ἀκρίβεια καὶ τὸ ἀνέλεγκτον ὑπάρχει τοῖς λόγοις;

219 Proclus, *In Euclidem*, p. 13, l. 3–5: πῶς δὲ τοῖς ἀκριβέσι καὶ ἀνελέγκτοις εἴδεσιν ἀπὸ τῶν μὴ ἀκριβῶν τὴν ἀκρίβειαν προσθήσομεν;

220 Proclus, *In Euclidem*, pp. 49, l. 7–50, l. 9. See section 3.5.2.

221 Proclus, *In Euclidem*, p. 43, l. 11–13: θριγχὸς οὖν εἰκότως ἐστὶν ἡ διαλεκτικὴ τῶν μαθημάτων, πᾶν τὸ νοερὸν αὐτῶν τελειοῦσα καὶ τὸ ἀκριβὲς ἀνελεγκτότερον ἀπεργαζομένη […].

222 Proclus, *In Euclidem*, p. 206, l. 16–18.

2.3.4.2 Objectual Precision or Accuracy

The second type of *certitudo* refers not to mathematical propositions but to the objects of study of mathematical disciplines and the properties which allow mathematical irrefutability and necessity to be achieved. One of the main arguments of Piccolomini in chapter VII is that mathematical objects cannot be embodied by sensible matter. As we have seen, this is one of the meanings of the term ἀκρίβεια in Pseudo-Simplicius' *In De anima*.[223] Nevertheless, the same semantics are also present in Proclus' commentary and, most likely, Piccolomini felt that both sources agreed on this point. When elaborating on the deficiencies of mathematical-like sensible items, Proclus posits the issue as follows:

> But the circle in phantasy is divisible, formed, extended – not one only, but one and many, and not a form only, but a form in instances – whereas the circle in sensible things is inferior in precision (ἀκριβείας), infected with straightness, and falls short of the purity of immaterial circles.[224]

The circle, Proclus argues, is attained in phantasy through the unfolding of innate reasons and instantiated as one or many. However, the same cannot be achieved in sensible matter. In what sense is the sensible circle deficient? Proclus mentions its lack of ἀκρίβεια: the sensible circle fails to be an "exact", "accurate", or "precise" circle. The next two clauses in the enumeration are (as I read them) explanations of this lack of ἀκρίβεια: sensible circles are inaccurate because they are contaminated by straightness; they are inaccurate because they fall short of the purity of immateriality. As expounded above, Pseudo-Simplicius made a very similar statement: both sources agree that the ἀκριβές of mathematical objects was given by the soul to the mathematical objects produced in phantasy, whereas sensible representations of mathematical entities failed to achieve it. As we will see in part 3 (see 3.5.2), this shall be one of the main arguments in Piccolomini's account of the ontological status of geometrical objects.

This evinces that the Sienese, translating the Greek ἀκρίβεια and related terms as *certitudo*, also incorporated objectual accuracy into its semantics. This claim is further reinforced by the third sense of *certitudo* below.[225]

---

223 Simplicius, *In De anima*, p. 277, l. 24–33; p. 233, l. 13–25; p. 214, l. 5–11.

224 Morrow's translation (slightly modified). Proclus, *In Euclidem*, p. 54, l. 8–13: ὁ δ'ἐν φαντασίᾳ μεριστὸς ἐσχηματισμένος διάστατος, οὐχ εἷς μόνον, ἀλλ'εἷς καὶ πολύς, καὶ οὐκ εἶδος μόνον, ἀλλά κατατεταγμένον εἶδος, ὁ δ'ἐν τοῖς αἰσθητοῖς καὶ τῆς ἀκριβείας ὕφεσιν ἔχων καὶ ἀνάπλεως τῆς εὐθείας καὶ τῆς καθαρότητος τῶν ἀΰλων ἀπολειπόμενος.

225 Piccolomini also describes mathematical entities that possess this form of ἀκρίβεια (objectual precision) as being "totally exact" and "absolute". See Alessandro Piccolomini,

2.3.4.3 Epistemic Transparency

The third sense of *certitudo* refers to the epistemic agent and its relation to mathematical objects. Piccolomini holds that mathematical reasons are, in their openness, easier to grasp than other entities. This adds to mathematical items a remarkable epistemic advantage, namely that, considered as objects of knowledge, they are endowed with a high degree of epistemic transparency:

> but as they are co-eternal with bare matter, they [quantities] will be most abstractable and, therefore, easy to understand, certain (*certae*) and manifest.[226]

Thus, mathematical objects or quantities are easy to know, certain, and manifest because of their special ontological status: being co-eternal with pure matter, nothing is hidden or not clear in them; their properties can be easily grasped.[227] On the other hand, the essence of real substances, flowing from their own complex principles and having a material structure of their own, is hidden and can be grasped only with great effort.[228]

Phantastic quantity, because of its (quite literally) unnatural accuracy, can be grasped with the greatest *certitudo*. The statement is reiterated in closing chapter XII, whereupon, in interpreting Pseudo-Simplicius (see above) Piccolomini attributes a superior *certitudo* to mathematical cognition because of the "more perceptible" nature of quantity:

> Quantity, however, because it does not ascribe to itself this particular matter more than that one, but, rather, does not require any defined disposition of matter, it is not surprising that it is highly abstractable. Therefore, it generates a more certain (*certiorem*) and "more sensed" (*sensatiorem*) cognition.[229]

Therefore, Piccolomini understands *certitudo* not only as a trait of the mathematical process of reasoning or demonstrating, but also as a trait of the

---

De certitudine, f. LXXXXVII–LXXXXVIIv: [...] *cum non detur inter quanta naturalia quod trinam dimensionem non recipiat, nec vera aequalitas, nec tandem quaevis figura, circularis, triangularis, aut quaevis alia, exacta penitus et absoluta.*

226 Alessandro Piccolomini, *De certitudine*, f. CVIIIv: [...] *sed cum denudata materia coaeternae sunt, abstrahibiles erunt maxime, et idcirco faciles cognitu, certae ac manifestae.*

227 I will deal with the question of the status of quantity and certainty more thoroughly in section 3.5.3.

228 Alessandro Piccolomini, *De certitudine*, f. CIX: [...] *tunc cum oriatur et fluat ex principiis propriis, et materiam sibi propriam limitet, cognitionem reddet arduam et difficilem.*

229 Alessandro Piccolomini, *De certitudine*, f. CIXv: *Quantitas autem, quia non hanc sibi ascribit materiam magis quam aliam, sed nulla limitata dispositione materiae indiget, nil mirum si abstrahibilis est valde. Et idcirco certiorem et sensatiorem generat cognitionem.*

cognition of the properties of quantitative entities. In this passage, he argues following the section from Priscian's commentary analysed above (402b, l. 11) that quantity is more certain because it is a "more sensed", or "more perceptible" accident (αἰσθητικώτερον). Epistemic transparency is, allegedly, the trait which Nifo called *evidentia* when indexing the many meanings of *certitudo* in his scholium (see 2.3 above).

The three meanings of *certitudo* can be linked up to one another with consistency, providing an account of the senses of "certainty" in Piccolomini's opusculum: the primary sense or *certitudo* (a) of mathematical demonstrations, i.e., their irrefutability and total necessity, is due to *certitudo* (b), namely to the rank of representative accuracy of mathematical objects as staged in phantasy. These metaphysical and epistemological properties allow *certitudo* (c) to happen: the intellect can easily grasp the properties of mathematical objects, thereby producing solid and absolutely necessary demonstrations (a).

2.4 Anti-mathematical Attitudes in the 16th-Century Italian Context

I claim that another source motivating Piccolomini's philosophy of mathematics is the widespread quantitative approach discussed in the university milieu in the 15th- and early 16th-century Italian context: the debates surrounding the so-called Mertonian philosophy and the distrust of mathematics that followed it. In this section, I shall provide a summary of the most relevant aspects of the Mertonian philosophy as well as a succinct account of the polemics surrounding it, indicating how such questions touched upon some of the issues treated in *De certitudine*. In posterior sections (2.5.3, 3.6), evidence will be provided that traces of the anti-Mertonian philosophy are contained in Piccolomini's paraphrase of the *Quaestiones mechanicae* and the last chapter of *De certitudine*.

2.4.1 *The Mertonians*

In the early 14th century, the treatment of Aristotelian natural philosophy at Merton College, Oxford experienced a mathematical shift. This approach is often associated with a group of scholars known as the *calculatores* or Mertonians (I will use these two labels interchangeably), amongst whom are enumerated Walter Burley (*c.*1274–1344), Thomas Bradwardine (*c.*1290–1349), Richard Kilvington (*c.*1302–1361), William Heytesbury (*c.*1313–1372), Roger Swineshead, Richard Swineshead, and John Dumbleton.[230]

230 See D. A. Di Liscia, "Introduction", in D. A. Di Liscia, E. D. Sylla (eds), *Quantifying Aristotle: The Impact, Spread and Decline of the Calculatores Tradition* (Leiden Boston: Brill, 2022), pp. 1–2.

The basic principle of the Mertonian philosophy can be summarised as follows: quantification has a place in the study of natural realities. Concepts in natural philosophy such as motion, time, and qualitative alteration (augmentation and diminution) are susceptible to mathematical description and treatment. This treatment can ultimately reveal something about the non-mathematical reality. Therefore, it consists in using mathematics as a meta-language, that is to say, to apply mathematical quantifications to our scientific speech about qualitative and motive alteration.

The Mertonian approach was backed by Aristotle's authority, who had declared that some qualities are susceptible to "more-and-less" (*Categoriae*, 8, 10b, l. 27–29) and who had himself engaged in a mathematical treatment of continua (including motion, the division of matter, and qualitative alteration) in books VI–VIII of the *Physica*.[231] A second line of theoretical defence of this approach was based on other mathematised disciplines known as *scientiae mediae* introduced by Aristotle in *Physica*, II, 2, such as harmonics, mechanics, and optics.[232] Following this approach, the Calculators (and their epigones throughout all of Europe) produced a set of original kinematics, and systems of addition and abstraction of *latitudines* for measuring the presence of a given quality or motion in an entity.[233] The key Greek terms for intensification and

231 For an analysis of the passages dealing with continuum in Aristotle's *Physica*, books VI–VII see B. M. Sattler, *The Concept of Motion in Ancient Greek Thought, Foundations in Logic, Method, and Mathematics* (Cambridge: Cambridge University Press, 2020), chap. 7, especially section 7.3; D. Bostock, *Time, Matter, and Form Essays on Aristotle's Physics* (Oxford: Oxford University Press, 2006), chap. 10.

232 As Molland puts it, Bradwardine set the agenda for extending mathematics to other domains in his *Tractatus de proportionibus*. In this work Bradwardine defends the legitimacy of the mathematisation of non-mathematical disciplines by claiming that if there were no arithmetical proportions amongst non-mathematical entities, disciplines such as harmonics would not be possible at all. See A. G. Molland, "Colonizing the world for mathematics", in E. Grant, J. E. Murdoch (eds), *Mathematics and its applications to science and natural philosophy in the Middle Ages: essays in honor of Marshall Clagett* (Cambridge New York: Cambridge University Press, 1987), pp. 45–65.

233 For an account of the traits of Mertonian philosophy at Oxford and its variations in the methods of quantification see J. Thijssen, "The nature of change", in R. Pasnau, Ch. Van Dike (eds), *The Cambridge History of Medieval Philosophy* (Cambridge: Cambridge University Press, 2011), pp. 279–292; Biard, "Blasius of Parma on the Calculation of the Variation of Qualities and Aristotelian Physics", in D. A. Di Liscia, E. D. Sylla (eds), *Quantifying Aristotle*, pp. 232–254; E. D. Sylla, "Medieval Quantifications of Qualities: The "Merton School"", in *Archive for History of Exact Sciences* 30/12 (1971), pp. 9–39. For a general study of kinematics and dynamics in the 14th century, see E. J. Dijksterhuis, *The Mechanization Of The World Picture* (Oxford: Oxford University Press, 1961), part II, Chapter 5. For a recent and thorough study of the Latin debates on quantification in the 13th- and 14th-century, see the monograph of S. Roudaut, *La mesure de l'être, Le problème de la quantification*

attenuation of a form (ἐπίτασις and ἄνεσις) were rendered by Boethius as *intensio* and *remissio*, and so they become part of the Latin technical jargon.[234]

Prima facie, this appears as an intuitive position: our everyday experience reveals that some traits, for instance, warmth, can be more-or-less present in a given object at different times, say in a cup of tea. Equally, we see moving objects, but we also see that they do not all move at the same speed all the time. Therefore, the fact that, to some extent, mathematical degrees can be used for describing such phenomena seems to be an unproblematic approach.

Or is it? The Calculators not only introduced a new philosophical jargon dealing with the *latitudes* (i.e., degrees in a geometrical representation) and with the intensity and remission of qualities; they also raised questions regarding the extreme states (*maximum, infinitum, minimum*) of such qualities or concerning the reality of degrees in natural quantities.[235]

In the 15th and 16th centuries, Italian philosophers inherited the debates and conceptual problems regarding this approach. Thus, some of them began to ask questions, formulate doubts, and add some caveats when dealing with the relationship between quantifications and the realities they, supposedly, described: are we sure that these mathematical statements describe something real? Are there real degrees in qualities? Does qualitative addition happen in things as it happens in our quantifications? Are the expressions "5 degrees of warmness" or "infinite velocity" meaningful outside the realm of mathematical concepts? If degrees are arbitrary, have these quantifications or, more generally, mathematics anything to reveal about natural entities?

One approach available to the Latin "Peripatetic" tradition to dismiss any aspirations to gain fundamental knowledge of the natural world through quantifications was the view that natural entities possess an unquantifiable essence. This essence could manifest, exist, or be expressed in varying degrees, thereby allowing for quantification: the *esse* of a quality, but not its *essentia* is susceptible of "more and less". This viewpoint was endorsed, each in their unique way, by influential Latin authorities such as Albert, Aquinas, Giles, and

---

des formes au Moyen Âge (ca. 1250–1370) (Leiden: Brill, 2022). For the Aristotelian and Averroistic background in 13th-century theories of motion see C. Trifogli, "Oxford Physics in the Thirteenth Century (ca. 1250–1270) Motion, Infinity, Place and Time", in A. Speer, W. Goris, K. Emery, G. Kapriev (eds), *Studien und Texte zur Geistesgeschichte des Mittelalters, 72* (Leiden Boston: Brill, 2000); Trifogli, "Change, time, and place", in R. Pasnau, Ch. Van Dike (eds), *The Cambridge History of Medieval Philosophy*, pp. 267–278.

234 Roudaut, *La mesure de l'être*, p. 4.

235 Di Liscia, "Introduction", in D. A. Di Liscia, E. D. Sylla (eds), *Quantifying Aristotle*, pp. 1–2.

Scotus.[236] To put it clearly, if an entity X possesses the form, ex. gr., of heat in an arbitrarily established degree of 5, the intensification of that quality to the degree of 8 does not involve any alteration in the essence of heat, nor does the quantification itself say anything either about what-heat-is or about the essence of the entity X. Quantifying only informs us of the "more or less" presence or intensity of heat in the entity X, without providing knowledge about causes or definitions. To phrase it in Latin terminology, the *quantitas* of heat is epistemically irrelevant in attaining the definition or cause(s) of the *qualitas* of heat; quantity is just an accident of the *esse* of qualitative form. Therefore, quantity, as an accident, is irrelevant in doing natural philosophy. As we shall see, the indifference of quantity regarding the cognition of essences is an opinion also held by no less a Peripatetic authority than Averroes.[237]

From this standpoint, a further potential objection against quantification follows, sc., that of the illicit transition between categories or μετάβασις. The concept of μετάβασις, introduced by Aristotle in *Analytica posteriora* I, 7 (75a) forbids inferring conclusions in a categorically delimited scientific branch from another branch, categorically different.[238] In the practice of science, this means that from principles and arguments obtained in the discipline devoted to the analysis of the category of *qualitas* (natural philosophy) or the natural affection of *motus*, conclusions related to the realm of the category of *quantitas* (mathematical conclusions) should not be deduced.[239] Inversely, no legitimate

---

236 On Albert's approach to the intensity and remission of forms see Roudaut, *La mesure de l'être*, chap. 2.1. On Scotus and his view that the quiddity considered as a universal species remains unable to be quantified, although individuals may possess that quiddity with greater or less perfection see Roudaut, *La mesure de l'être*, pp. 48–49.

237 Averroes, *Commentaria in libros De physico auditu Aristotelis*, IV, 3, f. 79v. See section 3.4.

238 L. Judson, "Aristotle and Crossing the Boundaries between the Sciences", in *Archiv für Geschichte der Philosophie* 28 (2019), pp. 177–204; M. P. Adams, "Demarcating Aristotelian Rhetoric: Rhetoric, the Subalternate Sciences, and Boundary Crossing", in *Apeiron* 48 (2015), pp. 99–122; J. G. Lennox, "Aristotle, Galileo, and 'Mixed Sciences'", in W. A. Wallace (ed.), *Reinterpreting Galileo* (Washington, D.C.: Catholic University of America Press, 1986), pp. 29–51.

239 The reason adduced by Aristotle is that absurdities follow when geometrical problems are resolved arithmetically, namely in the resulting incalculable means or irrational numbers. See *Analytica posteriora*, ed. I. Bekker, vol. 1, I, 7, 75a: Οὐκ ἄρα ἔστιν ἐξ ἄλλου γένους μεταβάντα δεῖξαι, οἷον τὸ γεωμετρικὸν ἀριθμητικῇ. Translation by Barnes at Aristotle, *Complete Works of Aristotle*, Vol. 1: "One cannot, therefore, prove anything by crossing from another genus – e.g. something geometrical by arithmetic". Aristotle's position was that irrational means in doing science followed from the inappropriate application of principles and procedures fitted for a given object of science to the subject-matter of a different science, such as asking qualities from numbers, or integer arithmetic from geometry, or quantity from quality.

conclusions can be obtained in natural philosophy through speculations on quantity, i.e., through mathematical arguments. Thus, if we use quantification in dealing with natural realities and problems, are not we infringing upon Aristotle's prohibition?

The Mertonian approach was transmitted from Oxford to Paris and Central Europe during the 14th century, where Mertonian teachings were discussed by influential figures in central Europe such as Nicholas Oresme (*c.*1320–1382), Albert of Saxony (*c.*1320–1390), and Marsilius of Inghen (*c.*1340–1396).[240] During the 15th and 16th centuries, the heated controversies that surrounded the Mertonian approach would be transferred from Oxford and Paris to Spain[241] and universities in northern Italy such as that of Padua.

2.4.2 *Italo-Mertonian Philosophy and the Reactions against It*

In northern Italy, the *calculatores* were destined for a long life in the philosophical debate. Their works were mainly imported into Italy by members of the mendicant orders (mostly Augustinians and Franciscans) studying at Oxford in the last part of the 14th century and travelling back to Italy after being educated in new logical methods, as is attested by manuscripts and bibliographical

240 H. Berger, "On the Reception of English Logic at Universities of Central Europe: Helmoldus de Zoltwedel (Prague, Leipzig) on the Liar-Paradox", in D. A. Di Liscia, E. D. Sylla (eds), *Quantifying Aristotle*, pp. 215–231.

241 In Spain, the Mertonian approach would find fertile ground. Later intellectual figures such as Diego de Astudillo (*c.*1489–1576), Domingo de Soto (1494–1560), and Diego Diest (n.c.1511) would discuss and develop Mertonian kinematics. Soto's reception of the Calculators' methods in his work *Super octo libros Physicorum questiones* (first edition in 1545) was very influential. In his treatment of the proportion between acceleration and distance, when discussing Albert of Saxony's and Oresme's views on book VII of the Physics (q.3, *Utrum velocitas motus ab effectu attendatur penes quantitatem spatii quod pertransitur*), Soto provided the first mathematised account of uniform acceleration (*uniformiter difformis velocitas motus*) of falling bodies in relation to the distance/time of fall, almost ninety years before Galileo's treatment of the same problem. Soto's works were reedited in Venice in 1585 and samples of his mathematised kinetics seem to have become part of the philosophical education of Italian intellectuals through the Jesuits of the Collegium Romanum, such as Soto's student in Salamanca, Francisco de Toledo (1532–1596). Wallace has persuasively defended the view that, through Toledo and other Jesuits such as Paolo Valla (1561–1622), Muzio Vitelleschi (1563–1645), and Ludovicus Rugerius (also teaching at the Roman College from 1589 to 1592), Soto's formalisation of kinetics had greatly influenced Galileo's master, Benetti, and Galileo himself. See W. A. Wallace, *Domingo de Soto and the early Galileo Essays on intellectual history* (New York: Routledge, 2018); Wallace, "Duhem and Koyré on Domingo de Soto", in *Pierre Duhem: Historian and Philosopher of Science. Part I: Duhem as Historian of Science*, in *Synthese* 83 (1990), pp. 239–260; Wallace, "The "Calculatores" in Early Sixteenth-Century Physics", in *The British Journal for the History of Science* 4/3 (1969), pp. 221–232.

remarks found in the *studia* of Padua and Bologna.[242] Also relevant for a diagnosis of the period are the complaints of intellectual figures such as Francis Petrarch (1304–1374). In a letter to Thomas of Mesina, he denounced the petulance of the Italian *dialectici* and their barbaric adoption of the logical Latin jargon of the Mertonians. In Petrarch's words, the "idiosyncratic evils of the British" were now worsened by adding to the "cohort of British logicians" a "new line of cyclopes from Etna".[243] Those "new cyclopes" were allegedly the philosophers partaking in debates of the early Italo-Mertonian philosophy: figures such as Giovanni of Casali (fl. 1340–1390) who, independently of Oresme, developed some geometrised kinetics in his *De velocitate motus alterationis* (1346, Bologna).

At the turn of the century, the northern Italian scholars Blasius of Parma (*c.*1350–1416), Paul of Venice (*c.*1369–1429), and Cajetan of Thiene (1387–1465) played a central role in the dissemination of Oxonian thought and were at the origin of an Italo-Mertonian philosophy.[244]

Blasius of Parma is one of the first Italian scholars to lecture on some of the Mertonian themes in the universities of Padua, Bologna, and Pavia. His *Tractatus de ponderibus* (*c.*1400) dealt with the new mathematised kinetics. He also produced a book of *Quaestiones* specifically treating Bradwardine's celebrated treatise *De proportionibus velocitatum in motibus* (1328).[245]

The highly influential Paul of Venice was sent by the Augustinian order to study at Oxford from 1390 to 1393, and moved thereafter to Paris, from 1393 to 1395. During this period, he became familiar with the Mertonian doctrines which he taught back in northern Italy. A token of Paul of Venice's current

---

242 C. Wilson, "William Heytesbury: Medieval Logic and the Rise of Mathematical Physics", in *British Journal for the Philosophy of Science* 8/31 (1957), pp. 254–256; N. G. Siasiri, *Arts and sciences at Padua: the studium of Padua before 1350* (Toronto: Pontifical Institute of Mediaeval Studies, 1973), pp. 62–64; W. J. Courtenay, "The Early Stages in the Introduction of Oxford Logic into Italy", in A. Maierù (ed.), *English Logic in Italy in the 14th and 15th Centuries* (Naples: Bibliopolis, 1982), pp. 13–32.

243 Francesco Petrarca, *Familiarium rerum libri*, eds V. Rossi, U. Bosco (Turin: Aragno, 2004), I, 6: *Immo vero iam insularum particulare malum est, si dialecticorum agmini Britannico Aethnea nunc novorum Cyclopum acies.*

244 For an account of the changes in the teaching of logic and natural philosophy in northern Italy from the 14th- to the 16th-century see P. F. Grendler, *The Universities of the Italian Renaissance* (Baltimore: Johns Hopkins University Press, 2004), chap. 7, and 8.

245 See Di Liscia, "Biagio Pelacani da Parma's Geometrisation of Latitudes and the Problems of the Mean Degree Theorem", in Biard, A. Robert (eds), *La philosophie de Blaise de Parme: Physique, psychologie, éthique* (Florence: Micrologus Library, 2019), pp. 129–152; G. F. Vescovini, "Blaise de Parme et l'intelligibilité mathématique", in *La philosophie de Blaise de Parme*, pp. 107–128; Biard, "Blasius of Parma on the Calculation".

knowledge of the mathematised kinetics can be found in his popular *Summa naturalium*, in which one finds a precise report of the arithmetical kinetics of Albert of Saxony.[246]

Cajetan of Thiene, a noteworthy disciple of Paul of Venice, dedicated a minimum of three of his expositions to expounding upon the doctrines put forth by the Mertonians. His commentaries played a pivotal role in fostering and shaping the late 15th- and early 16th-century reception of Mertonian philosophy within Italy, for they were notably published as guide texts alongside the works of the *calculatores* Heytesbury (*Regularum et sophismatum expositio*, Venice, 1483 and 1494), Ralph Strode (*Consequentiae Strodi cum commento Alexandri Sermonete et declarationes Gaetani in easdem Consequentias et obligationes eiusdem Strodi*, Venice, 1507), and Ferrybridge (*Consequentiae Ricardi de Ferabrich et expositio Gaetani super easdem*, Venice, 1507).

Cajetan's expositions and elucidations resulted in him being labelled as *calculator* himself. As a token of this, the Bolognese master Ludovico Boccadiferro (see below) referred to Cajetan as the quintessential *calculator*.[247] Hence it is clear that, when found in the Italian writers, the name *calculator* does not necessarily refer to the original 14th-century philosophers surrounding Merton College, but also may include any given epigone who sympathised with and applied their methods.

Regarding the philosophy of mathematics, there is a significant shift in Cajetan of Thiene's approach to and representation of Mertonian philosophy. As Wallace noted, with Cajetan the quantified statements of the Mertonians underwent a distinctly realist reading, aiming for potential application to the real world – a shift that, Wallace suggests, was pivotal for the development of early modern science.[248] This is particularly evident in Cajetan's examples.

---

246 Grendler, *The Universities of the Italian Renaissance*, pp. 250–253; Di Liscia, "Perfections and Latitudes: The Development of the Calculators' Tradition and the Geometrisation of Metaphysics and Theology", in D. A. Di Liscia, E. D. Sylla (eds), *Quantifying Aristotle*, pp. 278–328.

247 Ludovico Boccadiferro, *Lectiones in Aristotelis Stagiritae libros quos vocant Parva naturalia, De sensu et sensili* (Venice: Hieronymus Scotus, 1560), ff. 52v–53v.

248 See Wallace, *Prelude to Galileo: Essays on Medieval and Sixteenth-Century Sources of Galileo's Thought* (Dordrecht Boston: Springer, 1981), part II, chap. 3, 4, and 5. For the Mertonian use of mathematical imagination and some notes regarding Cajetan of Thiene see Sylla, "Mathematical physics and imagination in the work of the Oxford Calculatores: Roger Swinehead's *On natural Motion*", in E. Grant, J. E. Murdoch (eds), *Mathematics and its applications to science and natural philosophy in the Middle Ages: essays in honor of Marshall Clagett* (Cambridge New York: Cambridge University Press, 1987), pp. 69–102. For a wider analysis of Paul of Venice's metaphysical realism and his theory of universals see A. D. Conti, "Opinions on Universals and Predication in Late Middle Ages: Sharpe's

These are illustrative mental experiments of Cajetan's own invention, absent in his sources, intended to be taken as applicable descriptions of the Mertonian rules. For instance, in his influential commentary on Heytesbury's *Regulae*, in addressing the uniform velocity of a magnitude, Cajetan exemplifies the rule with well-defined mechanical cases: an expanding and contracting rotating wheel (*rota*) made of ice, put in an oven where it is melted while spinning or turned against the edge of a sword (*gladius*) continuously reducing it, or as a potting wheel to which parts are progressively and uniformly added.[249]

Cajetan's mechanical turn stands in contrast to the preceding Mertonian tradition, where the mechanical reality of quantified statements seemed secondary to the logical dimensions of the problems raised (or solved) by them. Regarding their metaphysics, most approaches were *nominales* (conceived as *conceptus mentis*) or softly realist.[250] Cajetan's plain realist reading of the Mertonian conceptual apparatus and his eloquent mechanical exemplifications would be decisive in establishing an implicit epistemology and metaphysics of the Mertonian methods in Italy. That is to say, Cajetan's plastically and practically fleshed out approach to quantification arguably had the consequence that, as his works and imagery spread as an introduction to the Mertonian philosophy, the default, tacitly agreed understanding of quantified meta-language amongst students was that such propositions and rules corresponded to the real world. In other words, what I call a **full-correspondence** reading of quantification thrived in the intellectual milieu surrounding the University of Padua.

The full-correspondence approach to quantified statements can be summarily put in two points:

(1) It holds that quantified statements about natural realities correspond, i.e., fit the real or possible extra-mathematical realities of which they are formalisations. In other words, there is an isomorphism between quantified statements and (actual or possible) quantified realities.

(2) It follows from the isomorphy granted in (1) that quantified statements are also epistemically valuable, i.e., they (can) convey or reveal something about the isomorphically quantified natural realities.

As we shall see, this reading of the Italo-Mertonian philosophy triggered a long-held debate amongst the Italian *litterati*, a controversy that reached Piccolomini and whose traces can be found in his 1547 publication (see section 2.5.3).

and Paul of Venice's Theories Compared", in *Documenti e Studi sulla Tradizione Filosofica Medievale* 18 (2007), pp. 483–500.

249 William Heytesbury, Cajetan of Thiene, *Regule eiusdem cum sophismatibus. Declaratio Gaetani supra easdem* (Venice: Bonetti, 1494), f. 38rb.

250 See Wallace, *Prelude to Galileo*, pp. 51–54; p. 61, note 9; pp. 66–68.

The nascent printing press was a great boom for the diffusion of Mertonian ideas in Italy: Swineshead's *Calculationes* was amongst the first printed books (*c.*1480), closely followed by the editions of Heytesbury's *Regulae* (1483, 1494, with Cajetan's comments), Oresmes' *De latitudinibus formarum* (1482), Albert of Saxony's *Tractatus de proportionibus* (1482), Thomas Bradwardine's *De proportionibus velocitatum in motibus* (1495), and Walter Burley's *De intensione et remissione formarum* (1496), to name a few. As Schmitt has pointed out, further evidence for the diffusion of Mertonian ideas in the northern Italian philosophy of mathematics from 1475 to (at least) 1520 can be inferred from the immense literary production on the topic on the eve of its introduction.[251]

The Mertonian approach was also institutionally welcomed by Italian universities. To propose a relevant case, in the 1496 curriculum of the University of Padua we find included Heytesbury's *Regulae*, Strode's *Consequentiae*, and Paul of Venice's *Logica parva*.[252] It is noteworthy that the Mertonians remained for the most part excluded from the natural philosophy syllabus and were taught as part of the logic syllabus.

Another token of the influence of the Mertonian philosophy in Italy is, as Luca Bianchi points out,[253] Ermolao Barbaro's widely read *Compendium scientiae naturalis ex Aristotele* (1484). This work was conceived without any innovative purpose: it consisted of a simplified explanation of the main tenets of Aristotelian natural philosophy as perceived and discussed at the time, together with the exposition of some famously debated problems. The *Compendium* was intended as an introductory manual, and it was reissued no fewer than six times in the years following its publication. One revealing trait of the *Compendium* is that it contains two sets of philosophical questions: the first regards the definition, unity, and immortality of the soul; the second (and the most detailed exposition in the entire work) concerns the quantification methods and puzzles of the Calculators. Amongst the problems more conspicuously considered we find kinetical questions and those of *intensio et remissio formarum*.[254]

---

251 See Ch. B. Schmitt, "Hieronymus Picus, Renaissance Platonism and the Calculator", in *The Aristotelian Tradition and Renaissance Universities* (London: Variorum Reprints, 1984), pp. 64–65. Schmitt analyses the works of Hieronymus Picus *De motu gravium et levium*, *Proportiones* (MS Venice, Marciana Lat VIII, 83, ff. 4–15r and 16–40v respectively), and his *Tractatus de intensione et remissione formarum* (MS. Bibl. Vat. Lat. 1905, ff. 5–36v).

252 J. Facciolati, *Fasti Gymnasii Patavini* (Sala Bolognese: Forni, 1978), Vol. 2, p. 114, pp. 118–119.

253 Bianchi, "Fra Ermolao Barbaro e Ludovico Boccadiferro: qualche considerazione sulle trasformazioni della "fisica medievale" nel Rinascimento italiano", in *Medioevo* 29 (2004), pp. 341–378.

254 In this second set of problems, two well-known natural questions are addressed in the Mertonian fashion. These questions are (1) the relation between movement, velocity, and

That such qualitative and kinetic quantifications were used in a book intended as a manual of natural philosophy evinces that an unavoidable tendency to consider the Mertonian meta-language as a part of natural philosophy providing real solutions and descriptions of reality had arisen by the end of the 15th century in northern Italy.

As expected, though, not every 15th- and early 16th-century northern Italian scholar gladly welcomed the Mertonian approach and Cajetan's full-correspondence reading of it. A trend of thought furiously opposed to the Mertonian quantification swiftly followed in Italy. Evidence of this open hostility is found with (i) Pietro Pomponazzi (1462–1525) and (ii) Agostino Nifo. Although these authors' philosophy of mathematics tackled a wide collection of problems, the question *de intensione et remissione formarum* will be revealed as the core issue of their disputes. Through this topic, further relevant positions on the nature and the place of mathematics amongst the sciences will be expounded.[255]

2.4.2.1 Pietro Pomponazzi

Of particular significance in an overview of the late 15th- and early 16th-century philosophy of mathematics is the case of the renowned Pietro Pomponazzi. Likely spurred by his direct engagement with the intellectual successors of Paul of Venice and Cajetan of Thiene, Pomponazzi expressed a distinct lack of enthusiasm for the Mertonian methods in addressing questions within natural

---

the resistance of the medium (*de motus velocitate quae sumitur ex ea proportione quae est inter proportiones moventium ad resistentia impedimenta*), and (2) the analysis of opposite qualities in a mixture of elements (*quomodo vocentur mixta ex iis, quae in ipsis qualitates insunt*). In treating the first, Barbaro reproduces an account of the proportionality involving the three relevant variables of the (amount of) movement, velocity, and resistance (*motus*, *velocitas*, and *resistentia*). He approaches the question arithmetically, claiming that a given amount of *motus/impetus* results in a different *velocitas* in a given body in a ratio inversely proportional to the *resistentia* of the *medium* through which the object moves. In the second case, the *Compendium* tackles the problem of qualitative mixture following the Calculators' procedure (sc. assuming the *intensio et remissio formarum*) of quantifying two related qualitative terms co-existing in the same entity, namely *caliditas* and *frigiditas*. Those two properties, warmness and coldness, are held to share the same *genus*, and, the one being the opposite of the other (they are related by abstraction), a numerically formalised account of their presence/absence in a given mixture is produced. The general arithmetic rule inferred is that the introduction of the amount *n* of *caliditas* into a mixture implies that the same amount *n* of *frigiditas* will abandon the mixture: *tantum vero deficit caliditatis quantum frigiditas*; ff. 27v–28v, quoted by Bianchi in "Fra Ermolao Barbaro e Ludovico Boccadiferro".

255 I will also deal with Genua De' Passeri's "Simplician" anti-mathematical thought in part 3 (3.4.1).

philosophy. This sentiment is evident in his extant lectures on *Physica* VII (Arezzo, MS 390, f. 179r), where he declared:

> And where Aristotle was reserved in this matter (the geometrical formalisation of the continuum), Heytesbury in his treatise and the Calculator (Swineshead) made extensive discussions of it. Aristotle dismissed these things because those compositions and tricks pertain to mathematics; and the Calculators surreptitiously beat philosophers, for they interpose geometrical things. But a philosopher, as a philosopher, does not involve himself in these matters. And these Calculators are called sophists; therefore, they should not involve themselves in philosophy, but in geometry.[256]

There are two revealing points in this paragraph for understanding what was at stake in the inclusion of Mertonian approaches within the domain of natural philosophy and as an official part of the university curriculum. The first problem is a philosophical one, the second is an institutional one, related to disciplinary boundaries.

Firstly, (1) Pomponazzi expresses the contempt felt by some natural philosophers in the Paduan context when confronted with what they perceived as discussions filled with alien quantitative terms within the field of natural philosophy. It is worth remembering that Pomponazzi considered the metaphysical misconceptions to which quantification could lead to be so momentous that he wrote three treatises intended to criticise the Mertonian approach: *Tractatus de maximo et de minimo* (between 1496 and 1500), *De intensione et remissione formarum ac parvitate et magnitudine*, and *De reactione* (between 1514 and 1515).

256 Arezzo, 390 f. 179r: *Et ubi Aristoteles in hoc loco fuit parcus, Entisbery in suo tractatu et calculator fecerunt de hoc magnos tractatus. Aristoteles enim dimisit haec, quia illae compositiones et illae truffae spectant ad mathematicam; et calculatores latenter vincunt philosophos, interponunt enim geometricalia. Sed philosophus, ut philosophus est, non se intromittit ad haec. Et isti calculatores sophisti apellantur, quare non se debent intromittere in philosophia, sed in geometria.* Quoted by Nardi, *Saggi sull'Aristotelismo padovano*, p. 112. (My translation). For another, late 15th-century, instance of hostility toward the Mertonian methods, see Giovanni Pico della Mirandola, *Conclusiones nongentae*, ed. S. A. Farmer, *Syncretism in the West: Pico's 900 Theses (1486): The Evolution of Traditional Religious and Philosophical Systems* (Tempe AZ: Arizona Center for Medieval and Renaissance Studies, 1998), p. 466: *Sicut dictum Aristotelis de antiquis dicentis quod ideo errarunt in physica contemplatione, quia mathematice res physicas tractarunt, verum esset si illi materialiter mathematica, non formaliter, accepisent, ita est verissimum modernos, qui de naturalibus mathematice disputant, naturalis philosophiae fundamenta destruere.*

Pomponazzi's theoretical hostility towards the Mertonian doctrines is deployed on three intertwined fronts: logical, physical, and metaphysical.[257] One of Pomponazzi's most relevant arguments relies on the paradoxes following the full-correspondence reading of the problematic notion of *infinita latitudo* – a notion that had caused uneasiness even amongst the Calculators themselves.[258] To expound on one of Pomponazzi's examples, let us suppose that we establish degrees in the quality of velocity. The Calculator claims that degrees can be considered by addition and subtraction *ad infinitum* and that infinite degrees of remission (= 0) or intension are meaningful.[259] Now, whereas the infinite remission of a quality is less problematic (this item is not P, or possesses the quality P in infinite remission) the same does not hold for infinite intensity. To begin with, the infinite *latitudo* of a form is never actualised in nature. Thus, a *maximum* or naturally possible highest degree of a given quality must be acknowledged (say, the highest naturally possible velocity). So, what is the exact meaning of "infinite velocity" if such a thing will never be actualised? Even worse, from the introduction of an operational infinite degree, two paradoxical conclusions follow: (a) that the would-be *maximum* naturally possible degree of a quality is an infinitely defective one compared with its *infinitum* degree, and (b) that the naturally possible *maximum* and the *minimum* degrees of a given quality are equally deficient (i.e., infinitely small) when compared to the *infinitum* degree of the same quality. Thus, the

257 F. P. Raimondi, "Pomponazzi e la polemica anticalculatoria", in *Pietro Pomponazzi, Tutti i trattati peripatetici* (Milan: Bompiani, 2013), pp. 9–86, see p. 23.

258 Pietro Pomponazzi, *De intensione et remissione formarum ac parvitate et magnitudine*, eds F. P. Raimondi, J. M. García Valverde (Milan, Bompiani, 2013), see I, cap. VI, pp. 270–282. For a complete enumeration of the arguments of Pomponazzi in this treatise see Wilson, "Pomponazzi's Criticism of Calculator", in *Isis* 44/4 (1953), pp. 355–362 and Raimondi's introductory essay to Pietro Pomponazzi, *De intensione et remissione*, "Pomponazzi e la polemica anticalculatoria", p. 23.

259 Pomponazzi puts the paradox in the following terms in Pietro Pomponazzi, *De intensione et remissione*, sec. V, cap. VII, 17, p. 518: *Quia quodlibet tempus finitum, comparatum aeternitati, est infinite parvum; etiam* ***quilibet gradus finitae velocitatis, comparatus toti latitudini, est infinite remissus.*** *Quare non sequitur: si est infinite remissus respectu totius latitudinis, quod respectu cuiuscumque sit infinite remissus, vel respectu cuiuscumque est remissus sit infinite remissus. Causa autem quare infinite intensum nullam remissionem habere potest, infinite autem remissum, quamcumque intensionem, dummodo sit finita, habere potest, est* ***quia infinite intensum respectu alicuius, quicquid sit illud, de necessitate habet qualitatem infinitam.*** On 13th- to 15th-century controversies on infinitum, continuum, and their reality see J. E. Murdoch, "Infinity and continuity", in N. Kretzmann, A. Kenny, J. Pinborg, E. Stump (eds), *The Cambridge History of Later Medieval Philosophy* (Cambridge: Cambridge University Press, 1982), from VII – Natural philosophy, pp. 564–592.

acceptance of meaningful infinite degrees is, in Pomponazzi's own saying, "at the very end of abomination and ignorance of a man unable not only of philosophy, but of any good at all".[260]

Thus, Pomponazzi teaches that the quantifications of *latitudines*, useful as they might be, risk establishing an illegitimate bridge between mathematics and the non-mathematical leading us to operate with meaningless quantifications and then inferring conclusions which supposedly describe real qualities and movements. The core philosophical problem of the Calculators is that, through their ill-conceived method, they confuse mathematical fictions with real entities:

> Therefore, the error of these [Calculators] arises because they equate natural things to mathematical things, thinking that everything said in mathematics fits natural things, which is utterly false.[261]

Nevertheless, the tradition initiated by Paul of Venice seems to have been so firmly settled that, when Pomponazzi lectured on the Mertonian views (presumably around 1513–1514), he was mocked by most of the students and professors alike, as he himself recounts in a letter to the nobleman Alberto III Pio (1475–1531).[262]

Secondly, (2) in the quotation from his lectures above, Pomponazzi confesses not only a feeling of methodological distrust regarding the Calculators' procedures but also a strong sense of professional intrusion.

The key sentence is the following: "the Calculators surreptitiously beat philosophers, for they interpose geometrical things". From this observation, it can be inferred that Pomponazzi felt that the Mertonian approach was a disciplinary transgression on the part of mathematicians and logicians, gradually taking over natural philosophy. In that sense, Pomponazzi's invectives against the Calculators must be considered not just as the fruit of his theoretical, but also of his professional zeal.

---

260 Pietro Pomponazzi, *De intensione et remissione formarum*, sec. V, cap. II, p. 440: *Quare dicamus quod dicere summum, qua summum est, esse remissum, est in ultimo abominationis et hominis ignorantis terminos neque capacis philosophiae, immo nullius boni.*

261 Pietro Pomponazzi, *De reactione*, sec. II, cap. VIII, p. 786: *Error igitur istorum provenit quia comparant naturalia mathematicis putantque quod omnia dicta in mathematicis conveniant in naturalibus, quod est falsissimum.*

262 *Sed cum in disputationibus calculatoribus transactis de hoc sermo haberetur dixissemque Calculatorem et a vero et ab antiquis philosophis longe remotum esse, quasi ab omnibus ridiculus habitus sum.* Quoted by Ramondi and Valverde in "Pomponazzi e la polemica anticalculatoria", p. 14.

In Italian universities like Padua, the same problems in natural philosophy might be discussed twice: the first time by the professor of logic, who dealt with it as one of the *sophismata* to be resolved in the Mertonian way; and the second time by the professor of natural philosophy, who would restate the same problem, assuring his students that quantification was not ontologically relevant. Such a situation seems ripe for two diverse philosophical views to develop the fiercest hostility against each other: the natural philosopher would show that solutions to the puzzles raised by Aristotle's texts like those proposed by the Mertonian logician were not relevant; meanwhile, in another room of the same building, the logician would instruct his audience on how to resolve some puzzles in Aristotle's *Physica* by adopting Mertonian methods.[263]

2.4.2.2 Agostino Nifo

Coming back to the Mertonian controversies, Di Liscia shows that a remarkable reception of the science of *latitudines* in 16th-century Italy is that of the highly influential Agostino Nifo.[264] A lengthy discussion on principles and rules of quantification is contained in his *Metaphysicarum disputationum dilucidarium* (1511) where the name *calculator* is invoked 42 times (far more abundantly than in his commentary on the *Physica*).

Nifo's account of the Mertonian questions of the "mean degree theorem" and the problematic quantification of the *perfectio specierum* as found in

263 See A. D. Lines, "Natural Philosophy in Renaissance Italy: The University of Bologna and the Beginnings of Specialization", in *Early Science and Medicine* 6/4 (2001), pp. 267–323. Lines' analysis of the gradual development of natural philosophy as a distinct academic specialization at the University of Bologna in the 16th century provides valuable insights. According to Lines, between 1400 and 1550, the average tenure of professors teaching natural philosophy increased from 10 years to 17.5–19 years. Moreover, the proportion of professors who taught natural philosophy only as a stepping stone to other positions decreased from two-thirds to one-third. This suggests that natural philosophy became, for many scholars, a definitive field of expertise or a final academic destination. Furthermore, data indicates that between 1469 and 1600, the budget allocated to natural philosophy at Bologna doubled, rising from 14% to 30% of the total budget. This shift points to the intellectual prominence of natural philosophy and the emergence of a highly specialized group of academic natural philosophers. With this in mind, it becomes evident that increasingly specialized Bolognese natural philosophers would have had a vested professional interest in "rescuing" natural philosophy by countering the views of the *Calculatores*.

264 Di Liscia, "Perfections and Latitudes". For a study of Nifo's logical thought see E. De Bellis, *Il pensiero logico di Agostino Nifo* (Galatina: Congedo, 1998), especially chapter 4: *La teoria del* regressus demonstrativus. For an account on Nifo's philosophy and intellectual profile see E. P. Mahoney, *Two Aristotelians of the Italian Renaissance: Nicoletto Vernia and Agostino Nifo* (Aldershot Burlington: Ashgate, 2000).

book VIII of the *Dilucidarium* do not seem to contain great doctrinal innovations beyond his source: it can be considered as a restatement of some of the contents of Blasius of Parma's *Questiones super tractatu de latitudinibus formarum* (Padua, 1486). As Di Liscia shows, regardless of the great space given to these questions in the *Dilucidarium*, Nifo's anti-Mertonian stance is openly revealed only at the end of his analysis. There he closes by saying:

> These are from the twelfth disputation, where I wish to excuse myself, for I debated questions that have nothing to do with metaphysics. I did this because of a single external question that touched upon all these matters. Therefore, to clarify the truth reached in those matters, I wanted to debate these questions. Furthermore, through these questions, we have clarified many obscure statements in natural philosophy, so please be lenient with me. And if God grants me life, I will compile a compendium of calculatorial disputations, in which we will object to the importunities of all Calculators, refuting their errors. For we know that these [questions] are neither Aristotelian nor necessary for knowledge, but rather displays of human vanity.[265]

Nifo firstly excuses himself for the treatment of the Mertonian questions in dealing with Aristotle's *Metaphysica*, confessing that, to his mind, such topics are alien to the discipline. He explains that he wanted to scrutinise them to test to what extent they have touched the truth. His overall judgment regarding this former point is shown: Nifo held that through the Mertonian means many things were confusedly stated in natural philosophy. He expresses his intent to compose a treatise (most likely never written) aimed at refuting the Calculators' assertions. He argues that their methods are neither genuinely Peripatetic nor essential for science, but rather a mere exercise in self-aggrandizement of the human mind.

Regardless of his suspicions, Nifo did not entertain overly hostile views against mathematics as such. It does not seem that he found that the use of

265 Agostino Nifo, *Metaphysicarum disputationum dilucidarium* (Naples: Sigismundus Mayr Alemannus, 1511), VIII, f. 328v: *Haec de disputatione duodecima, ubi volo me excusare, nam quaestiones disputavi quae nihil attinent ad metaphysicum. Haec enim feci propter unam exterorum quaestionem, quae tetigit omnia haec. Ut igitur in his quae tetigerunt veritas dilucidetur, volui quaestiones hasce disputare. Praeterea per has quaestiones multa confuse dicta in naturalibus illustravimus, quare parcatur mihi. Et si Deus vitam producet, conficiam dilucidarium disputationum calculatoriarum, in quo omnibus calculatoribus obiiciemus eorum importunitates refellentes: scimus enim illas nec esse Peripateticas, nec necessarias ad scientiam, sed ostentationes hominum.*

imaginatio in mathematics (or in any other science) was a source of distrust. He sustains that, whenever the aspect of a real object is isolated by a discipline, this is done by some form of imagination. For instance, a human body might be studied as a natural object by natural philosophy and as a quantity by mathematics. The same goes for time or movement. In this way, as Nifo puts it, we can consider things by separating different traits according to *physicam imaginationem* or *mathematicam imaginationem*.[266] By these kinds of imagination, he arguably means different ways of abstraction. Thus, a scientific consideration based on the imaginative disjunction of aspects is the (only) way to do science. Nevertheless, in the case of the Calculators, there is a tendency to overstep the boundaries of mathematical imagination, leading to misconceptions – such as the notion that the *physicus* and the *mathematicus* are one and the same. This Italo-Mertonian idea is addressed and refuted by Piccolomini in chapter XII of *De Certitudine* (see 3.6).

Regarding the mathematical demonstration, Nifo produced some comments which also predate some of the contents of *De certitudine*.[267] Much as his contemporary Zimara, Nifo endorsed Averroes' view that *demonstratio potissima* was abundantly found in mathematics. Moreover, siding with Themistius in the ἀκρίβεια/*certitudo* controversy, Nifo also held that it is by its formal means of proof that mathematics possesses the highest certainty.[268] This would be an opinion targeted by Piccolomini, who names Nifo as his adversary regarding this point (see section 2.6.3).

266 Agostino Nifo, *Metaphysicarum disputationum dilucidarium*, III, f. 127. Here, discussing the ways in which a quality "white" or "sweet" can be understood, he asserts: [...] *potest adhuc intelligi secundum imaginationem mathematicam, aut secundum imaginationem physicam: quod quia in positionibus dictum est, omitto*. See also VIII, f. 295v, where he uses the term *imaginatio* to describe the ways in which different disciplines establish degrees – pointing out the vice of the calculators in thinking that the natural philosopher and the mathematician are the same: *Igitur de gradu formae aliter loqui possumus secundum imaginationem astrologi, aliter secundum imaginationem medici, aliter etiam secundum imaginationem physici, aliter secundum imaginationem mathematici sive calculatoris, qui physicus est mathematicus*.

267 Nifo claims that *regressus demonstrativus*, the Paduan research and justification methodology (see section 2.6.1 below for an elaboration on the topic), is not possible in mathematical sciences. See Agostino Nifo, *Expositio super octo Aristotelis Stagiritae libros De physico auditu* (Venice: Hieronymus Scotus, 1559), p. 13b: *Magna fieri digraduatio de regressu solet in naturalibus demonstrationibus; dico in naturalibus demonstrationibus quia in mathematicis non fit regressus*.

268 Agostino Nifo, *Super libros De anima* (Venice: Giunti, 1544), f. 4b: *Ergo demonstratio mathematica sola est simpliciter et per consequens in primo gradu* [*certitudinis*] *et preferenda*.

2.4.3 *Dismantling Mathematics: Ludovico Boccadiferro*

Ludovico Boccadiferro was a disciple of Alessandro Achillini (1461–1512) and Pomponazzi. He earned a doctorate in arts and medicine in 1516 at Bologna and, after three years of teaching logic, he was promoted to the chair of natural philosophy and lectured on that discipline for the rest of his life. In other words, Boccadiferro was a natural philosopher by profession – and one of the most influential of his time. As we have learned above, Piccolomini was one of his most enthusiastic students. In this section, I will provide an account of Boccadiferro's views on the philosophy of mathematics and akribology. Some of these views would be echoed in Piccolomini's opusculum.

One of the texts containing some of Boccadiferro's lectures is the *Explanatio libri Physicorum Aristotelis* (posth. 1558). It contains some of the master's views on the status of mathematics as a discipline. In the introductory lectures, the scientific status, the place amongst other disciplines, and the methods of natural philosophy are discussed. The first two *lectiones* include a discussion of the definitions, distinctions, and the order of the sciences. Topics such as sub-alternation and the *ordo scientiarum* regarding the three contemplative sciences (sc. metaphysics, physics, and mathematics) are also tackled. In these chapters, Boccadiferro advances the following relevant positions concerning mathematical items and their epistemological traits:

(1) He posits the ontological dependency of mathematical upon natural objects: mathematical objects are the same as natural quantities but conceived in themselves in the reason's use of imagination.[269] Thus, mathematical objects are posterior to natural entities.

(2) Boccadiferro holds that traits which are common in many entities and perceptible by many senses are more easily grasped by the intellect than properties which belong only to one entity, or which are perceived only through one of the senses. This refers to the so-called question of *sensibilia communia*, mainly emanating from Aristotle's *De anima* (see III, 425a, l. 15–28). The rationale here would be that because quantity is common to every corporeal entity and is perceptible through more than one human sense (vision and touch, at least), this property is grasped with greater certainty/accuracy (*certius*) than properties which are perceptible only through one sense or through no sense.[270]

269 Ludovico Boccadiferro, *Explanatio libri Physicorum*, lec. 1, f. 3v: *Dico quod est falsum quod res mathematicae solo intellectu ac imaginatione, non sensu, quia eadem est quantitas naturalis et mathematica.* The classical *loci* for sustaining the inseparability of mathematical objects are found in Aristotle, *Metaphysica*, XIII, 3, 1077b, l. 12–34 and *Physica*, II, 2, 194a, l. 8–11.

270 Ludovico Boccadiferro, *Explanatio libri Physicorum*, lec. 1, f. 3: *Dico quod sensibile commune licet apprehendatur obscurius a singulo sensu particulariter quam proprium sensibile,*

(3) Therefore, mathematics is easier to understand and "more known to us" (*notior nobis*) than natural philosophy because of the reason above: its subject-matter or *quantus simpliciter* is the commonest accident in the sensible world.[271] Most things encountered in ordinary life are three-dimensional, possess size, and can be numbered. Thus, knowing quantity is a relatively easy matter. This view is further defended by the Aristotelian example found in the *Ethica Nicomachea* VI where the Stagirite establishes that a child can easily become a mathematician, but not a natural philosopher.[272] This exact argument and *locus* will be adopted by Piccolomini in *De certitudine* (see section 3.5.3.2).

(4) In *lectio* 34 of his *Lectiones super tres libros De anima* (posth. 1556), Boccadiferro attacks Thomas Aquinas' view (as the Bolognese master understands it) that mathematical definitions involve some sort of matter.[273] This attack is motivated by Boccadiferro's theory of definition according to which

tamen ab homine ipso certius apprehenditur sensibile commune quam proprium, quia sensibile commune pluribus sensibus percipitur.

271 Ludovico Boccadiferro, *Explanatio libri Physicorum*, lec. 1, f. 3: *Sensibilia enim communia eodem modo a naturali et mathematico percipiuntur, tantum ratione different, quia naturalis considerat ea ut termini corporis naturalis quanti, mathematicus ut quantus simpliciter.*

272 Aristotle, *Ethica Nicomachea*, VI, 9, 1142a, l. 6–19: σημεῖον δ'ἐστὶ τοῦ εἰρημένου καὶ διότι γεωμετρικοὶ μὲν νέοι καὶ μαθηματικοὶ γίνονται καὶ σοφοὶ τὰ τοιαῦτα, φρόνιμος δ'οὐ δοκεῖ γίνεσθαι. αἴτιον δ' ὅτι καὶ τῶν καθ' ἕκαστά ἐστιν ἡ φρόνησις, ἃ γίνεται γνώριμα ἐξ ἐμπειρίας, νέος δ' ἔμπειρος οὐκ ἔστιν· πλῆθος γὰρ χρόνου ποιεῖ τὴν ἐμπειρίαν· ἐπεὶ καὶ τοῦτ' ἄν τις σκέψαιτο, διὰ τί δὴ μαθηματικὸς μὲν παῖς γένοιτ'ἄν, σοφὸς δ' ἢ φυσικὸς οὔ. Ludovico Boccadiferro, *Explanatio libri Physicorum*, lec. 1, f. 2v: *Ideo* VI *Ethicorum ait Aristoteles: puer potest esse mathematicus, naturalis non, quia res certior est mathematica quam naturalis.*

273 The passages scrutinized by Boccadiferro, from which he polemically infers that Aquinas asserts the inclusion of intelligible matter in mathematical definitions, are found in Thomas Aquinas, *In Physicam*, II, lec. 3, n. 5–6: *Et ideo huiusmodi quantitates et quae eis accidunt, sunt secundum intellectum abstracta a motu et a materia sensibili, non autem a materia intelligibili, ut dicitur in* VII *Metaphysicorum. Quia igitur sic sunt abstracta a motu secundum intellectum, quod non claudunt in suo intellectu materiam sensibilem subiectam motui; ideo mathematicus potest ea abstrahere a materia sensibili.* [...] *Deinde cum dicit: fiet autem utique manifestum etc., manifestat positam solutionem dupliciter: primo quidem per differentiam definitionum quas assignat mathematicus et naturalis; secundo per scientias medias, ibi: demonstrant autem et quae magis et cetera. Dicit ergo primo quod hoc quod dictum est de diverso modo considerationis mathematici et physici, fiet manifestum si quis tentaverit dicere definitiones naturalium et mathematicorum, et accidentium eorum: quia mathematica, ut par et impar, et rectum et curvum, et numerus et linea et figura, definiuntur sine motu et materia; non autem caro et os et homo: sed horum definitio est sicut definitio simi, in cuius definitione ponitur subiectum sensibile, scilicet nasus; non autem sicut definitio curvi, in cuius definitione non ponitur aliquod subiectum sensibile. Et sic ex ipsis definitionibus naturalium et mathematicorum apparet quod supra dictum est de differentia mathematici et naturalis.*

perfect definitions must refer to matter. Thus, he engages in a discussion concerning *materia intelligibilis* or ὕλη νοητή to show that mathematical definitions are imperfect.

The first reason alleged by Boccadiferro to reject material definitions in mathematics is that Euclid does not make any reference to matter in his definitions.[274] Hence, when Aquinas points out that some specific geometrical definitions and entities (like that of the semicircle) are "material", he is wrongly taking the so-called *materia intelligibilis* as real matter.[275]

Boccadiferro proceeds then to refute Aquinas' view that the intelligible matter of mathematical items can be equated with prime matter.[276] He argues that by *materia intelligibilis* we should understand the formal, discernible parts of abstracted geometrical items. As such, intelligible matter is said to be "material" only equivocally, but in no other way is it either ontologically or epistemically comparable to real matter. Therefore, mathematics, as it deals with items abstracted from matter, lacks definitional perfection as defined

274 Ludovico Boccadiferro, *Lectiones super tres libros De anima Aristotelis* (Venice: Iohannes Baptista Somaschus et fratres, 1566), lec. 34, pp. 57–58: *Sed, domini, quod dicit iste vir* [*Aquinas*] *multum difficile est et falsum videtur, quia nullibi ipse Euclides definit cum materia, quia dicit ipse: superficies est longitudo et latitudo absque profunditate. Item dicit: circulus est figura plana et cet., et facit in omni alia definitione, quare et cet. Secundo errat quia abutitur nomine materiae intelligibilis Divus Thomas quia Metaphysicorum II, 37 et XV, 8 Aristoteles vocat materiam intelligibilem ipsas partes integrales totius, ut est semicirculus respectu circuli et cet.* ***Non ergo materia intelligibilis est materia prima,*** *ut dicit vir iste Divus Thomas, quare etc.*

275 Boccadiferro targets Aquinas' notion of intelligible matter in a different manner in *Explanatio libri Physicorum*, lec. 16, pp. 40–41. There he completely rules out the notion of intelligible matter in mathematics, pointing out that mathematical items exist solely in cognition without any kind of matter: *Divus Thomas propter hoc argumentum dixit, quod mathematicus considerare materiam sine quantitate revera non potest, nec etiam considerare quantitatem sine materia, quia posterius non potest separari a suo priori. Sed quantitas posterior est materia prima, ergo non potest ab ea separari. Et sic eam considerat, quia ita res se habet ad cognosci sicut se habet ad esse, ut in II Metaphysicorum et cet. Ideo dixit divus Thomas quod duplex est materia, sensibilis et intelligibilis. Materia sensibilis est materia prima qualitatibus affecta, materia autem intelligibilis est materia sine sensibilibus conditionibus. Et mathematicus considerat materiam intelligibilem in conditione quantitatis, non autem materiam sensibilem. Sed hoc quod divus Thomas non est verum, quia si mathematicus potest considerare quantitatem sine materia sensibili, potest etiam ipsam considerare absque alia materia, quia vel quantitas habet esse subsistens in re, vel habet tantum esse in cognitione. Si primum non potest considerari absque materia sensibili, si secundum nullam materiam requirit in eius cognitione, quia si aliqua requireretur, maxime esset sensibilis, quia ab ipsa dependet in esse, sed ipsa non requiritur, ergo nec intelligibilis.*

276 For an account of Aquinas' reception of the concept of *materia intelligibilis* see G. Galluzzo, *The Medieval Reception of Book Zeta of Aristotle's Metaphysics* (*Studien Und Texte Zur Geistesgeschichte Des Mittelalters*) (Leiden Boston: Brill, 2001), pp. 435–436.

by the Bolognese master. Consequently, perfect definitions are only found in natural philosophy, the only science where the defined objects are considered without separating them from real (i.e., sensible) matter.[277] Yet again, an argument found in Boccadiferro's lectures would be echoed by Piccolomini: the claim that *materia intelligibilis* cannot be understood as real matter would be adopted by the Sienese in his denial of mathematical causation (see 3.3.3).

In the *lectiones* 14 and 15 on *De anima*, Boccadiferro explores the rival opinions regarding the superiority of mathematical demonstrations: according to some authors (he most often names Averroes) mathematical demonstration, unlike demonstration in natural philosophy, provides the cause and the effect at the same time. Because of this particular epistemic trait, mathematics would be, as they claim, a more certain science.[278]

The Bolognese master is compelled to acknowledge that mathematics is more certain than natural philosophy, but he utterly refuses to admit that this happens because of any special trait of mathematical demonstrative means. Rather, he claims, the superior degree of certainty of mathematics stems from both our cognitive familiarity with quantity as the commonest accident (see above), and the utter simplicity of *quantus simpliciter*. Mathematical items are stripped via abstraction from matter and movement, sources of imprecision and error and, hence, they are more stable and epistemically diaphanous.[279] Thus, certainty is transferred from mathematical objects to mathematical demonstrative procedures – which, formally, do not differ from demonstration in natural philosophy.

277 Ludovico Boccadiferro, *Lectiones super tres libros De anima*, lec. 34, f. 58: *Ideo aliter dicemus cum Graecis expositoribus ut Ioanne Grammatico, et Simplicio, et Themistio, quod mathematicus citra omnem materiam definit. O, tu dices: omnis definitio debet declarare rem sicuti est ipsa res; dico: domini, quod definitio illa mathematica imperfecta est. O, vos dicetis: quae igitur perfecta est definitio? Dico: illa quae traditur philosopho naturali, mathematicus autem imperfecte definit, quia sine materia.*

278 Ludovico Boccadiferro, *Lectiones super tres libros De anima*, lec. 14, f. 22b: *Tum quia illa scientia est certior quae dicit quia et propter quid, sed mathematica dat utrumque, naturalis autem tantum dat quia, vel rarissime propter quid, nunquam autem simul quia et propter quid, quia scientia naturalis primo procedit ab effectu ad causam, postea si ipsa debeat generare scientiam, fit regressus a causa ad effectum, ergo scientia mathematica est certior quam scientia de anima* [...].

279 Ludovico Boccadiferro, *Lectiones super tres libros De anima*, lec. 14, f. 22a–b: *Praeterea I Posteriorum, cap. 17, probavit Aristoteles demonstrationes mathematicas esse certiores demonstrationibus naturalibus, quia naturales sunt in materia et motu, materia autem est causa omnis erroris et incertitudinis, quia est causa transmutationis et motus. Mathematica autem non sunt in materia et motu, quare etiam certiores.*

This concession notwithstanding, Boccadiferro swiftly introduces a caveat to save the privileged epistemic position of natural philosophy vis-à-vis mathematics. He adopts a double standard of *certitudo*: scientific principles and the demonstrations derived from them are certain, we are told, either *per se* or *ad nos*. Hence, the special ontological properties of *quantus simpliciter* make mathematics the most certain science, but only *to us* (*nobis*). Nonetheless, if the scientific principles of each science are considered in themselves (i.e., regarding their ontological independence, reality, and substantiality), mathematics is the least certain of the three contemplative sciences: the most certain one is metaphysics, followed by natural philosophy; the third and last place is conceded to mathematics, which deals with entities which have no subsistence of their own, but only subsist as accidents parasitic of natural substances.[280]

2.5 Pseudo-Aristotle's *Quaestiones mechanicae*

Another of the themes which influences Piccolomini's views in *De certitudine* is the 16th-century recovery of Pseudo-Aristotle's *Quaestiones mechanicae*. In this section, I will deal with this work and, more generally, with mechanics as a discipline. Once the necessary context is provided, I will show how Piccolomini's philosophy of mathematics in *De certitudine* continues and completes his philosophy of mechanics as held in the Latin paraphrase of the *Quaestiones mechanicae* with which it was published.

Although the pseudo-Aristotelian *Quaestiones mechanicae* held significant importance in the history of early modern science and technology, its influence was short-lived. Within a century of its publication in the Aldine edition, its contents had been superseded and critiqued by Tartaglia, Benedetti, and Galileo.[281]

The *Quaestiones mechanicae* is structured as a collection of 35 problems and their solutions, primarily focused on the behaviour of devices like the balance, lever, ship's rudder, sling missile, forceps, and wheel. The work provides

280 Ludovico Boccadiferro, *Lectiones super tres libros De anima*, lec. 14, f. 17a: *Quia primo est certior metaphysica, postea naturalis, ultimo mathematica. Metaphysica enim est de rebus certissimis, quae a nullo dependent et est de rebus, ut dixi, actualioribus et certissimis, et quod hoc sit verum patet, quia substantia, quae ad naturam in se, est certior accidente, quia prior natura in se.*

281 I follow Rose, S. Drake, "The Pseudo-Aristotelian Questions of Mechanics in Renaissance Culture", in *Studies in the Renaissance* 18 (1971), pp. 65–104. See W. R. Laird, *The Renaissance of Mechanics: Ancient Science in the Age of Humanism* (Cham, Springer, 2024), pp. 157–163; R. Suter, "The Scientifc Work of Alessandro Piccolomini", in *Isis* 60 (1969), pp. 210–222.

mathematised explanations for questions such as: Why does a stone thrown with a sling travel farther than one cast by hand (problem XII)? How can a small rudder steer a large ship (problem V)? Why are pebbles dispersed by the constant action of ocean waves (problem XV)? Why are moving objects easier to move than those at rest (problem XXXI)?

Despite the controversy surrounding its authorship from the outset, the contents of the work captivated mathematical practitioners, technically minded scholars, and an erudite readership interested in advancing human endeavours and practical life through knowledge – particularly mathematical insights – revived from antiquity. Notably, it drew the interest of Piccolomini's patron, Diego de Hurtado y Mendoza.

The *Quaestiones mechanicae* were included in the second volume (1497) of the Aldine *editio princeps* of Aristotle's works. Aldus took the text from Bessarion's manuscript Marciana Z. Gr. 200.[282] The first Latin translation of the *Quaestiones* was published in Paris in 1517.[283] Vittore Fausto (1480–*c.*1551), its translator, was a Venetian humanist, some of whose interests lay in technical innovation and ship construction.

The second translation of the *Quaestiones mechanicae* was published in 1525 by Niccolò Leonico Tomeo (1456–1531). Tomeo, the son of a Greek émigré and a friend of culturally relevant figures such as Bessarion and Pietro Bembo (1470–1547), taught at the University of Padua from 1497 to 1506, encouraging the direct use of Greek sources. His Latin version of the mechanics was published in a collection of his translated pieces which also included a partial Latin version of Proclus' *In Timaeum*.[284] His Latin rendition of the mechanics was reprinted several times.[285]

Piccolomini's paraphrase of the mechanics (preserved in manuscript at Biblioteca Nazionale Centrale di Firenze, II, IV 214) appeared under the name *In mechanicas quaestiones Aristotelis paraphrasis paulo quidem plenior*. A second edition of the text was published in 1565 in Venice by Traianus Curtius. Rose and Drake have argued that Piccolomini most likely used at least two

---

282 Rose, Drake, "The Pseudo-Aristotelian Questions of Mechanics".

283 Pseudo-Aristotle, *Mechanica*, ed. Vittore Fausto (Paris: Josse Bade, 1517).

284 Niccolò Leonico Tomeo, *Opuscula nuper in lucem edita*.

285 As reported by Rose's and Drake's study of the print-history of the work, although deprived of the diagrams and the author's orienting notes, Tomeo's Latin translation of the *Quaestiones* was included in several editions: Bekker's edition of Aristotle's works (III, 409–415); the Giunti *Opera Aristotelis* (Venice, 1552, 1562); Laemarius' (Geneva, 1597); LaRoviere's (1606); Duvalle's consecutive reprintings (1619–1629, 1639, 1654); it is also contained in Michael Psellus' *Compendium mathematicum* (Leyden: Elzevirii, 1647). See Rose, Drake, "The Pseudo-Aristotelian Questions of Mechanics", p. 79.

known manuscripts, namely the 12th-century Marciana, Z. Gr. 214, and the 15th-century Laurenziana, Pl. XXVIII, 45. Yet, the possibility that Piccolomini had access to other sources during his period in Padua and under the patronage of Diego Mendoza cannot be discarded.[286]

Four traits of Piccolomini's attitude towards the *Quaestiones mechanicae* are especially noteworthy for our ends.

The first is that, concerning the authorship controversy, he considered it an authentic work of Aristotle, which he already mentioned in his 1540 reference to the *Quaestiones mechanicae*.[287] In the *Praefatio* to the paraphrase, Piccolomini tackled the question through the linguistical resemblance of the demonstrative and mathematical language present in the *Quaestiones mechanicae* to that of other Aristotelian works, in which mathematical means are used in the resolution of non-mathematical problems.[288]

Secondly, Piccolomini believed that the work required both a more precise Latin translation and a clearer exposition, as he expressed dissatisfaction with Tomeo's exiguous annotations to his edition.[289] The Sienese further argues that the perceived obscurity of the text is not solely due to the poor condition of the Greek text, but also to the lack of mathematical knowledge amongst contemporary readers, which hinders their ability to grasp the arguments presented.

The third mark of Piccolomini's approach, akin to and most likely influenced by Mendoza, is an emphasis on the necessity for a wider understanding and public knowledge of the useful contents of the book.[290] Years later, this very same persuasion would move the Sienese to ask his student, the architect Oreste Vannocci Biringucci (1558–1585), to translate his paraphrase into Italian, so that a wider readership of technicians and artisans could benefit from it. In his preface to the Italian version, Biringucci recollects some of Piccolomini's views on the work about the *Quaestiones mechanicae* and, generally, about mechanical practice:

---

286 Rose, Drake, "The Pseudo-Aristotelian Questions of Mechanics", p. 84.

287 Alessandro Piccolomini, *De la institutione di tutta la vita*, III, 17, p. 72.

288 Alessandro Piccolomini, *In mechanicas quaestiones*, f. IIII: *Si nos phrasim atque elocutionem huius libelli et aliorum Aristotelis locorum, in quibus mathematicis utitur demonstrationibus, simul invicem contulerimus, sicut in disputatione de iride et aliis locis, hanc esse Aristotelis peculiarem phrasim manifestum erit.*

289 Alessandro Piccolomini, *In mechanicas quaestiones*, f. IIII: *Quem quidem libellum, propter eius, ut dicunt, obscuritatem, nullus fuit quem ego viderim qui ad hunc usque diem exposuerit, praeter unum Leocinum, cuius quaedam extant brevissimae annotationes.*

290 See Hobson, *Renaissance Book Collecting*, pp. 79–80; Keller, "Machines as Mathematical Instruments".

> Similarly, he showed me how much utility there would be to bring, especially to engineers and architects, among whom there are many who work and construct some beautiful machines with subtle inventiveness. However, lacking that second element of their science, called "ratiocination" (*ratiocinatione*) by Vitruvius, they are unable to give any reason (*ragione*), contrary to the precept of Aristotle, who says that it is not enough to assert one's opinion, but proofs and demonstrations must be adduced.[291]

Biringucci reveals in this passage one of Piccolomini's motivations: he felt that it was not sufficient for technicians to have mere know-how of their endeavours, a practical ability handed down through generations of artisans and the result of cumulated trial-and-error experience. Piccolomini was convinced that offering demonstrations, or "ratiocinations" in Vitruvian terms, of their craft could enhance not only the artisans' knowledge and technical skill but also elevate the overall quality of the arts. As we will see, Piccolomini also asserts in the *Praefatio* that a certain type of understanding arises from the practice of mechanics.

But what are the proofs, reasonings, or demonstrations in the mechanical arts? What exactly do they teach and how?

Such questions bring us to the fourth and final aspect of Piccolomini's approach to mechanics. This is also the most relevant aspect to our examination of Piccolomini's philosophical stance. Mechanical proofs, such as those articulated by Pseudo-Aristotle and highly appreciated by Diego de Hurtado y Mendoza and embraced by mathematician-technicians like Niccolò Tartaglia, raised significant philosophical questions which required a fresh philosophy of mechanics.

2.5.1 *The Philosophy of Mechanics*

In the controversy which followed Piccolomini's and Barozzi's confrontation, the debate always revolved around pure geometry. Although it is undeniably true that Piccolomini's opusculum uses purely geometrical examples, an

---

291 Alessandro Piccolomini, *Parafrasi di monsignor Alessandro Piccolomini arciuescouo di Patras, sopra le Mecaniche d'Aristotile, tradotta da Oreste Vannocci Biringucci, gentilomo senese*, transl. Oreste Vannocci Biringucci (Rome: Francesco Zanetti, 1582), p. 5: *Mostrommi parimente quanta utilità fusse per apportare, principalmente à gl'ingenieri e architetti, tra i quali si trovano molti che operano e fabricano con sottile inventione qualche bella machina, ma poi non havendo quel secondo membro della scienza loro, chiamato da Vitruvio 'ratiocinatione' non sanno rendere ragione alcuna, contra il precetto d'Aristotele, il qual dice que non basta affermar il parer suo, mà sene debe addur prove e demonstrationi.*

account of *De certitudine* which omits the author's concerns with mechanics misses (in my view) one of the main points that Piccolomini ultimately meant with his work.

For a better interpretation of Piccolomini's aims, one should first ask oneself: what did he understand by *scientiae mathematicae*?

The answer to this question is to be found in the *Praefatio* to the paraphrase and the first chapter of the *In mechanicas quaestiones Aristotelis paraphrasis*. As mentioned, whereas the paraphrase was finished by 1540, both the *Praefatio* and *De certitudine* were composed and added by 1546: Piccolomini mentions in the *Praefatio* to the paraphrase the questions of *demonstratio potissima*, his little treatise on the certainty of mathematics, as well as the name of Proclus as the main authority in the opusculum.[292]

In the same *Praefatio*, Piccolomini explains what he means by *mathematicae*. Here he offers a list of subalternate or mixed sciences, declaring firstly that disciplines such as astronomy, stereometry, perspective, and harmonics derive both their principles and demonstrative procedures from mathematics (that view will be swiftly qualified). Then, he affirms that those disciplines are called "mathematical" for a very specific reason:[293] disciplines that adopt demonstrative procedures or the "instrument" (i.e., the logic) of the mathematicians must be called "mathematical" (*mathematicae dici debent*). Every science, Piccolomini goes on, receives its name from the mode and power of its demonstrations (*ex modo et vi demonstrandi*). Therefore, all subalternate disciplines are mathematical, as they employ the demonstrative means of mathematics, adopt middle-terms which express mathematical properties, and share in the principles of geometry and arithmetic.

Piccolomini apparently (but only apparently) assumes here the traditional account of *scientiae mediae* as *mathematicae*. This view ultimately proceeds

292 Proclus is only mentioned once more in the paraphrase as an authority for reminding the reader that the same theorem can be demonstrated through many, varied proofs. In this way, Piccolomini defends his understanding and conclusion of the 30th problem, which diverges in Piccolomini's proof from that of Pseudo-Aristotle. See Alessandro Piccolomini, *In mechanicas quaestiones*, f. LVIII. See Alessandro Piccolomini, *In mechanicas quaestiones*, f. III: *Verumenimvero, et nos alias disputantes sustentavimus, et nunc constanter asserimus, demonstrationes quibus utuntur mathematici non esse potissimas illas ac praecipuas quas Aristoteles in libros De demonstratione omni diligentia inquirit, quo fit ut alia sit causa qua in primo certitudinis gradu mathematicae scientiae reponantur. Sicut nos abunde in libello quem fecimus De certitudine mathematicaum scientiarum opinionem hanc nostram.*

293 Alessandro Piccolomini, *In mechanicas quaestiones*, f. IIV: *Verum etiam, etsi eius [Aristotelis] testimonium et auctoritas non adesset, nihilominus, cum mathematicorum instrumento pertractentur, mathematicae dici debent.* [...] *ex modo et vi demonstrandi quaelibet scientia recte nuncupabitur.*

from Aristotle's accounts of the so-called subalternate sciences (θάτερον ὑπὸ θάτερον):[294] a special kind of sciences which speak mathematically of non-mathematical objects. Examples of such sciences are optics (which treat sight), music (sound), astronomy (heavenly bodies), and mechanics (artificial movements). According to the Stagirite, these sciences treat a fact (ὅτι) physically and provide an explicative account (διότι) through mathematical means and demonstrations.[295]

Philosophically speaking, the Aristotelian account of how subalternate sciences are possible consists of adopting a double use of the *qua*-operator: optics treats the sight-space *qua* geometrical lines, and geometry is adopted in optics *qua* sight (i.e., it isolates and adopts those parts of geometry that fit sight).[296] As Aristotle summarily affirms in *Physica* II, 2 "geometry researches the physical line, but non *qua* physical, whereas optics researches the mathematical line, but non *qua* mathematical, but *qua* physical".[297]

Because of this, Aristotle described these sciences as "the more physical of the mathematical sciences".[298] Distelzweig has remarked that the text of the *Aristoteles Latinus* had an interesting translation of this phrase. The genitive τῶν μαθημάτων was rendered in Latin in a comparative sense, resulting in the problematic description of subalternate sciences as "more physical than mathematical" ([...] *magis naturales quam mathematicae* [...]).[299]

Finally, it is to be pointed out that Aristotle declared these disciplines as forms of mathematical sciences, something that, as we shall see, was exploited by Piccolomini in his account of the scientific status of mechanics.[300]

Engaging in the philosophy of mechanics, Piccolomini finally turns to the question of the status of mechanics itself as a discipline. Firstly, he declares, in consonance with the Aristotelian common places just expounded, that

---

294 The common *loci* devoted to the so-called subalternate mathematical sciences and their methodology and demonstrative idiosyncrasy are *Analytica posteriora*, I, 27, 87a, l. 33; I, 7, 75b, l. 14–17; I, 9, 76a, l. 10–16; I, 13, 78b, l. 36–8 and 79a, l. 10–17; *Physica*, ed. I. Bekker, vol. 1, II, 2, 194a, l. 9–11; *Metaphysica*, III, 2, 997b, l. 14–24; XIII, 2, 1077a, l. 1–8.

295 Ex. gr. in *Analytica posteriora*, 79a Aristotle claims the following about optics: τὸ μὲν γὰρ ὅτι φυσικοῦ εἰδέναι, τὸ δὲ διότι ὀπτικοῦ, ἢ ἁπλῶς ἢ τοῦ κατὰ τὸ μάθημα.

296 P. M. Distelzweig, "The intersection of the mathematical and natural sciences: The subordinate sciences in Aristotle", in *Apeiron* 46/2 (2013), pp. 85–105.

297 *Physica*, II, 2, 194a, l. 8–11: ἡ μὲν γὰρ γεωμετρία περὶ γραμμῆς φυσικῆς σκοπεῖ, ἀλλ' οὐχ ᾗ φυσική, ἡ δ' ὀπτικὴ μαθηματικὴν μὲν γραμμήν, ἀλλ' οὐχ ᾗ μαθηματικὴ ἀλλ' ᾗ φυσική.

298 *Physica*, II, 2, 194a, l. 7: [...] τὰ φυσικώτερα τῶν μαθημάτων.

299 Distelzweig, "The intersection of the mathematical and natural sciences", p. 90.

300 Aristotle, *Metaphysica*, VI, 1, 1026a, l. 24–27; I, 8, 989b, l. 30–34; and XII, 8, 1073b, l. 2–7.

mechanics is beyond any doubt a science subalternate to geometry, and hence a mathematical discipline.[301]

In the first chapter of the paraphrase, the Sienese insists on the fittingness of calling mechanics a science (*scientia*) rather than an art (*ars*). He argues that Aristotle himself used such a denomination to refer to mechanics.[302] In addition, he claims that occasional references to mechanics as an art in Aristotle's writings do not diminish its status, for in the Greek of the Stagirite, Piccolomini assures the reader, the terms ἐπιστήμη and τέχνη are interchangeably adopted.[303]

Once he has determined that mechanics is a mathematical science, accounting for the legitimacy of both terms *scientia* and *mathematica*, Piccolomini proceeds to clarify the specific place mechanics occupies in the hierarchy of disciplines. And then he introduces one of his major twists. In the first chapter of the *Quaestiones mechanicae*, he explains:

> The art of mechanics is to be considered only that through which, by inventing many machines and constructions, we discover things useful to various practical arts. There should be no doubt that it should be counted amongst the contemplative parts of philosophy (*philosophiae contemplatricis*), for it speculates on the principles of many arts, as it is subjected to geometry. And indeed, although these principles are directed towards practical work, the contemplative faculty (*facultas contemplatrix*) of these principles will not be disregarded for this reason.[304]

According to the Sienese philosopher, mechanics is one of the contemplative parts of philosophy (*philosophia contemplatrix*). In other words, Piccolomini

301 Alessandro Piccolomini, *In mechanicas quaestiones*, f. III: *Mechanica igitur scientia, cum mathematico medio instrumento ac medio pertractetur, nemini dubium esse debet quin sub geometria comodissime collocetur.*

302 I will elaborate on what Piccolomini meant when he applied the term *scientia* to mathematical disciplines in section 3.4.2.

303 Alessandro Piccolomini, *In mechanicas quaestiones*, f. VI: *Scientia igitur magis quam ars facultas haec mechanica merito nuncupabitur, praesertim cum Aristoteles ipse non solum ab initio mechanicarum quaestionum, sed etiam in Generatione animalium, in Metaphysicis, et quam pluribus aliis in locis de scientiis loquens, nomine artis abutatur.*

304 Alessandro Piccolomini, *In mechanicas quaestiones*, ff. VV–VI: *Mechanica enim ars illa tantum existimanda est qua excogitantes quamplures machinas, constructiones ceteris artibus sellularibus invenimus, quam philosophiae contemplatricis partibus annumerandam esse nemini dubium esse debet, quippe quae, geometriae subiecta, multarum artium principia excogitata speculatur. Quae quidem principia, etsi dirigantur ad opus, non ob id tamen horum principiorum facultas contemplatrix non existimabitur.*

claims that the very discipline affording many practical applications related to the banausic or servile arts (*sellulariae, banausicae, serviles*) and inquiring into how these sorts of machines work or even discovering new ones must be considered as a contemplative science.

This claim presupposes a crucial shift in the understanding of mechanics in comparison with the commonly held views within the Latin context.[305] To understand the momentum of such a statement, some historical considerations are in order.

At first sight, the leap from being a mathematical science to contemplative philosophy seems an all-natural consequence of the Aristotelian position just explained: much like astronomy is a *scientia mixta* eminently enumerated amongst the lofty contemplative sciences, by the same token, mechanics should be legitimately taken not just as a mathematical, but also as a contemplative science. Nevertheless, the history of this discipline in the Latin West has shown otherwise.

Mechanics had occupied, to say the least, an ambiguous place amongst the sciences and arts during the Latin Middle Ages. To offer some examples, Isidore of Seville (*c.*560–636) excludes mechanics from mathematical sciences in book III of his *Etymologiae*; instead, he introduces *mechanica* as a craft in his *De differentiis rerum*.[306] Eriugena (*c.*800–*c.*877), in his commentary on Martianus Capella's *De nuptiis Philologiae et Mercurii* – the first known appearance in Latin of the syntagm *artes mechanicae* – places mechanics along with other seven humble arts, including tailoring, agriculture, and cooking.[307] William Wheatley (fl. 1309–1316) in his exposition of Pseudo-Boethius' *De scholarium disciplina* affirms that the very name of mechanics is related to the Latin verb *mechari/moechari* (to commit adultery). The argument provided for this etymology is that the intellect, when humans are devoted to such plebeian arts, is excluded or used (if any use is made) for merely practical means. Hence, either

305 For a history of mechanical arts see Laird, *The Renaissance of Mechanics*, especially pp. 160–163 where Piccolomini is addressed; E. Whitney, *Paradise Restored: The Mechanical Arts from Antiquity through the Thirteenth Century* (Philadelphia: American Philosophical Society, 1990); S. A. Walton, *An Introduction to the Mechanical Arts in the Middle Ages* (Toronto: Association Villard de Honnecourt for Interdisciplinary Study of Medieval Technology, University of Toronto, 2003).

306 Isiodore of Seville, *Etymologiarum sive originum libri XX*, ed. W. M. Lindsay (Oxford: Clarendon, 1911), III: "De mathematica"; Isiodore of Seville, *Liber differentiarum*, ed. M. A. Andrés Sanz (Turnhout: Brepols, 2006), II, 29. See also Whitney, *Paradise Restored*, p. 42 and 59.

307 John Scot Eriugena, *Annotationes in Marcianum*, ed. I. Ramelli (Milan: Bompiani, 2006), II, 79, 12: [...] *Postquam Mercurius dederit septem liberales artes, tunc Virgo dabit septem mechanicas.*

neglected or defiled by banausic affairs and the knowledge of particulars, the intellect commits adultery.[308]

Regarding the scholarly works within the field of mechanics proper, there did not exist an erudite and continuously cultivated corpus of mechanics in the university context, for (as seen) mechanics was not numbered amongst the seven liberal arts. The only exceptions were short and disconnected works such as those of Jordan of Nemore's (fl. *c.*1250) *Scientia de ponderibus*, Pseudo-Euclid's *De ponderoso et levi*, and Archimedes' *De insidentibus in humidum*, *Liber de canonio*, and *Liber karastonis*.[309] The overall conclusion is that mechanics, unlike its more fortunate subalternate siblings such as astronomy, music or optics, was generally held in low esteem by Latin scholars: erudite elaborations on the field were rare, its dignity was called into question given its practical and low-class implications, and it was excluded from the curricula as a servile art or a mere craft.

As Whitney has pointed out in her seminal work,[310] the 11th- to 13th-century infusion of Islamic philosophy in the Latin world together with a more perspicuous notion of the Aristotelian "productive" (ποιητικαί) disciplines provided the occasion for some Latin authors to regard mechanics in a slightly better light. Some cases of this revalorisation of mechanics in the 13th century are Robert Kilwardby (1215–1279) and Roger Bacon (*c.*1219/20–1292), who considered that mechanics could be connected to higher branches of knowledge through mathematics.[311]

Nevertheless, those are isolated, rare cases: the general appreciation of mechanics remained that of a servile craft, loosely connected with mathematics in some of its parts, but philosophically low-ranking. Those are the opinions, for instance, of Thomas Aquinas and Albert the Great.[312] Thus, although the theoretical elements and assumptions for considering mechanics as a

308 William Wheatley, *Expositio in Boethii De scholarium disciplina*, in Thomas Aquinas, *Opera omnia*, vol. 24 (Parma: Pietro Fiaccadori, 1869), cap. 5: *Notandum, artes mechanicae in sola manuum exercitatione consistunt omni operatione intellectus penitus aut pro majori parte seclusa. Intellectus enim in suis operationibus quietus est; ipsae autem artes plebeiae intellectum moechari cogunt, quare non immerito obtusioribus qui solum singularia comprehendere valent, ascribendae sunt.*

309 For an edition and translation of the mentioned works see M. Clagget, *The Science of Mechanics in the Middle Ages* (Madison WI: University of Wisconsin, 1959). For the Renaissance shifting in the social view and dignity of artistic and mechanical endeavours see P. Rossi, *I filosofi e le macchine 1400–1700* (Milan: Feltrinelli, 2017), pp. 37–71.

310 Whitney, *Paradise Restored*, pp. 129–134.

311 For a full historical account of these figures and their views on crafts and mechanics see Whitney, *Paradise Restored*, chapter v.

312 Whitney, *Paradise Restored*, pp. 137–141.

mathematical, and hence contemplative philosophy, together with astronomy and optics, were already present at least since the 11th century in the Latin West, this position would have to wait for more convenient times to be recast and widely accepted.

From the middle of the 15th century onwards, the enthusiasm for the benefits of mechanical arts applied to architecture, sculpture, or general machine construction was remarkably noticeable in circles of *litterati* in Italy. An extraordinary flourishing of associated literary genres followed. To name only a few, Ghiberti's (1378-1455) *Commentarii* (1447) displayed deep familiarity with Euclid's and Alhazen's optics applied to the fine arts, inviting artists to become more familiar with the mathematical techniques employed by the ancients and the Arabs in their artistic endeavours. In 1450 Pope Nicholas V (1397–1455) commissioned a bilingual edition of the Archimedean corpus to Iacobus Cremonensis (1395/1410–1454) (*Archimedem Grecum et Latinum*) to improve current knowledge about statics and dynamics, an edition enthusiastically received by Nicholas of Cusa who made use of it in his work *De mathematicis complementis* (1453); moreover, Regiomontanus (1436–1476) used Iacobus' edition for his lectures at Padua in 1464.[313] By 1450 Leon Battista Alberti (1404–1472) published his *De re aedificatoria*. This was the first ever printed book on building, followed by Sulpitius' (fl. *c.*1470–1490) edition of Vitruvius' ten books on *De architectura* (1486): all those works provided mathematically justified solutions to several problems related to architecture and mechanical engineering; toys and fountains were designed under the inspiration of Hero's *Pneumatica* and *Automata* and usually found amongst the other delights of the villas of the nobility.[314] By 1492 the humanist Giorgio Valla was lecturing on Vitruvius at Venice.[315]

The social prestige and success achieved by mechanical practitioners in the fields of arts, ballistics, and fortification in Piccolomini's time, the inclusion of the *Quaestiones mechanicae* in the Aldine edition, as well as the fostering patronship of renowned public figures such as his patron Diego de Mendoza

313 See Rose, "Patrons, Collectors and Translators: Humanist Origins of the Mathematical Renaissance", in *The Italian Renaissance of Mathematics*, p. 40. For the humanists' and courtiers' attitudes towards mechanics and machinery see J. L. Wolfe, *Humanism, Machinery, and Renaissance Literature* (Cambridge and New York: Cambridge University Press, 2004).

314 Rose, Drake, "The Pseudo-Aristotelian Questions of Mechanics", pp. 70–71.

315 Rose, "Patrons, Collectors and Translators", p. 49. For an account of the theoretical changes surrounding the flourishing of early modern mechanics see Keller, "Mathematics, Mechanics and the Origins of the Culture of Mechanical Invention", in *Minerva* 23/3 (1985), pp. 348–361.

or the Venetian *Infiamatto* Daniele Barbaro (who translated Vitruvius into Italian and commented upon it), seem to have impelled Piccolomini to his (still extravagant) opinions regarding the scientific status of mechanics. The study of mechanics, a trade successfully cultivated by artists, engineers, and architects, a discipline favoured by the patronage of distinguished noblemen and enjoyed in its public beauty and usefulness, vindicated an intellectual revalorisation. Piccolomini avowedly attempted to accomplish this.[316] In this sense, Piccolomini can be considered as one of the final steps in a process of intellectually positive revaluation of mechanics, marginally initiated in the Latin world in the 11th century.

Therefore, when Piccolomini employed the syntagm *mathematicae scientiae*, both in the title and in the body of *De certitudine*, he not only referred just to pure mathematics or the traditional mathematical sciences of the *quadrivium*, but also included amongst mathematical and contemplative parts of philosophy the discipline of mechanics and even the activity of mechanical invention. This detail will be significant when considering the problems which he discusses along the *De certitudine* and the reason why this treatise was published as an attachment to the paraphrase of the *Quaestiones mechanicae*.

To conclude, what is then Piccolomini's position regarding the place of mechanics amongst other disciplines? Is it a craft, a contemplative discipline, or a combination of both?

My interpretation of Piccolomini's standpoint runs as follows: the Sienese held that mechanical disciplines have two moments, a contemplative one,[317] and a technical one. In the contemplative one, an epistemic agent searches for the mathematical principles and explanations behind mechanical behaviour or produces a new pattern from them. In such a manner, by grasping mechanical principles and deducing particular consequences from them, mechanics can be deemed to grant some sort of (mathematical) cognition to the practitioner.[318]

In the practical moment, such principles are applied in craftsmanship, but with a great degree of inaccuracy. Therefore, although the abstract search for and speculation upon mechanical principles is oriented to practice, nonetheless, the very mathematical inquiry and the resulting understanding are contemplative in nature. The same applies in the case of pure geometry: a

316 Laird, "The Scope of Renaissance Mechanics", in *Osiris* 2 (1986), pp. 43–68.

317 See, for instance, Alessandro Piccolomini, *In mechanicas quaestiones*, p. 111v: *Mechanicus tamen artifex, ea ratione qua mechanicus est, solummodo causas et principia earum consyderans, in contemplatione ipsa quiescit et sistit.*

318 I will elaborate on this question in 3.4.2.

carpenter or a farmer contemplates geometrical truths with the sole aim of applying them to obtain the right shape in table-making or the equal division of the land: their practical purposes notwithstanding, the carpenter or the farmer would be engaging in contemplation at the very moment in which they perceive the geometrical truth as such, as a step in their practical goals.

It is to be pointed out that not only purely Aristotelian considerations might have directed Piccolomini's statement that mechanics is a contemplative science. Indeed, Proclus' *In Euclidem* might have also played a part in the Sienese's views. For, when it comes to the subalternate sciences and mechanics, Proclus describes them as follows:

> In its lowest applications, therefore, it [mathematics] projects all of mechanics, as well as optics and catoptrics and many other theories.[319]

According to Proclus, not only the more abstract sciences of arithmetic and geometry are unfolded and projected onto phantasy. The inferior mathematical sciences are also cast from the soul (προύβαλεν), including, quite literally in Proclus' statement above, "the entirety of mechanics" (μηχανικὴν ὅλην). By this, Proclus' passage seems to suggest that all projected mathematical sciences obtain the rank of theoretical sciences (θεωρίαν). From passages like this, it is easy to see how Piccolomini could find further support and inspiration for enhancing the status of mechanics in Proclus' commentary.

2.5.2 *Mechanics and Natural Philosophy: Do They Study the Same Objects?*

The ever-increasing public role of a mathematised body of knowledge applied to the domestic, martial, and plastic arts during the Renaissance raised philosophical questions. This is a problem that was tackled by Piccolomini both in the 1540 paraphrase and in the 1547 *De certitudine*, although with slightly different approaches. The main issue can be outlined as follows: when mechanical demonstrations are used for establishing the principles ruling the balance, the sling, the wheel, or the movement of some bodies, what kind of knowledge is produced? Is ballistics natural or mathematical knowledge? Is it both physical and mathematical knowledge, or perhaps neither? Or is it, in a Mertonian-like manner, mathematical knowledge of a physical object?

---

319 Morrow's translation (slightly modified). Original Greek in Proclus, *In Euclidem*, p. 19, l. 25–26: διὸ δὴ καὶ ἐν ταῖς ἀποπερατώσεσιν ἑαυτῆς τήν τε μηχανικὴν ὅλην προύβαλεν καὶ τὴν ὀπτικὴν καὶ κατοπτρικὴν θεωρίαν.

One answer to this challenge is already given in the very text of Pseudo-Aristotle's *Quaestiones mechanicae*, I, 5:

> These [mechanical problems] are not altogether identical with physical problems, nor are they entirely separate from them, but they have a share in both mathematical and physical speculations, for "how it happens" is demonstrated by mathematics, but the "about which" belongs to physics.[320]

According to the author of the *Quaestiones mechanicae*, this discipline partakes of both natural philosophy and mathematics: natural philosophy concerns the phenomenon the science is occupied with (τὸ περί), whereas mathematics shows how the phenomenon occurs (τὸ ὥς).

This meagre line could be interpreted in many ways. One of them is that the analysis of an object in movement (say, a balance) belongs to natural philosophy – what the balance is *qua* material-bronze object (about which), whereas mathematics would explain how the balance works in such-and-such a way when such-and-such weights are applied. In other words, the answer to the question "What is this?" would be answered by natural philosophy, whereas the response to "How or why does it work as it does?" would be a mathematical explanation.

The former paragraph is susceptible to a far more controversial reading: what the text of the *Quaestiones mechanicae* is referring to with the two sorts of explanations (τὸ περί and τὸ ὥς) are the two sorts of demonstration (τὸ ὅτι and τὸ διότι; *quia* and *propter quid*). Indeed, considering the possible causative meaning of ὥς in ancient Greek, it can be easily interpreted that what Pseudo-Aristotle affirms is that, in mechanics, the demonstrations of effect-to-cause are conceived in terms of natural philosophy. Then, once a natural cause is reached, a cause-to-effect explanation is offered in mathematical terms. Therefore, when we explain how the balance works, we first reason from the effect to the cause by attending to the material nature of the artefact (the weights are attracted to their natural place on the ground because of such-and-such cause, pushing down the plates of the balance, etc). Once we

320 Translation from *Mechanica sometimes attributed to Aristotle* (Loeb Classical Library: Cambridge (Mass.) and London, 1936). See original at Pseudo-Aristotle, *Quaestiones mechanicae*, ed. J. P. van Capelle (Amsterdam: Petrus Den Hengst et filius, 1812), p. 3: Ἔτι δὲ ταῦτα τοῖς φυσικοῖς προβλήμασιν οὔτε τ᾽αὐτὰ πάμπαν, οὔτε κεχωρισμένα λίαν, ἀλλὰ κοινὰ τῶν τε μαθηματικῶν θεωρημάτων καὶ τῶν φυσικῶν. τὸ μὲν γὰρ ὣς διὰ τῶν μαθηματικῶν δῆλον, τὸ δὲ περὶ ὅ, διὰ τῶν φυσικῶν.

have found a natural cause of the behaviour of the balance, the cause-to-effect account is produced by appealing to the mathematical principles ruling over the balance.[321] In this reading, the *Quaestiones mechanicae* delivers a mathematical discipline (for it uses geometrical demonstrations and middle terms) whose demonstrations explain the behaviour of sensible entities.

As shown in section 2.4.2 above, the latter reading was understood as deleterious to the status of natural philosophy by some philosophers. It implies that quantification is a legitimate approach to scientifically explaining sensible phenomena. As we will see, Piccolomini sought to address this problem in several ways.

2.5.3 *Piccolomini's 1540 Anti-Mertonian Standpoint*

Piccolomini's early answer to the puzzle of the scope of mechanics reveals a reluctance towards the idea that mathematical meta-language can be adopted to conclude truths about natural objects – as the author of the *Quaestiones mechanicae* seems to suggest. This approach is singled out for criticism in his Latin paraphrase. In question I, chap. VII, Piccolomini proposes to himself a momentous doubt: whether the mathematical principles of mechanics (and, generally, of all subordinate sciences) hold when applied to sensible realities:

---

321 It is worth mentioning that, although the Pseudo-Aristotle establishes the difference between movement according to nature and movement against nature (κατὰ φύσιν, παρὰ φύσιν, *Quaestiones mechanicae*, I), and declares that mechanics deals with the second case, in the pieces of dynamics, kinetics, and statics such as those produced by Tartaglia, Benedetti, and Guidobaldo, this distinction is increasingly blurry. This ever-increasing mathematical neutrality or indifference to the kinds of movement seems to be a legacy of the Mertonian approach to kinematics, in which the mathematical descriptions of movement and motive-powers, as well as imagined counterfactuals, were undertaken without consideration of their being natural or violent, or even naturally possible cases. On the history of early modern mechanics and the gradual intertwinement of mathematics and natural philosophy see A. Gabbey, "Between Ars and Philosophia Naturalis: Reflections on the Historiography of Early Modern Mechanics", in J. V. Field, F. A. J. L. James (eds), *Renaissance and Revolution: Humanists, Scholars, Craftsman, and Natural Philosophers in Early Modern Europe* (Cambridge New York: Cambridge University Press, 1993), pp. 133–145; on this topic see also Keller, "Mathematics, Mechanics and the Origins of the Culture of Mechanical Invention". Also noteworthy is the fact that a categorical distinction between different scientific fields was not respected by Aristotle himself: as De Groot has shown, some mechanical principles in the pseudo-Aristotelian work (lever-balance and their theorems related to the circle) were adopted by Aristotle in some natural explanations, biological and astronomical. See J. De Groot, "*Dunamis* and the Science of Mechanics: Aristotle on Animal Motion", in *Journal of the History of Philosophy* 46/1 (2008), pp. 43–67. For some general remarks on the epistemological problem of artifacts as a source of natural knowledge in late Renaissance, see the Appendix "Il rapporto natura-arte e la macchina del mondo", in Rossi, *I filosofi e le macchine 1400–1700*.

[1] However, if anyone finds absurd what we have just said, namely that a small weight can be suspended on a small balance, which not only will not be visibly moved by this action but indeed will not be moved at all, especially since it can be reasoned that, because something has been added to the plates that was not there before, it must either be that [what has been added] has no weight (which is false by concession) or that that weight must exert no force, which leaves an absurdity in nature.
[2] In response to someone doubting in this way, it must be said that many things are often concluded to be true through mathematical demonstration and imagination, which cannot exist in the nature of things. For it is necessary that, for example, if we abstract it by imagination from natural matter, a straight line must not touch a sphere except at a point. Nevertheless, this cannot be found in natural things. For, although celestial bodies are immune from any blemish, perfectly round and turned, they cannot be touched by a straight line in such a way that they do not touch at an array (of points). I say, therefore, that the reasoning employed by those who doubt in this manner holds true but only in a mathematical mode. However, in natural things, it does not hold true due to the interposition of natural matter.
[3] Therefore, this is the cause why many pseudo-philosophers are deliriously led astray while engaging in extensive and intricate debates on various questions such as the maximum and minimum and similar topics, spending entire days on them. This is particularly done, as they say, concerning what is written in book VI of *Natural Principles* [i.e., *Physica*], without noticing that Aristotle himself, in that book, adopts mathematical, not natural arguments.[322]

322 Alessandro Piccolomini, *In mechanicas quaestiones*, f. XX–XXV: [1] *Si cui tamen absurdum videbitur, quod modo diximus, aliquod scilicet modici momenti pondus in quapiam parva libra pendi posse, quae non solum ob id mota sensui non apparebit, sed vere etiam non movebitur, praesertim cum ratione concludi possit quod, quia lancibus illis aliquid appositum est quod ibi prius non erat, necesse est, vel illud nullius ponderis esse (quod falsum est ex concessu), vel pondus illud nutum aliquem non habere, quod absurdum relinquitur in natura.* [2] *Hoc pacto dubitanti respondendum est, quod multa saepenumero concluduntur vera ex demonstratione atque imaginatione mathematica, quae tamen in rerum natura dari nequeunt. Necesse enim est, exempli gratia, si abstrahimus imaginantes a materia naturali, lineam rectam non contingere globum nisi in puncto, nihilominus hoc in rebus naturalibus reperiri nequit. Siquidem, quamvis corpora caelestia ab omni labe immunia sint, perfecte rotunda ac tornata, non ob id tamen a recta linea possent ita contingi, ut quadam intercapidine non contingantur. Dico igitur quod ratio quam adducunt qui eo modo dubitant verum concludit, sed modo mathematico. In naturalibus tamen, ob interpositam materiam naturalem, non concludit.* [3] *Quare id est in causa quod multi pseudo-philosophi hallucinantur dum multis*

The problem triggering Piccolomini's philosophical argument is posed in paragraph [1]: How is it that, according to the principles of mechanics, a balance should be moved even by a minimum weight located in one of the plates, whereas this is empirically false? Why does not an imperceptible speck of dust (which, nonetheless, must possess some weight) unbalance the balance?

Piccolomini's answer is delivered in paragraph [2]. He tells the reader that, although mathematical statements are true (following Aristotle) they are just mathematically, but not physically true.[323] In other words, mechanical principles and theorems, as they are about entities abstracted by imagining, do not invariably hold if applied to natural objects. Thus, subordinate sciences are not really concerned with natural objects: they describe mathematical objects, entities belonging to a completely different kind. Piccolomini then adapts the example taken from Aristotle's discussion of abstraction in *De Anima* I, 1 to affirm that, although that "a straight line, qua straight, touches a sphere only at one point" is true in mathematics, it cannot be naturally actualised – not even in the celestial region.[324]

It is worth noting that, in this early passage, Piccolomini uses (and only here) the syntagm *imaginatio mathematica*. Six years later, in *De certitudine*, he would invariably adopt *phantasia* and *quantum phantasiatum*. In addition, although he invokes the doctrine of abstraction, he does not provide any clue as to how this process is supposed to work, assuming that the reader would understand what he means. We are only told that, by virtue of this mathematical-imaginative maneuverer, the abstracted quantitative object of study of mechanics and its treatment are fully mathematical, and not natural anymore.

In paragraph [3] Piccolomini scornfully identifies some philosophers who hold the refuted position, i.e., that quantified statements can hold exactly true in natural objects and, that they can provide knowledge of the sensible world. Piccolomini noticeably mentions the terminology of the Calculators: those who discuss *de maximo et de minimo* in the quantified analysis of movement by speculating upon Aristotle's mathematisations in book VI of the *Physica*, where Aristotle concludes through geometrical reasoning that time and movement,

---

ac prolixis quaestionibus, veluti de maximo et minimo et huiusmodi, integros disputant dies. Et hoc praesertim, ut dicunt, occasione sumpta ab iis quae scribitur ab Aristotele in VI de Principiis naturalibus, non animadvertentes Aristotelem ipsum in illo libro disputantem rationibus mathematicis, non naturalibus. The division into paragraphs and numeration is mine. I will refer to these paragraphs by the assigned numbers.

323 *Physica*, II, 2, 193b, l. 35; *Metaphysica*, 1078a, l. 18–19.

324 Aristotle, *De anima*, I, 1, 403a, l. 13–15.

as continua, cannot be composed of indivisible units.[325] The anti-Mertonian streak of philosophers such as Pomponazzi is noteworthy in this passage: Piccolomini refers to those who think that they can legitimately use quantification in natural philosophy as "pseudo-philosophers" (*pseudo-philosophi*) and emphasises that their misguided and obscure speculations are nothing more than hallucinations (*hallucinantur*).

What is behind Piccolomini's position in this early section?

The reasons provided in this citation and the passages surrounding it are:

(1) Mathematical statements are *ex demonstratione et imaginatione mathematica* and

(2) The pseudo-philosophers (the Calculators) do not understand or even perceive the epistemological difference between mathematics and natural philosophy.

Piccolomini points out that the Calculators misinterpret Aristotle's use of mathematical procedures in *Physica* VI. They overlook the fact that, whenever Aristotle quantifies a physical entity in his writings to address certain problems, he is no longer dealing with physics, but with mathematics. Using mathematics requires abstraction, and abstracted mathematical objects are, by definition, non-natural. The reason argued to explain why mathematical statements cannot be verified by physical items is the obstacle of natural matter itself: only mathematical objects, separated from sensible matter, stand as the genuine truth-makers of mathematical propositions.[326] In this manner, Piccolomini opens an unsurmountable epistemic and ontological gap between mathematics and physics, between natural items and abstracted mathematical beings.[327]

325 See, for instance, Aristotle, *Physica*, VI, 5, 231b ff.

326 A very similar account is found in Tartaglia's philosophical explanations added to his Italian version of the *Elementa* of Euclid. Mathematical constructions, Tartaglia claims, cannot exactly be reproduced in matter (paper, earth, or whatever other matter), but only as an approximation. See Euclid, Niccolò Tartaglia, *Euclide Megarense philosopho, solo introduttore delle Scientie Mathematice* (Venice: Venturino Ruffinelli, 1543), f. IIIIv: *Onde per resolver questo dubbio rispondo (come habbiamo detto nel principio del Prohemio) che tutte le operationi e costruttioni fatte dal operante in materia, cioe, in carta, over in terra, over in qual si voglia altra materia mai puonno esser cosi vere e precise che le non possano esser piu vere e piu precise.* On Piccolomini's 'anti-realist' philosophy of mathematics, as he labels it, see Higashi, *Penser les mathématiques*, p. 136.

327 Another passage where the unrelatedness and independence of mathematics from sensible matter is stressed is Alessandro Piccolomini, *In mechanicas quaestiones*, p. II: *Caeterum mathematica, quae contemplatricis philosophiae tertia restat pars, quia quantum, magnitudinem scilicet et multitudinem, easque nullo habito respectu ad materiam in qua firmentur, consyderat, idcirco duae sunt eius partes, alteram numerum contemplans,*

This evinces that by 1540 Piccolomini saw a full-correspondence interpretation of the *Quaestiones mechanicae* as an actual threat: should the Mertonian approach take over as the default philosophy in interpreting Pseudo-Aristotle's *Quaestiones mechanicae*, it would follow that mechanics draws conclusions epistemically relevant in natural philosophy.

Piccolomini will return to this precise argument six years later, in the last chapter of *De certitudine* (XII). As we shall see, the Sienese repeats there almost *ad litteram* his statement that the quantified passages in *Physica* VI do not hold isomorphically true in natural items. Nevertheless, by that time Piccolomini would have matured in his thought and designed a full-fleshed philosophy of mathematics inspired by Proclus and Pseudo-Simplicius. This philosophy would allow him to provide an ameliorated answer to the same problem by tackling the question of the precise ontological status of mathematical objects.[328]

Returning to our discussion of Piccolomini's motivations, it has already been noted that subalternate sciences were viewed as disciplines that employ mathematical language to describe non-mathematical entities. For example, the traditional view holds that the geometrical sphere reveals something about this bronze sphere, inasmuch the bronze sphere is a token of the geometrical sphere. These arguments prompted the philosophical questions raised by Piccolomini: How far does this alleged "inasmuch" actually go? To what extent do sensible objects and processes match their mathematical counterparts? Piccolomini's answer to this question can be inferred from the preceding sections. We need to consider that:

(a) Piccolomini sought to establish the dignity of mechanics and promote this discipline as a form of contemplative philosophy.

(b) At the same time, Piccolomini shared in the anti-Mertonian view that mathematics has (and must have) no say on natural realities *qua* natural realities. This task belongs to the natural philosopher or the metaphysician, not to the mathematician, let alone to the technician.

---

arithmetica scilicet, reliqua vero continuam magnitudinem speculans, quam geometriam nuncupamus. This statement should be understood in the context of Piccolomini's philosophy of mathematics as outlined in chapter VII of *De certitudine*. From this, it becomes clear that the matter not considered "in any respect" (*nullo habito respectu*) in mathematics refers exclusively to sensible matter, which is the occasional cause of phantasy's triggering input. But this does not apply to the so-called "intelligible matter", or the sensible-like traits of phantastic quantity. For further clarification, see sections 3.5.2 and following.

328 I shall offer a detailed analysis of the implications of Piccolomini's 1546 philosophy of mathematics vis-à-vis a full-correspondence reading of mathematised kinetics in section 3.6.

Following the preceding Italian, anti-Mertonian tradition, Piccolomini seeks to prevent any full-correspondence reading of the mechanics as a mathematical account of sensible movement. Hence the philosophical concerns which motivated him: he needed a philosophy of mathematics able to dignify mechanics and, at the same time, prevent a full-correspondence reading of it: an ontology avowedly designed to promote mechanics without undermining natural philosophy.

Piccolomini developed an earlier solution to both (a) and (b). It leads us back to the troublesome paragraph of the *Quaestiones mechanicae* discussed in the last section, i.e., Pseudo-Aristotle's statement regarding the "about which" and the "how" of the discipline. In this paragraph lies a not insignificant hindrance for Piccolomini's project, since the very author of the *Quaestiones mechanicae* declares that, although diverse in their scientific procedures and idioms, natural philosophy and mechanics overlap in their object of study, namely, sensible kinetic processes depending on natural principles.[329] The Sienese's strategy in tackling this problem was omitting and reformulating some parts of the text in the Latin paraphrase. Indeed, when in Chapter I of the paraphrase he came across the problematic lines, he avoided any explicit mention of a would-be natural object which mechanics explores. Instead, the reader gets the following rephrasing:

> Indeed, these [mechanical problems], because they cannot be entirely called natural or wholly mathematical, but they have the flavour of both natures, are nonetheless more closely related to mathematical inquiries.[330]

---

329 As noted earlier, Piccolomini tends to blur the line between natural and non-natural movement. He believes that machines rely on natural principles to create effects that nature, in its simplicity, does not produce on its own, but permits. Therefore, according to Piccolomini, the behaviour of machines *qua* sensible entities is ultimately governed by natural principles. See ex. gr., Alessandro Piccolomini, *In mechanicas quaestiones*, f. Vv: *Ars enim quamvis naturam imitetur, ac illam adiuvet, multa tamen, ut usui nobis esse possint, aliter operatur ac illa facit. Neque minus ob id naturae imitatrix existimari debet, quippe quae ex eo modo efficit quidquid facit quo illamet efficeret, si illud faceret.* See section 3.6.

330 Alessandro Piccolomini, *In mechanicas quaestiones*, f. VI: *Quae quidem, quamvis non penitus naturales nec omnino mathematicae dici possint, sed utrorumque naturam sapiant, magis tamen mathematicis quaestionibus propinquae sunt.* See also Alessandro Piccolomini, *In mechanicas quaestiones*, f. IIv: [...] *quae tamen omnes, etsi purae ac syncerae mathematicae dici nequeant, cum materiam modo quodam respiciant, convenientius tamen mathematicae quam naturales appellantur.*

Piccolomini's paraphrase concedes to the original Greek that mechanics is not *prima facie* a completely natural discipline, nor a wholly mathematical one: mechanics possesses the "taste" of both. He adopts the Latin phrase *sapere* + accusative,[331] a common Latin trope amongst humanists to signal stylistic resemblances, to indicate a similarity between natural questions and mechanical problems. But just after indicating this resemblance, Piccolomini swiftly concludes his own saying, sc., that mechanics is nearer to mathematics.

As I understand it, Piccolomini can be deemed a subtle reformer of the views contained in the original text of the *Quaestiones mechanicae*. He posits that strictly speaking natural philosophy and mechanics do not overlap. The mathematical balance and the sensible balance share a resemblance as the former is obtained after observing and paying attention to the behaviour of the latter. Yet the sensible balance and the mathematical balance (whose ontology is allegedly linked to the power of imagination) belong to starkly discrete realms: the principles and variables analysed by mechanics and real movement and weights do not symmetrically track each other.[332]

Consequently, mechanical demonstrations produce theorems which are unqualifiedly true only when stated mathematically, not physically.[333] If a mechanical conclusion happens to apply to sensible matter, it will do so only accidentally and with the inherent lack of precision of sensible matter, serving as an approximation. This approximate, yet effective, application of mechanical principles in dealing with real machines is not enough to postulate that mechanics can provide anything more than mathematical truths. Therefore, physical truth remains untouched by mechanical demonstration.

2.5.3.1 *Coda*: Tartaglia's Akribology and the Gap between Mathematics and Physics

It is noteworthy that a similar position was voiced by Tartaglia when addressing the first question posed by Piccolomini's patron, Diego de Hurtado y Mendoza in book VII of his work *Quesiti et inventioni diverse* (1538). Tartaglia's claim is even more revealing given that (1) he offers his view when addressing the very same question of the *Quaestiones mechanicae* regarding the balance. Secondly

331 See C. T. Lewis, C. Short, *A Latin Dictionary* (Oxford: Clarendon Press, 1879), Sapio, īvi or ĭi II. Trop. 1. To taste or smell of, savour of, i.e., a. **to resemble** (late Lat.): "patruos", Pers. 1, 11.

332 See section 3.5.3, ii for an elaboration on this point.

333 *Nota bene* that according to Piccolomini the sensible movement present in the artefacts of mechanics also depends on natural principles and, thus, its ultimate explanation belongs to natural philosophy. I will deal with Piccolomini's blurring of the distinction of motion *secundum naturam* and *praeter naturam* in section 3.6.

(2), in the sketchy philosophy of mechanics that he provides as an answer, the concept of certainty plays a central role.

The debate concerns an error that Tartaglia points out in Pseudo-Aristotle's account of one of the problems tackled in the discussion on the balance: that bigger balances are more precise (i.e., more sensitive) than small ones. This statement follows the principle that the further a point lies from the centre of the balance, the quicker it moves when pushed by the same force. Since in bigger balances, the points upon which the force is applied are more distant, it follows that they are also more sensitive. Tartaglia swiftly points out that this claim is falsified by experience: if one is to measure a light portion of cinnamon, one would use a small balance and the resulting measure would surely be far more accurate than if the sprinkle of cinnamon were measured by using a balance designed, say, for weighting ships. Why is this so? Tartaglia answers that Pseudo-Aristotle's view holds only in mathematics: the material bulk of which the ship-balance is made hinders the precise application of mathematical principles. Then he adds:

> And for this and other reasons the mathematician does not accept or consent to proofs and demonstrations made on the strength and authority of the senses in matter, but only to those made by demonstrations and arguments abstracted from all matter. Consequently, the mathematical disciplines are considered by the wise not only to be more certain than the physical ones but even to be in the highest degree of certainty (*nel primo grado di certezza*). And, thus, the questions which can be demonstrated through mathematical arguments cannot be suitably proved by physical arguments. Likewise, those [questions] which have already been demonstrated by mathematical arguments (which are the most certain) should not be subjected to attempts to certify (*certificarle*) them further by physical arguments, which are less certain.[334]

334 Translation from R. Pisano, D. Capecchi, *Tartaglia's Science of Weights and Mechanics in the Sixteenth Century, Selections from Quesiti et inventioni diverse: Books* VII–VIII, pp. 200–201. Original at Niccoló Tartaglia, *Quesiti et inventioni diverse* (Venice: Venturino Ruffinelli, 1546), p. 77: *Et per questo et altri simili respetti, el mathematico non accetta, ne consente, alle demostrationi, over probationi, fatte per vigor et auttoriá di sensi in materia, ma solamente à quelle fatte per demostrationi et argomenti astrati da ogni materia. Et per questa causa, le discipline mathematice non solamente sono giudicate dalli sapienti esser piu certe delle naturale, ma quelle esser anchora nel primo grado di certezza. E pero quelle questioni che con argomenti mathematici se possono dimostrare non è cosa conveniente ad approbarle con argomenti naturali. Et finalmente quelle, che sono gia dimostrate con argomenti mathematici (che sono piu certi) non è da tentare, ne da persuadersi de certificarle*

First, Tartaglia introduces a similar gap to the one postulated by Piccolomini: mechanics holds true only if abstractly considered; material realities do not conform to the principles established in mathematical speculation. Hence, we do not reason in mathematics by arguing from sensible experience as we do not expect a perfect instantiation of mathematical principles in matter. Admittedly, these lines deliver the view that mechanics, as a fully mathematical discipline, is practically functional in matter only by approximation.

Secondly, Tartaglia invokes a familiar, akribological view: that mathematics is in the highest and first degree of *certezza* – more than natural philosophy. Doubtlessly, the term *certezza* possesses here the meaning (ii) of objectual precision or accuracy (see 2.3.4). Mathematics deals with principles and entities endowed with such exactness as to be unattainable in sensible matter. Moreover, Tartaglia uses the (for him granted) different degrees of certainty of mathematics and physics as an argument for their disciplinary isolation. His akribological principle is that it is illicit to relate propositions coming from sciences with different degrees of certainty to ensure their truth: judging the truth of mathematical propositions by confronting them with empirical propositions ignores the insurmountably distant degrees of *certezza* with which each disciple speaks.

Tartaglia's testimony also reveals that outside universities theoretical problems related to the applicability and certainty of mathematics were part of the debate amongst practically-oriented mathematicians. The question of the epistemic and ontological bridges (or the absence thereof) linking natural philosophy and mechanics was discussed in the years that Piccolomini wrote his opusculum and the *Praefatio* to the paraphrase. As seen, concerns were raised regarding the same work that he had paraphrased by 1540, and commenting on the very same question: the accuracy of the balance and the fitting realisability of mathematical propositions in natural objects.

No influence of Tartaglia on Piccolomini (or *vice versa*) can be attested, although it would be strikingly odd if, concerned with the same topics and working for the same patron, they were totally unknown to each other. What is clear is that a comparison of Tartaglia's few lines on the philosophy of mathematics and Piccolomini's paraphrase together with his *Praefatio* to it (and *De certitudine*, as we shall see) confirms a doctrinal core shared by both authors. This evinces that Piccolomini was mindfully taking part in a discussion triggered by the divulgation of the *Quaestiones mechanicae*.[335]

---

meglio con argomenti naturali, li quali sono men certi. See also De Pace, *Le matematiche e il mondo*, pp. 242–260.

335 For an elaboration on certainty in the mechanical tradition ever since Leonardo da Vinci see Rose, "Certitudo Mathematicarum from Leonardo to Galileo", pp. 43–49.

Nevertheless, Piccolomini would address such questions in the *Praefatio* and *De certitudine* with the tools of his erudition, i.e., by backing his view with an elaborated philosophy of mathematics. The resulting outlook was surely more deeply founded than Tartaglia's piecemeal exposition of an idea alike.

2.6 *Demonstratio potissima*

One of the most evasive, yet pivotal, concepts of *De certitudine* is that of *demonstratio potissima*, a notion inextricably linked to the Latin elaborations on Aristotle's *Analytica posteriora* and its secular commentary tradition. This logical arrangement is of the utmost importance for this study, as it is the scientific structure that Piccolomini famously denied to mathematics by invoking the authority of Proclus, consequently causing the famous *quaestio de certitudine mathematicarum* (see section 1.3 above).

In this section, I shall provide a background for better grasping this logical and epistemic concept and its link to the notion of *certitudo*, considering its historical and formal aspects as well as the sources quoted by the Sienese, either concordantly or antagonistically, in his discussion and elaboration of it.

2.6.1 *The Most Scientific Demonstration: a Historical Introduction*

As known, Piccolomini engages in a lengthy discussion concerning the scientific status of the logical procedures in mathematics. Mathematical proofs were considered of a superior rank by authors of such great influence as the Averroist Marco Antonio Zimara (*c.*1470–1532), one of the most relevant voices in the 16th-century Paduan context and a figure targeted in *De certitudine*.[336] Such demonstrative superiority was guaranteed by the capacity of mathematics to instantiate the scientifically paradigmatic, demonstrative arrangement: the so-called *demonstratio potissima*.

This section is divided into three parts. It delivers (i) an account of the origin of *demonstratio potissima* in Averroes' interpretation of Themistius, (ii) of

336 Zimara was highly influential in the philosophy of mathematics of Piccolomini's time. In his encyclopaedic work *Theoremata*, Zimara endorses the opinion that mathematical demonstrations are cases of *demonstratio potissima*, i.e., cases of the most scientific logical procedure. See Marco Antonio Zimara, *Theoremata seu memorabilium propositionum limitationes* (Venice: Hieronymus Scotus, 1543), f. 31v: *Simpliciter etiam non est dubium quod demonstrationes mathematicae sunt demonstrationes simpliciter dantes causam et esse, procedentes a notioribus naturae a nobis simpliciter*. For an excellent account of Marco Antonio Zimara's philosophy of mathematics see Higashi, *Penser les mathématiques*, part 1: "Zimara et l'encyclopedisme averroïste". I will deal with the Latin debates on mathematical demonstration in section 2.6.

its 13th-century Latin reception, and (iii) its later re-elaborations (15th century and beyond).

When one speaks of a mathematical demonstration as being a supreme demonstration or *demonstratio potissima*, one does not mean a proof like those found in the Euclidean corpus. Within the logical and scientific framework of the 13th to 16th centuries, a mathematical *demonstratio potissima* would consist of the syllogistic formalisation of a mathematical proof.[337] That means that using syllogistic models of deduction as a meta-language, a given mathematical proof is rephrased so that it fits the syllogistic form. As an example, let us consider the pattern of mathematical *demonstratio potissima* designed by Aquinas as applied to Euclid's theorem I.32:[338]

> *Major premise:* Every three-sided rectilinear figure has its exterior angles equal to the sum of the opposite interior angles (i.e., to two right angles).
> *Minor premise:* Every triangle is three-sided rectilinear figure.
> *Conclusion:* Therefore, every triangle has its exterior angles equal to the sum of the opposite interior angles.

One of the main *loci* providing some background knowledge for *demonstratio potissima* is *Analytica posteriora*, II, 13 (78a, l. 22–78b, l. 13).[339] There Aristotle

337 See Mancosu, Mugnai, *Syllogistic Logic*, Introduction, chapter 1, and chapter 2.

338 Thomas Aquinas, *Expositio in libros Posteriorum analyticorum*, ed. Leonina, vol. 1.2 (Paris: Commissio Leonina, 1989), II, lec. 1: *Si velimus de triangulo demonstrare quod habet tres angulos aequales duobus rectis, accipiamus primo pro medio quod est figura habens angulum extrinsecum aequalem duobus intrinsecis sibi oppositis, quod est quasi definitio passionis. Quod iterum demonstrare oportet per definitionem subiecti, ut dicamus: omnis figura tribus rectis lineis contenta habet angulum exteriorem aequalem duobus interioribus sibi oppositis; sed triangulus est huiusmodi; ergo etc.*

339 Aristotle, *Analytica posteriora*, II, 13, 78a, l. 22–78b, l. 13: τὸ δ' ὅτι διαφέρει καὶ τὸ διότι ἐπίστασθαι, πρῶτον μὲν ἐν τῇ αὐτῇ ἐπιστήμῃ, καὶ ἐν ταύτῃ διχῶς, ἕνα μὲν τρόπον ἐὰν μὴ δι'ἀμέσων γίνηται ὁ συλλογισμός (οὐ γὰρ λαμβάνεται τὸ πρῶτον αἴτιον, ἡ δὲ τοῦ διότι ἐπιστήμη κατὰ τὸ πρῶτον αἴτιον), ἄλλον δὲ εἰ δι' ἀμέσων μέν, ἀλλὰ μὴ διὰ τοῦ αἰτίου ἀλλὰ τῶν ἀντιστρεφόντων διὰ τοῦ γνωριμωτέρου. κωλύει γὰρ οὐδὲν τῶν ἀντικατηγορουμένων γνωριμώτερον εἶναι ἐνίοτε τὸ μὴ αἴτιον, ὥστ'ἔσται διὰ τούτου ἡ ἀπόδειξις, οἷον ὅτι ἐγγὺς οἱ πλάνητες διὰ τοῦ μὴ στίλβειν. ἔστω ἐφ' Γ πλάνητες. ἐφ' ᾧ Β τὸ μὴ στίλβειν, ἐφ' ᾧ Α τὸ ἐγγὺς εἶναι. ἀληθὲς δὴ τὸ Β κατὰ τοῦ Γ εἰπεῖν· οἱ γὰρ πλάνητες οὐ στίλβουσιν. ἀλλὰ καὶ τὸ Α κατὰ τοῦ Β· τὸ γὰρ μὴ στίλβον ἐγγύς ἐστι· τοῦτο δ' εἰλήφθω δι' ἐπαγωγῆς ἢ δι' αἰσθήσεως. ἀνάγκη οὖν τὸ Α τῷ Γ ὑπάρχειν, ὥστ'ἀποδέδεικται ὅτι οἱ πλάνητες ἐγγύς εἰσιν. οὗτος οὖν ὁ συλλογισμὸς οὐ τοῦ διότι ἀλλὰ τοῦ ὅτι ἐστίν· οὐ γὰρ διὰ τὸ μὴ στίλβειν ἐγγύς εἰσιν, ἀλλὰ διὰ τὸ ἐγγὺς εἶναι οὐ στίλβουσιν. ἐγχωρεῖ δὲ καὶ διὰ θατέρου θάτερον δειχθῆναι, καὶ ἔσται τοῦ διότι ἡ ἀπόδειξις, οἷον ἔστω τὸ Γ πλάνητες, ἐφ' ᾧ Β τὸ ἐγγὺς εἶναι, τὸ Α τὸ μὴ στίλβειν· ὑπάρχει δὴ καὶ τὸ Β τῳ Γ καὶ τὸ Α τῳ Β, ὥστε καὶ τῷ Γ τὸ Α τὸ μὴ στίλβειν. καὶ ἔστι τοῦ διότι ὁ συλλογισμός· εἴληπται γὰρ τὸ πρῶτον αἴτιον. πάλιν ὡς τὴν σελήντην δεικνύουσιν ὅτι σφαιροειδής, διὰ τῶν αὐξήσεων – εἰ γὰρ τὸ αὐξανόμενον οὕτω σφαιροειδές, αὐξάνει δ'ἡ

arranges two syllogisms in the field of astronomy which depict two different relationships amongst three terms: the twinkling (τὸ στίλβειν), the wandering celestial bodies (οἱ πλάνητες), and the planets' proximity to the earth (τὸ ἐγγὺς εἶναι). Two kinds of proof are generated. The first one is a syllogism called τὸ ὅτι: this is a proof leading from a given effect to its cause. We know that the planets are near to the Earth (cause) because they twinkle (effect).

This sort of demonstration was rendered into Latin as Latin *quia, esse*, or *signi*. Whereas *esse* was associated with the answer to the question *an sit* (εἴ ἐστιν), *signi* was the Latin equivalent to the Themistian syntagm "διὰ τοῦ σημείου".[340] As illustrated by De Bellis, these three aspects of the Aristotelian heuristics were destined to be conjoined and assimilated into the Latin tradition.[341]

The second proof arrangement deduces an effect from its cause: the planets twinkle (effect) because they are near the Earth (cause). This second arrangement is called τὸ διότι. It was translated as *demonstratio propter quid*; also

σελήνη, φανερὸν ὅτι σφαιροειδής – οὕτω μὲν οὖν τοῦ ὅτι γέγονεν ὁ συλλογισμός, ἀνάπαλιν δὲ τεθέντος· τοῦ μέσου τοῦ διότι· οὐ γὰρ διὰ τὰς αὐξήσεις σφαιροειδής ἐστιν, ἀλλὰ διὰ τὸ σφαιροειδὴς εἶναι λαμβάνει τὰς αὐξήσεις τοιαύτας. σελήνη ἐφ' ᾧ Γ, σφαιροειδὴς ἐφ' ᾧ B, αὔξησις ἐφ'ᾧ A. ἐφ'ὧν δὲ τὰ μέσα μὴ ἀντιστρέφει καὶ ἔστι γνωριμώτερον τὸ ἀναίτιον, τὸ ὅτι μὲν δείκνυται, τὸ διότι δ'οὔ. Translation in Aristotle, *Complete Works of Aristotle*, vol. 1 (slightly modified): "Understanding the fact and the reason why differ, first in the same science – and in that in two ways: in one way, if the deduction does not come about through immediates (for the primitive explanation is not assumed, but understanding of the reason why occurs in virtue of the primitive explanation); in another, if it is through immediates but not through the explanation but through the more familiar of the converting terms. For nothing prevents the non-explanatory one of the counter-predicated terms from sometimes being more familiar, so that the demonstration will occur through this. E.g., that the planets are near, through their not twinkling: let C be the planets, B not twinkling, A being near. Thus, it is true to say B of C; for the planets do not twinkle. But also to say A of B; for what does not twinkle is near (let this be got through induction or through perception). So it is necessary that A belongs to C; so that it has been demonstrated that the planets are near. Now this deduction is not of the reason why but of the fact; for it is not because they do not twinkle that they are near, but because they are near they do not twinkle. But it is also possible for the latter to be proved through the former, and the demonstration will be of the reason why – e.g., let C be the planets, B being near, A not twinkling. Thus, B belongs to C and A to B; so that A belongs to C. And the deduction is of the reason why; for the primitive explanation has been assumed".

340 Themistius, *Analyticorum posteriorum paraphrasis*, ed. M. Wallies (Berlin: Academiae litterarum regiae Borusicae, 1900), p. 45, l. 15: ἕτερος δὲ τρόπος, ὅταν δι' ἀμέσων μὲν ἄμφω, ἀλλ'ὁ μὲν διὰ τῆς αἰτίας, ὁ δὲ διὰ τοῦ σημείου. Translation: "Another way is when both are immediate, but one is through the cause and the other through the sign".

341 See De Bellis, "La trattazione della teoria aristotelica della demonstratio nel Ms. Gal. 27 di Galileo Galilei", in *Rivista di Filosofia Neo-Scolastica* 109/3 (2017), pp. 725–736.

following Themistius' expression "διὰ τῆς αἰτίας", it was rendered into Latin as *demonstratio causae*.

2.6.1.1 Averroes and Themistius

The concept of a would-be *demonstratio potissima* appears as an elaboration of Averroes on some of Themistius' comments in his *Analyticorum posteriorum paraphrasis*. The epistemological remark that might have inspired Averroes to introduce a third kind of demonstration runs as follows:

> Many times it happens indeed that the cause and the sign (τὸ αἴτιον καὶ τὸ σημεῖον) reciprocate and that each of them is demonstrated through the other: through the sign as "the fact" (ὡς τὸ ὅτι), through the latter as the "why" (ὡς τὸ διότι); through the former [it is demonstrated] from what is better known than the conclusion, through the latter from the cause of the fact.[342]

Themistius stresses that the cause/explanation (τὸ αἴτιον) and the effect or, as he puts it in Greek, the sign (τὸ σημεῖον), meaning here the sign following a given causal event, can be in some cases inferred from the other. The remote ancestor of this logical formula was mathematical analysis (ἀνάλυσις, see sections 2.1.2 and 3.2), paramount in the practice of ancient mathematics.[343] As Dubouclez has shown, the notion of mathematical analysis is the source of the Aristotelian distinction between demonstration τὸ ὅτι and τὸ διότι.[344] Thus, this logical scheme is also at the origin of a tradition of commentators such as Galen or Themistius, who uphold that the highest form of scientific account establishes some explanative reciprocity between analysed departing points (facts, effects, things to be explained) and their explanations (causes, or first principles). It is also important to recall that the Latin readership had access to Themistius' paraphrase through Gerard of Cremona (ob. c. 1187).[345]

Averroes posits in the *Prooemium* to his *Commentarium magnum* on the *Physica* and on *Analytica posteriora* that there is a third, more accomplished

---

342 Themistius, *Analyticorum posteriorum paraphrasis*, p. 45, 19: πολλάκις μὲν οὖν συμβαίνει καὶ ἀντιστρέφειν ἀλλήλοις τὸ αἴτιον καὶ τὸ σημεῖον καὶ ἄμφω δείκνυσθαι δι' ἀλλήλων, διὰ τοῦ σημείου μὲν ὡς τὸ ὅτι, διὰ θατέρου δὲ ὡς τὸ διότι, δι'οὗ μὲν ὡς γνωριμωτέρου τοῦ συμπεράσματος, δι'οὗ δὲ ὡς αἰτίου τοῦ πράγματος.

343 For some details on the history and traits of this mathematical technique see Hintikka, Remes, *The Method of Analysis*. See also Hintikka, "Method of Analysis".

344 Dubouclez, *Descartes et la voie de l'analyse*, part 1, *passim*.

345 Sorabji, "Introduction", in *Aristotle Transformed*, p. 28.

sort of demonstration in addition to the two kinds stated by Aristotle.[346] This new kind of demonstration would grant both the effect-to-cause and the cause-to-effect knowledge at the same time. It is called by Averroes البرهان المطلق (*al-burhān al-muṭlaq*) literally "absolute proof". Averroes' syntagm *al-burhān al-muṭlaq* was rendered into Latin as *demonstratio simpliciter* or *absoluta*; it was also known as *demonstratio potissima* by the 13th-century tradition of commentary on the *Analytica posteriora* at least since Robert Grosseteste.[347] This is a nomenclature which, as we shall see, would reach the debates of the Paduan philosophers in the 16th century.

This unqualified demonstration is intended to be the paradigmatic scientific procedure, a demonstration *par excellence*, or a *sensu stricto* demonstration, considering all the other forms of demonstration to be weaker variations of this genuine model. Nonetheless, the exact details of how such a double-account demonstration is produced are not provided by Averroes, and they would become the object of controversy. Amongst later Paduan authors, this logical arrangement was supposed to answer all the questions *an sit, quid sit, quia sit* and *propter quid sit*.[348]

---

346 Averroes, *Commentaria in libros De physico auditu Aristotelis*, in *Aristotelis opera cum Averrois commentariis*, transl. Mantini, vol. 4, Proemium, E2–F3: *Modus autem doctrinae qui in eo sit est ex generibus doctrinae quae in hac scientia fieri solent, qui quidem sunt modi omnium doctrinarum trium generum demonstrationum, nempe signorum seu demonstrationum quod est, et demonstrationum propter quid, necnon demonstrationum absolutarum vel simpliciter, licet in hac arte* [*in physica*] *maior pars demonstrationum quae in ea fiunt sint demonstrationes quod est et propter quid; raro enim fiunt in ea demonstrationes absolutae, contra id quod fit in ipsis mathematicis, in quibus semper fiunt ut plurimum demonstrationes simpliciter et absolutae*. See also Averroes' comments from the section I.2 in Averroes, *Sharh al-Burhān li-Aristū wa-Talkhīs al-Burhān, al-Silsila al-turāthiyya*, vol. 12 (Al-Kuwayt: Al-Majlis al-Watanī li-l-Thaqāfa wa-l-Funūn wa-l-Ādāb, 1984): ولكون البرهان المطلق الذي يعطى سبب الشيء القريب هو البرهان الذي مقدماته موجودة بهذه الشروط التي تقدمت كلها شرح البرهان، فصل ١٠ مختلف المبادئ [...]. Transl. "And since absolute demonstration which gives the proximate cause of a thing is the one whose premises are found according to all the aforementioned conditions [...]". I cordially thank al-Walīd al-Saqqāf for his assistance in dealing with the Arabic text.

347 Robert Grosseteste, *Commentarius in Posteriorum analyticorum libros*, eds L. S. Olschki, P. Rossi (Florence: Testi e Studi 1981), I, 9: *Et non est intelligendum universaliter quod medium vel minor extremitas sint semper talia, quia medium erit in potissima demonstratione convertibile cum maiori vel cum minori, erit enim alterius diffinitio*. See also Albert the Great, *In libros Posteriorum analyticorum*, ed. A. Borgnet, vol. 2 (Paris: Ludovicus Vives, 1890), I, T. III, chap. 1: *Ex his enim ut primis et immediatis ut indemonstrabilibus demonstrant in potissima demonstratione, quae est demonstratio ostensiva dicens propter quid et quid in medio per quod demonstrat*.

348 De Bellis, "La trattazione della teoria aristotelica della demonstratio nel Ms. Gal. 27 di Galileo Galilei", p. 726. As Biard has pointed out in his study of the types of demonstration,

In the *Prooemium*, Averroes also makes a remark which would be of the utmost importance for our research: that *demonstratio potissima* is mostly found in mathematics, whereas it is rarely obtained (if produced at all) in natural philosophy (*raro enim fiunt in ea demonstrationes absolutae*).

2.6.1.2 The Latin Fortune

This notion of a syllogistic procedure which provides scientific knowledge *par excellence* and that it is somehow linked to mathematics and certainty (*certitudo*, اليقين / *al-yaqīn*) reached the Latin West through the translations of Averroes' *Prooemium* and of his comments on *Metaphysica* II, 3 (where *demonstratio simpliciter* also occurs, see 2.3.2). It was, thereafter, explored by Latin authors.

Authors after Grosseteste (the first known Latin philosopher who produced a commentary on *Analytica posteriora*) would debate the possibility and precise traits of an unqualified demonstration providing genuine science.[349] The main questions in the early Latin reception can be summarily stated as follows:

(1) What is the most scientific syllogistic structure?

(2) What is the most scientific middle-term in a demonstration?

Following the textual evidence provided by Burnett and Mendelsohn,[350] we learn that not every author assumed, as a starting point, that (1) and (2) were

---

there was not complete agreement amongst the Latin authors regarding the exact meaning of demonstration *quia* or *esse* and its exact relation with the *propter quid* account. See Biard, "Quante tipologie di dimostrazione? Origini incrociate nella tradizione degli Analitici Secondi", in A. Musco, G. Musotto (eds), *Traduzioni e Tradizioni: Il pensiero medievale nell'incontro tra le culture mediterranee. Siracusa, 26–29 Settembre 2011* (Palermo: Officina Studi Medievali, 2014), pp. 61–78. For the sake of this discussion, I shall adopt the commonest and most agreed meaning of the terms and those leading to Piccolomini's understanding of the question. For further reading on Averroes' influence on the Paduan context see K. Richardson, "Averroism", in H. Lagerlund, B. Hill (eds), *The Routledge Companion to the Sixteenth Century Philosophy*, pp. 137–155; C. Martin, "Humanism and the Assessment of Averroes in the Renaissance", in A. Akasoy, G. Giglioni (eds), *Renaissance Averroism and Its Aftermath: Arabic Philosophy in Early Modern Europe* (New York, London: Springer, 2012), pp. 65–81; C. Burnett, "Revisiting the 1552–1550 and 1562 Aristotle-Averroes Editions", in *Renaissance Averroism and Its Aftermath*, pp. 55–64; E. Troilo, *Averroismo e aristotelismo Padovano* (Padova: Olschki, 1939).

349 For an account of the early reception of *Analytica posteriora* and the logical and epistemological questions it raised since the 12th century, including demonstration, see S. Ebbesen, "The Posterior Analytics 1100–1400 in East and West", in J. Biard (ed.), *Raison et démonstration*, pp. 11–21.

350 C. Burnett, A. Mendelsohn, "Aristotle and Averroes on Method in the Middle Ages and Renaissance: The "Oxford Gloss" to the Physics and Pietro D'Afeltro's *Expositio Proemii Averroys*", in D. Di Liscia, A. R. Kessler, Ch. Methien (eds), *Method and Order in the Renaissance Philosophy of Nature: The Aristotle Commentary Tradition* (Aldershot, Hants:

two forms of the same question. Some examples of attempts of an account of *demonstratio potissima* trying to answer (1) can be found in John of Jandun's (*ob.* 1328) (widely read and respected in Padua) *Quaestiones in duodecim libros Metaphysicae*, or in the Oxonian Walter Burley's *Expositio in octo volumina Aristotelis De physico auditu*.

Amongst many others, Albert the Great, Thomas Aquinas, Giles of Rome, William of Ockham, John Buridan, and Paul of Venice wrote specific comments on question (2). The early Latin debates will help to clarify some aspects of Piccolomini's position. Although several points of disagreement would emerge, there were some widely diffused *loci* which were commonly assumed as the starting points of the debate. Those are:

(1) The syllogistic form of *demonstratio potissima* should be that of the first figure (σχῆμα πρῶτον, *prima figura*), namely *Barbara*, *Celarent*, *Darii*, or *Ferio*. This assumption was based on Aristotle's account as found in *Analytica priora* of what a perfect deduction (συλλογισμὸς τέλειος) is.[351] *Barbara* fulfils Aristotle's

---

Ashgate, 1997). See Jean de Jandun's (of great influence at Padua) exposition in Jean de Jandun, *Quaestiones perspicacissimi peripatetici Ioannis de Ianduno in duodecim libros Metaphysicae* (Venice: Hieronymus Scotus, 1553), lib. XII, q. 3, f. 129r: *Notandum ulterius propter solutionem rationum: sicut Commentator tangit II Physicorum, et aliqualiter in isto XII quod tres sunt modi demonstrationum. Quaedam est demonstratio simpliciter, quaedam propter quid, et quaedam quia. Sed Aristoteles tantum in libro Posteriorum de duabus demonstrationibus dixit, scilicet de demonstratione propter quid et quia. Sed forte comprehendit demonstrationem simpliciter sub demonstratione propter quid. Unde Commentator forte intellexit per demonstrationem simpliciter demonstrationem quae est per causas remotas, seu per demonstrationem propter quid, intelligit demonstrationem quae est per causas propinquas. Sed per demonstrationem quia intelligit demonstrationem quae est per effectus posteriores et signa.*

351 *Analytica priora*, I, 4, 25b, l. 30–26a, l. 4: ὅταν οὖν ὅροι τρεῖς οὕτως ἔχωσι πρὸς ἀλλήλους ὥστε τὸν ἔσχατον ἐν ὅλῳ εἶναι τῷ μέσῳ καὶ τὸν μέσον ἐν ὅλῳ τῷ πρώτῳ ἢ εἶναι ἢ μὴ εἶναι, ἀνάγκη τῶν ἄκρων εἶναι συλλογισμὸν τέλειον. καλῶ δὲ μέσον μὲν ὃ καὶ αὐτὸ ἐν ἄλλω καὶ ἄλλο ἐν τούτῳ ἐστίν, ὃ καὶ τῇ θέσει γίνεται μέσον· ἄκρα δὲ τὸ αὐτό τε ἐν ἄλλῳ ὂν καὶ ἐν ᾧ ἄλλο ἐστίν. εἰ γὰρ τὸ Α κατὰ παντὸς τοῦ Β καὶ τὸ Β κατὰ παντὸς τοῦ Γ, ἀνάγκη τὸ Α κατὰ παντός τοῦ Γ κατηγορεῖσθαι· πρότερον γὰρ εἴρηται πῶς τὸ κατὰ παντὸς λέγομεν. ὁμοίως δὲ καὶ εἰ τὸ μὲν Α κατὰ μηδενὸς τοῦ Β, τὸ δὲ Β κατὰ παντὸς τοῦ Γ, ὅτι τὸ Α οὐδενὶ τῷ Γ ὑπάρξει. Translation by A. J. Jenkinson at Aristotle, *Complete Works of Aristotle*, Vol. 1: "Whenever three terms are so related to one another that the last is in the middle as in a whole, and the middle is either in, or not in, the first as in a whole, the extremes must be related by a perfect deduction. I call that term middle which both is itself in another and contains another in itself: in position also this comes in the middle. By extremes I mean both that term which is itself in another and that in which another is contained. If A is predicated of every B, and B of every C, A must be predicated of every C: we have already explained what we mean by 'predicated of every'. Similarly also, if A is predicated of no B, and B of every C, it is necessary that A will belong to no C". See also *Analytica priora*, ed. I. Bekker, vol. 1, I, 4, 26b, l. 25–32: φανερὸν οὖν ἐκ τῶν εἰρημένων ὡς ἐὰν ᾖ συλλογισμός ἐν τούτῳ τῷ σχήματι κατὰ μέρος, ὅτι ἀνάγκη τοὺς ὅρους

requirement of knowledge as being about universals (κατὰ παντός).[352] *Celarent* and *Ferio* are excluded, since Aristotle declared that proofs resulting in a negated statement (as *reductio ad absurdum* or *ad impossibile* in mathematics) provided knowledge of what-is-not-the-case instead of knowledge of what-is-the-case, the latter being preferable.[353] Since *demonstratio potissima* produces by definition *scire simpliciter*, it must provide positive knowledge of a real entity. Thus, its conclusion must be affirmative. It was controversial whether a first figure proof containing an affirmative particular in the minor premise (*Darii*) could be considered as a *demonstratio potissima*, for a conclusion about a singular would fall short of the universal status of scientific knowledge postulated by Aristotle.[354]

(2) Following Aristotle's requirements of scientific demonstration in *Analytica posteriora* I, 2,[355] it was agreed that the premises of *demonstratio potissima* must fulfil the intrinsic criteria (considered in themselves) of being (i) true

---

οὕτως ἔχειν ὡς εἴπομεν· ἄλλως γὰρ ἐχόντων οὐδαμῶς γίνεται. δῆλον δὲ καὶ ὅτι πάντες οἱ ἐν αὐτῷ συλλογισμοὶ τέλειοί εἰσι· (πάντες γὰρ ἐπιτελοῦνται διὰ τῶν ἐξ ἀρχῆς ληφθέντων), καὶ ὅτι πάντα τὰ προβλήματα δείκνυται διὰ τούτου τοῦ σχήματος· καὶ γὰρ τὸ παντὶ καὶ τὸ μηδενὶ καὶ τὸ τινὶ καὶ τὸ μή τινι ὑπάρχειν. καλῶ δὲ τὸ τοιοῦτον σχῆμα πρῶτον. Translation by Jenkinson in Aristotle, *Complete Works of Aristotle*, Vol. 1: "It is evident also that all the deductions in this figure are perfect (for they are all completed by means of what was originally assumed) and that all conclusions are proved by this figure, viz. universal and particular, affirmative and negative. Such a figure I call the first".

352 *Analytica priora*, I, 4, 26a, l. 4; *Analytica posteriora*, I, 4, 73b, l. 26–27; *Metaphysica*, VII, 10, 1036a, l. 2–11 and 15, 1039b, l. 28–1040a, l. 5.

353 *Analytica posteriora*, I, 26, 87a, l. 1–2: ἐπεὶ δ' ἡ κατηγορικὴ τῆς στερητικῆς βελτίων, δῆλον ὅτι καὶ τῆς εἰς τὸ ἀδύνατον ἀγούσης. Translation by Barnes in Aristotle, *Complete Works of Aristotle*, Vol. 1: "Since affirmative demonstration is better than negative, it is clear that it is also better than demonstration leading to the impossible". See M. Malink, "Demonstration by reductio ad impossibile in Posterior Analytics 1.26", in Victor Caston (ed.), *Oxford Studies in Ancient Philosophy*, Vol. LVIII. (Oxford: Oxford University Press, 2020), pp. 91–155.

354 On the possibility of science about particulars in the early Latin reception, see A. Corbini, *La teoria della scienza nel XIII secolo. I commenti agli Analitici Secondi* (Florence: Testi e Studi, 2006), chap. I, especially pp. 4–8.

355 Aristotle, *Analytica posteriora* I, 2, 71b, l. 20–22: εἰ τοίνυν ἐστὶ τὸ ἐπίστασθαι οἷον ἔθεμεν, ἀνάγκη καὶ τὴν ἀποδεικτικὴν ἐπιστήμην ἐξ ἀληθῶν τ' εἶναι καὶ πρώτων καὶ ἀμέσων καὶ γνωριμωτέρων καὶ προτέρων καὶ αἰτίων τοῦ συμπεράσματος· οὕτω γὰρ ἔσονται καὶ αἱ ἀρχαὶ οἰκεῖαι τοῦ δεικνυμένου. συλλογισμὸς μὲν γὰρ ἔσται καὶ ἄνευ τούτων, ἀπόδειξις δ' οὐκ ἔσται. Translation in Aristotle, *Complete Works of Aristotle*, vol. 1: "If, then, understanding is as we posited, it is necessary for demonstrative understanding in particular to depend on things which are true and primitive and immediate and more familiar than and prior to and explanatory of the conclusion (for in this way the principles will also be appropriate to what is being proved). For there will be deduction even without these conditions, but there will not be demonstration; for it will not produce understanding".

(ἐξ ἀληθῶν, *ex veris*), (ii) primary ([ἐξ] πρώτων, *ex primis*), and (iii) immediate ([ἐξ] ἀμέσων, *ex immediatis*), and the extrinsic criteria (considered in relation to other propositions in the syllogistic form) of being (iv) better known than ([ἐξ] γνωριμωτέρων, *ex notioribus*), (v) prior to ([ἐξ] προτέρων, ex *prioribus*), and (vi) explanatory of the conclusion ([ἐξ] αἰτιῶν τοῦ συμπεράσματος, *ex causis conclusionis*).[356]

(3) The premises of *demonstratio potissima* must incorporate what I shall call "three-term definitions" or "complex definitions".[357] Whether such definition operates as the middle-term (μέσον, *medium*) and whether it is the definition of the *subiectum* (ὑποκείμενον) or that of the *passio* (τὸ πάθος) is one of the issues at stake in the 13th- and 14th-century Latin discussion (see below).

(4) The last commonly held view was Averroes' opinion in the *Prooemium* and in *Metaphysica* II, 3 that *demonstratio potissima* is prominently obtained in mathematics, as it is the science endowed with the greatest degree of certainty. Therefore, the certainty of mathematics was commonly held as the reason for its superior demonstrative status – the view *certissimae ergo potissimae* that Piccolomini would eventually subvert.

2.6.1.3 The "Paduan Doctrine" of *regressus demonstrativus*

From the 14th to the 15th century, *demonstratio potissima* received new treatment in the Paduan circles with a renewed interest in Averroes' original proposal of a third, superior kind of demonstration. Averroes' teachings on *demonstratio potissima*, together with some other passages from Galen's highly influential *Ars parva*, would be interpreted in a particular way by Paduan authors. They saw the logical operation introduced by Averroes providing both the *quia* and the *propter quid* not as a single syllogism, but as a procedural arrangement of interrelated syllogisms. Thus, Averroes' sketch of *demonstratio potissima* was destined to be transformed into the celebrated Paduan scientific methodology, the so-called *regressus*, a specific Latin coinage which does not possess a Greek counterpart.

Peter of Abano (*c.*1257–1316) can be considered amongst the first authors to promote the notion that a double explanative process (effect-to-cause

---

356 In the following sections, I will follow Bronstein's terminology and formalisation in order to analyse the Medieval and Renaissance reception of some Aristotelian logical concepts contained in *Analytica posteriora*. The difference between intrinsic and extrinsic criteria regarding the premises of a syllogism is referred to by D. Bronstein in *Aristotle on Knowledge and Learning: The Posterior Analytics* (Oxford: Oxford University Press, 2016), pp. 61–63.

357 *Analytica posteriora*, II, 2, where Aristotle uses the three-term definitions of the lunar eclipse. On this topic see section 2.6.2.

and cause-to-effect) was achieved by intertwining two syllogisms.[358] Peter of Abano's interpretation was highly influenced by the Paduan reception of Galen's methodological teachings. Galen's own application in medicine of the terms ἀνάλυσις and σύνθεσις for describing an effect-to-cause and cause-to-effect logical entanglement can be found described and employed in his *Ars medica, quae et ars parva dicitur*.[359] Galen adopted the medical names of "signs" (σημεῖα) of disease to mean the departing point of the medical reasoning, whereas the αἴτιον stands in Galen's text as the cause to be found by researching from the signs.[360] Peter of Abano interprets the *demonstratio quia* as equivalent to Galen's *doctrina resolutiva* or ἀνάλυσις and the *demonstratio propter quid* as the *doctrina compositiva* or σύνθεσις.[361] Hence, a long tradition of what I shall call "Galenic" analysis was started: this position identifies the resolutive and compositive processes of Galen (and of other sources) with Aristotle's two demonstrations (τὸ ὅτι, τὸ διότι). Such a theoretical approach led to the elaborations and debates on *regressus*. The precise arrangement of *regressus* was

358 Wallace, "Circularity and the *Paduan Regressus*: From Pietro D'Abano to Galileo Galilei", in *Vivarium* 33/1 (1995), pp. 76–97; Dubouclez, *Descartes et la voie de l'analyse*, pp. 137–161. For a general account of the Paduan philosophers since Peter of Abano see Nardi, *Saggi sull'Aristotelismo Padovano*, chap. I and II.

359 Galen, *Ars parva*, Preface, 1. 51–54: τρεῖς εἰσὶν αἱ πᾶσαι διδασκαλίαι τάξεως ἐχόμεναι. πρώτη μὲν ἡ ἐκ τῆς τοῦ τέλους ἐννοίας κατ' ἀνάλυσιν γινομένη. δευτέρα δὲ ἡ ἐκ συνθέσεως τῶν κατὰ τὴν ἀνάλυσιν εὑρεθέντων. τρίτη δὲ ἡ ἐξ ὅρου διαλύσεως, ἣν νῦν ἐνιστάμεθα. See Galen, *Ars medica quae et ars parva dicitur* (Venice: Guillielmus Rovillius, 1548). English translation: "All the teachings have an order in which there are three components. The first is from the concept of the end and arises from analysis. The second is from the synthesis of those things discovered by analysis [i.e. synthesis]. The third is from dialysis of a definition which is what we are now setting up" in Galen, *The Art of Medicine*, ed. I. Jonhson (Cambridge: Loeb Classical Library, 2016). For Galen's methodological teachings and epistemology see R. J. Hankinson, M. Havrda (eds), *Galen's Epistemology: Experience, Reason, and Method in Ancient Medicine* (Cambridge: Cambridge University Press, 2021); T. Tieleman, "Methodology", in R. J. Hankinson (ed.), *The Cambridge Companion to Galen* (Cambridge: Cambridge University Press, 2008), chapter 5, pp. 49–65; B. Morison, "Logic", in R. J. Hankinson (ed.), *The Cambridge Companion to Galen* (Cambridge: Cambridge University Press, 2008) chapter 4, pp. 66–115.

360 See, for instance, Galen, *Ars parva*, I, 1. 5–9.

361 Peter of Abano, *Conciliator differentiarum, quae inter philosophos et medicos versantur* (Venice: Giunti, 1496), dif. 3, I: *Etenim cum effectibus per se causis suis inhaerentibus ordine priori e converso proximis et immediatis secundum consequentiam pervenitur in causam quae intenditur venari, aut causis remotioribus intermissis quibusdam proprinquioribus deducitur conclusio in effectum, scientia per demonstrationem elicitur quia seu doctrina dicta resolutiva*. See also dif. 8.

also discussed by authors such as Paul of Venice, Cajetan of Thiene, and Urban the Averroist (*ort.* 1274).[362]

Jacob of Forlì (*ob.* 1413), who held the chair of medicine and natural philosophy at Padua, followed this approach. As Randall has shown, Jacob of Forlì introduced some qualifications concerning the *doctrina resolutiva*.[363] This elaboration, mixing logical procedures and medical diagnosis, would make Peter of Abano's *viae* something akin to a strategy of inquiry. An example of its application is diagnosing a fever. The physician uses a trial-and-error approach to test various potential causes in the *via resolutiva*. Ultimately, he identifies the specific cause of the fever affecting the patient and reveals this etiological link from cause to effect in the *via compositiva*.[364]

Hugh Benzi (*c.*1376–1439), also known as Hugh of Siena, provides another token of the *regressus* methodology. In his *Expositio super libros Tegni Galeni*, Hugh follows Peter of Abano in declaring that the process of attaining scientific knowledge consists of two steps (*duplicem processum*): one from-effect-to-cause as an analysis of the effect which leads to the cause (*resolutivus, inventio*), and a second one which concludes the effect from the found cause and confirms the first process (*compositivus, notificatio*).[365] The understanding of the logical device that provides both the cause-to-effect and the effect-to-cause as a concatenation of syllogisms and non-deductive steps conforming to a scientific methodology was more clearly delineated by Hugh of Siena's times.

---

362 For an account of the views of those authors see J. H. Randall, *The School of Padua and the Emergence of the Modern Science* (Padova: Editrice Antenore, 1961), chap. 1, "The Development of Scientific Method". See also N. W. Gilbert, *Renaissance Concepts of Method* (New York Chichester: Columbia University Press, 1960), part 2 *passim* and Jardine, "Epistemology of the sciences". For an analysis of Agostino Nifo's doctrine of *regressus* see Jardine, "Galileo's road to truth and the demonstrative regress", in *Studies in History and Philosophy of Science* 7/A (1976), pp. 277–318. The most recent study on the topic, to which I shall return later, is Palmieri's, "On *scientia* and *regressus*".

363 Randall, *The School of Padua*, pp. 35–36.

364 Jacob of Forlì, *Super Tegni Galeni* (Padua: Giunti, 1475), comm. T. 1: *Ut si primo febriens comprehendas conceptum febris, habes generalem et confusum; deinde resolve febrem in suas causas; quia quaelibet aut est ex calefacto esse humoris, aut sanguinis aut flegmatis, etc. donec ad febris specificam ac distinctam causam et notitiam pervenias* [...]. *Tunc enim quae in demonstratione quia resolvendo inveniuntur, in demonstratione postea propter quid componuntur et iunguntur, donec ad causam immediate perveniamus et concludamus effectum.*

365 Hugh Benzi, *Super libros Tegni Galeni* (Venice: Octavianus Scotus, 1518), I, p. 3, c. A: *Ego autem video in inventione scientiae effectus per causam duplicem processum, et similiter in inventione scientiae causae per effectum: unum inventionis medii, alium autem notificationis. Processus inventionis in demonstratione per causam est resolutivus, alter compositivus.*

The early Latin question on the middle-term of *demonstratio potissima* was also inherited and discussed in the *regressus* debates. This continuity can be paradigmatically shown in the speculations of Nifo concerning the middle-term in the *regressus* arrangement:

> Recent thinkers propose four cognitive moments (*notitias*): the first is knowledge through the senses or observation of effects; the second is the discovery of the cause through the effect, which is obtained with a demonstration "of the sign" (*demonstratione signi habetur*); the third is knowledge of the same cause through intellectual inquiry (*negotiationem*), from which, with the first, the cognition of the cause increases to such an extent that it becomes worthy of being the middle-term of the unqualified demonstration (*demonstrationis simpliciter*); the fourth is cognition of the "why" (*propter quid*) of the same effect through such a certain cause that it becomes the middle-term.[366]

In this passage, Nifo engages in the controversy regarding the number and the description of each of the steps of *regressus*. While describing *regressus*, Nifo mentions that the *demonstratio signi* and the further "negotiation" would lead to the appropriate middle-term which, adopted in the second syllogism of the *regressus* (*propter quid*), produces a *demonstratio potissima*.[367]

In the late 15th and the early 16th centuries, the *regressus* methodology would be the object of further controversy amongst Italian authors such as Pomponazzi, Zimara, Gerolamo Balduino (fl. 1528), Bernardino Tomitano (Piccolomini's fellow at Padua), and the famous Paduan Jacob Zabarella. The elaborated logico-methodological *regressus* about which the later debate was produced would ideally be composed of the following steps:[368]

---

366 Agostino Nifo, *Expositio super octo Aristotelis Stagirite libros De physico auditu*, p. 12b: [...] *Recentiores volunt esse 4 notitias: prima est effectus per sensum, aut observationem; secunda inventio causae per effectum, quae cum demonstratione signi habetur; tertia est eiusdem causae per intellectus negotiationem, ex qua cum prima crescit notitia causae in tantum, ut digna sit effici medium demonstrationis simpliciter; quarta est eiusdem effectus notitia proper quid, per talem causam sic certam, ut sit medium.*

367 Mahoney, *Two Aristotelians of the Italian Renaissance.*

368 Jardine, "Epistemology of the sciences"; Wallace, "Circularity and the *Paduan Regressus*"; Dubouclez, *Descartes et la voie de l'analyse*, chap. IV, *passim*; Poppi, "Pietro Pomponazzi tra Averroismo e Galenismo sul Problema del "Regressus"", in *Rivista Critica di Storia della Filosofia* 24/3 (1969), pp. 243–266; De Bellis, *Il pensiero logico di Agostino Nifo*, chap. 4: "La teoria del *regressus demonstrativus*".

Step 1. The first step proceeds from signs to their alleged cause (there is smoke, therefore there is fire) once we have learned about the sign through sense experience (the initial sense experience is enumerated by authors such as Nifo as an independent, previous step,[369] resulting in a four-stage methodology). This first demonstrative stage corresponds to the relation σημεῖα-αἰτίαι or *demonstratio quia*.

Step 2. The second step is an examination of the cause called *negotiatio, consideratio*, or *examen mentale*.[370] This non-demonstrative step of the process somehow (for it was never agreed how) delves into the nature of the cause reached in step 1, clarifying it (fire is such and such kind of thing). In scrutinising the nature of the cause, some insight regarding the cause shall be discovered. This finding will function as the appropriate middle-term for the last syllogism (or *demonstratio potissima*, see below).[371] This step was introduced for shielding *regressus* against the accusation of circularity: through this process, the second syllogism provides knowledge which was inaccessible through just the syllogism *quia*. Formally speaking, it can be also taken as a re-definition step, in which one of the terms reached in step 1 is rephrased so that it does not appear again in the second syllogism.

Step 3. The third and last step goes back from the cause expanded upon in step 2 to its (now better understood) effect, providing thereby new insight into the effect through a more profound knowledge of its cause. The new insight is adopted as the middle-term, so producing a second syllogism. In this case, the syllogism will possess the αἰτίαι-σημεῖα form; it will be a *demonstratio propter quid*.[372]

This process would not be solely reserved for the medical field, but it was intended as a universal pathway to knowledge. To justify it, a tradition of authoritative *loci* in the Aristotelian corpus and Averroes', Themistius', and Galen's works was established.[373]

---

369 Jardine, "Epistemology of the sciences", p. 688.

370 *Negotiatio* is the name used by Nifo; Piccolomini calls it *consideratio*, and Zabarella *examen mentale*. See Jardine, "Epistemology of the sciences", *passim*; see also Alessandro Piccolomini, *De certitudine*, f. LXXXIII.

371 On this aspect, see Jardine's analysis of Nifo changing positions in "Galileo's road to truth and the demonstrative regress", sec. IV.

372 Peter of Abano, *Conciliator differentiarum, quae inter philosophos et medicos versantur*, dif. 3.

373 The main *loci* adduced for sustaining such a view were *Physica* I, 1; *Analytica priora*, XI, 5, 8, 20; *Analytica posteriora*, I, 2, 3, 13, 22 (I, 13 being the most important); *Topica*, ed. I. Bekker,

Regressus theorists discussed *demonstratio potissima* within the context of their methodology. Some of the questions they raised are: what is the relation of *regressus* with the unqualified demonstration that 13th- and 14th-century authors such as Albert the Great, Aquinas, Giles of Rome had discussed following Averroes? To this question, we can find (at least) two approaches:

One way was to identify the whole structure of *regressus* with *demonstratio potissima*: *regressus* is the only logical structure providing the *quia* and the *propter quid* accounts simultaneously; such was the case of Pomponazzi.[374]

Another was to identify *demonstratio potissima* with the compositive moment of *regressus*: an unqualified demonstration can only be produced as step 3 (*propter quid*) of a *regressus*.[375] A proponent of this position is Zabarella, the most celebrated *regressus* theorist, who in his treatise *De regressu* identifies the demonstration *propter quid* in the last step of the *regressus* with *demonstratio potissima*.[376]

The transmission of the idea of reciprocal scientific accounts, from mathematical analysis and Aristotle, through many of its historical forms and elaborations, up to the later Latin interpretation, could be roughly sketched as follows:[377]

---

vol. 1, II, 2; *Ethica Nicomachea*, I, 2, 3; and *Metaphysica*, V, 11. In the case of Averroes the *loci* are *Expositio magna in libros posteriorum resolutoriorum*, in *Aristotelis opera cum Averrois commentariis*, vol. 1.2, lib. I, text. comm. 95–102; *Epitome in primum librum posteriorum*, in *Aristotelis opera cum Averrois commentariis*, vol. 1.2, lib. I, comm. 2–4; the passage from which *demonstratio potissima* was generally quoted is Averroes, *Commentaria in libros De physico auditu Aristotelis*, lib. I, text. comm. 2–5; and Averroes, *Expositio media in libros Physicorum*, in *Aristotelis opera cum Averrois commentariis*, vol. 4, lib. I, 2–5. Of Themistius were adduced the already mentioned *loci* in *Analyticorum posteriorum paraphrasis*, pp. 2–3, pp. 20–22, and his well-read in the Paduan context *In libros De anima paraphrasis*, I, 7–9. See Wallace, "Circularity and the *Paduan Regressus*", pp. 77–79.

374 Pietro Pomponazzi, *Quaestio de regressu*, ed. A. Poppi, *Corsi inediti dell'insegnamento padovano*, vol. 1 (Padua: Antenore, 1970), pp. 153–176.

375 Jardine, "Epistemology of the sciences", p. 687; Wallace, "Circularity and the *Paduan Regressus*".

376 Jacopo Zabarella, *De methodis libri quattuor: Liber de Regressu*, ed. C. Vasoli (Bologna: Cooperativa Libraria Universitaria Editrice Bologna, 1985), p. 328: *Ex tribus igitur partibus necessario constat regressus: prima quidem est demonstratio quod, qua ex effectus cognitione confusa ducimur in confusam cognitionem causae; secunda est consideratio illa mentalis, qua ex confusa notitia causae distinctam eiusdem cognitionem acquirimus; tertia vero est* **demonstratio potissima**, *qua ex causa distincte cognita ad distinctam effectus cognitionem tandem perducimur.*

377 *Nota bene*: two Greek sources on mathematical analysis known to Piccolomini, namely Proclus' *In Euclidem* and Eustratius' *In Analyticorum posteriorum* are omitted from this scheme. I will deal with Piccolomini's specific use of them in section 3.2.

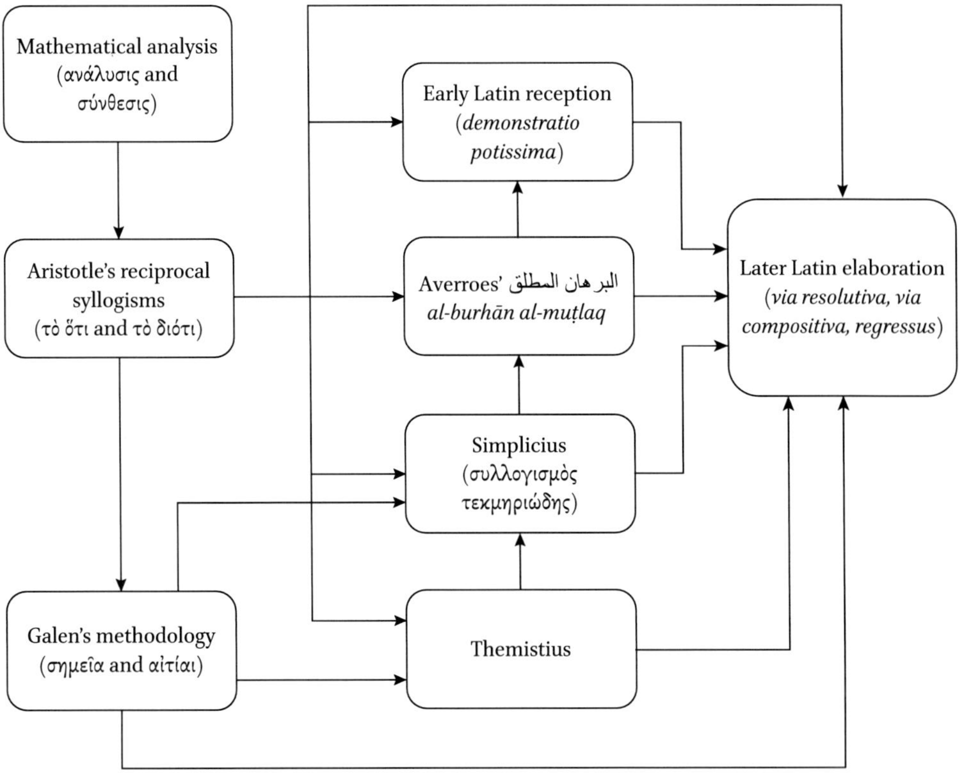

FIGURE 1 The history of *demonstratio potissima*

In what follows, I will provide a formal introduction to the concept of *demonstratio potissima* and an overview of the logical and akribological doctrines in the Latin tradition whose elaborations are relevant for grasping Piccolomini's reception of it. Ultimately, it shall be shown that Piccolomini's *demonstratio potissima* is a particular form of *regressus demonstrativus.*

2.6.2 Demonstratio potissima*: Formal Traits*

The initial 13th- and 14th-century debate surrounding *demonstratio potissima* tries to answer this question: what is the most scientific logical structure? In which science can it be produced? The 13th-century debate would have as its core the formal relation between the terms being the subject (S) and the predicate (P) in the syllogism.

A good starting point for getting an overview of the question is Aristotle's definition of universality, the main requirement of scientific propositions. In *Analytica posteriora* I, 4 Aristotle defines "universality" as follows: "I call universal whatever belongs to something both in every case and in itself (per se) and as such. It is evident, therefore, that whatever is universal belongs from necessity to its objects".[378] Aristotle states that having a true proposition of the form 'Every S is P' is not sufficient for having a truly universal proposition – as is required in science. To the requirement (a) of the universal syntactical form (κατὰ παντός), Aristotle adds two other requirements, namely (b) "in itself" (καθ'αὑτό), and (c) "as such" (ᾗ αὐτό).

Aristotle, and the tradition following him, distinguished several ways in which a predicate can be said (b) καθ'αὑτό of a subject. The various ways in which a predicate can belong "in itself" to a subject would be a question at the centre of the early Latin debate on *demonstratio potissima*. But before analysing this issue, it is necessary to introduce some additional notes regarding the kinds of definitions functioning either as premises or conclusions.

2.6.2.1 *Three-Term* Definitions

Aristotelian scientific definitions are of two kinds:

Firstly, we have statements such as "human being is a living being (or animal) which has reason (or is rational)", in which the *definiendum* is identified with a two-term predication. The first predicative term is a genus, indicating the position of a would-be taxon of entities within a given science; the second singles out the species, denoting what is unique to the *definiendum* if compared with other entities sharing the same genus.[379]

Secondly, we have three-term definitions (or complex definitions). They are introduced in *Analytica posteriora* II, 2 when Aristotle ponders the example of the lunar eclipse in his study of the role of the middle-term in a scientific demonstration. Three-term definitions capture the Aristotelian idea that knowing what something is means knowing what causes something to be what it is.[380]

---

378 Translation in Aristotle, *Complete Works of Aristotle*, vol. 1 (slightly modified). Original in Aristotle, *Analytica posteriora* I, 4, 73b, l. 26–27: καθόλου δὲ λέγω ὃ ἂν κατὰ παντός τε ὑπάρχῃ καὶ καθ' αὑτὸ καὶ ᾗ αὐτό. φανερὸν ἄρα ὅτι ὅσα καθόλου, ἐξ ἀνάγκης ὑπάρχει τοῖς πράγμασιν.

379 To explain the basics of Aristotle's theory of demonstrative knowledge in this section, I will mainly follow Bronstein's illuminating exposition and exemplifications in *Aristotle on Knowledge*. See also M. Deslauriers, *Aristotle on Definition* (Leiden: Brill, 2007), especially chapter 2; P. Pellegrin, "Definition in Aristotle's Posterior Analytics", in J. Lennox, R. Bolton (eds), *Being, Nature, and Life in Aristotle: Essays in Honor of Allan Gotthelf* (Cambridge: Cambridge University Press, 2010), pp. 122–146; A. Gómez-Lobo, "Definitions in Aristotle's *Posterior Analytics*", in D. J. O'Meara (ed.), *Studies in Aristotle* (Washington D.C.: Catholic University of America Press, 1981), pp. 25–46.

380 See *Analytica posteriora*, II, 3, 89b, l. 37–90a, l. 1 where 'what it is' (τὸ τί ἐστι) and 'why is it' (τὸ διὰ τί) are put as the later stages in scientific inquiry of the middle-term: ὅταν δὲ γνόντες ἢ τὸ ὅτι ἢ εἰ ἔστιν, ἢ τὸ ἐπὶ μέρους ἢ τὸ ἁπλῶς, πάλιν τὸ διὰ τί ζητῶμεν ἢ τὸ τί ἐστι, τότε ζητοῦμεν τί τὸ μέσον. In Bronstein's translation (slightly modified): "[…] having come to know either

In other words, knowing what X is means knowing what causes X to be X, where the cause of X must be understood as a formal cause deriving from X's essence. We will find that three-term definitions contain a definitional term and two explanative terms: one term linking the defined term with a subject, and another term providing a final or efficient cause of the defined term.[381] Thus, formally speaking, every complex definition will be composed of three terms. The formalisation that I will follow will be:

Term-1 reveals a related cause of the *definiendum* which, in principle, could be either the efficient or the final cause.

The cause given in Term-1 links the *definiendum* to the other two terms: Term-2 and Term-3 in a complex definition.

Term-2 is the definitional term: it is a descriptive statement of the *definiendum.*

Term-3 stands for the underlying reality (normally a substance) to which the *definiendum* is said to belong.

Therefore, three-term definitions possess the paradigmatic form:

> ***Definiendum*** is (described as) the content given by **Term-2**, which belongs to the substance given by **Term-3**, because of the cause given by **Term-1.**

A formalised example provided by Aristotle himself of a three-term definition is that of thunder:

> Thunder (***Definiendum*-Predicate**) is noise (**Term-2**) in the clouds (**Term-3**) because of fire extinguishing (**Term-1**).[382]

---

the fact that or whether it is either partially or without qualification, we seek in turn the reason why or what it is, we are then seeking what the middle term is". See Bronstein, *Aristotle on Knowledge*, pp. 89–90.

381 See *Metaphysica*, VIII, 4, 1044b, l. 9–15 for Aristotle's account of how causal explanation relates to the terms of complex definitions that I will now enumerate. On this topic see Bronstein, *Aristotle on Knowledge*, pp. 96–99. For the Aristotelian definition in relation to his metaphysics, see also M. Deslauriers, *Aristotle on Definition*, chap. 4.

382 *Analytica posteriora*, II, 8, 93b, l. 7–14: τί ἐστι βροντή; πυρὸς ἀπόσβεσις ἐν νέφει. διὰ τί βροντᾷ; διὰ τὸ ἀποσβέννυσθαι τὸ πῦρ ἐν τῷ νέφει. νέφος Γ, βροντὴ Α, ἀπόσβεσις πυρὸς τὸ Β. τῷ δὴ Γ τῷ νέφει ὑπάρχει τὸ Β (ἀποσβέννυται γὰρ ἐν αὐτῷ τὸ πῦρ), τούτῳ δὲ τὸ Α, ψόφος. Aristotle understands that thunder is not a substance, but an accident that exists or happens in another entity, namely a cloud. Thus, it is to be analysed as a predicate. See Deslauriers, *Aristotle on Definition*, pp. 60–61. Other three-term definitions adopted by Aristotle are found in *Analytica posteriora*, II, 2, 90a, l. 18–23; 8, 93b, l. 7–14; 12, 95a, l. 16–21; *De somno*, ed. I. Bekker, vol. 1, I, 454b, l. 25; II, 455b, l. 20.

Another example is provided in *De anima* I, 1. There Aristotle proposes the structure that the physical definition of "being angry" (τὸ ὀργίζεσθαι) would have without specifying "for the sake of what" (τοῦ ἕνεκα) anger occurs:

> Being angry (***Definiendum*-Predicate**) is a sort of motion (**Term-2**) of a body of such a sort, or of a part or capacity of a body (**Term-3**), brought about by this for the sake of that (**Term-1**).[383]

2.6.2.2 Καθ'αὑτό Is Said in Many Ways

Three meanings of καθ'αὑτό are found in *Analytica posteriora*, I, 4.[384] These are three ways of expressing the relation Subject-Predicate.[385] Two of them are qualifications of the accidental features occurring to a substance: (1) κατὰ συμβεβηκότα or simply "accident", and (2) καθ' ὑποκειμένου συμβεβηκότα, i.e., "accident according to the subject", "accident in itself" or, as Bronstein aptly labels it,[386] "demonstrable attribute". To the three ways of καθ' αὑτό-belonging recognised in *Analytica posteriora* a fourth, introduced in *Metaphysica* (V, 18, 1022a, l. 24–35), must be added. The syntagm καθ' αὑτό would be translated into Latin as *per se* and the four ways the relationship S-P can be obtained would be named as *per se modi*. Thus, the four modes of belonging treated in the 13th-century Latin controversy following the reception of *Analytica posteriora* are:[387]

---

383 Aristotle, *De anima* I, 1, 403a, l. 27–29: τὸ ὀργίζεσθαι κίνησίς τις τοῦ τοιουδὶ σώματος ἢ μέρους ἢ δυνάμεως ὑπὸ τοῦδε ἕνεκα τοῦδε.

384 Aristotle, *Analytica posteriora*, I, 4, 73a, l. 34–73b, l.5: καθ'αὑτὰ δ' ὅσα ὑπάρχει τε ἐν τῷ τί ἐστιν, οἷον τριγώνῳ γραμμὴ καὶ γραμμῇ στιγμή (ἡ γὰρ οὐσία αὐτῶν ἐκ τούτων ἐστί, καὶ ἐν τῷ λόγῳ τῷ λέγοντι τί ἐστιν ἐνυπάρχει), καὶ ὅσοις τῶν ὑπαρχόντων αὐτοῖς αὐτὰ ἐν τῷ λόγῳ ἐνυπάρχουσι τῷ τί ἐστι δηλοῦντι, οἷον τὸ εὐθὺ ὑπάρχει γραμμῇ καὶ τὸ περιφερές, καὶ τὸ περιττὸν καὶ ἄρτιον ἀριθμῷ, καὶ τὸ πρῶτον καὶ σύνθετον, καὶ ἰσόπλευρον καὶ ἑτερόμηκες· καὶ πᾶσι τούτοις ἐνυπάρχουσιν ἐν τῷ λόγῳ τῷ τί ἐστι λέγοντι ἔνθα μὲν γραμμὴ ἔνθα δ'ἀριθμός. ὁμοίως δὲ καὶ ἐπὶ τῶν ἄλλων τὰ τοιαῦθ' ἑκάστοις καθ'αὑτὰ λέγω, ὅσα δὲ μηδετέρως ὑπάρχει, συμβεβηκότα, οἷον τὸ μουσικὸν ἢ λευκὸν τῷ ζῴῳ. Translation in Aristotle, *Complete Works of Aristotle*, vol. 1 (slightly modified): "One thing belongs to another in itself both if it belongs to it in what it is – ex.gr. line to triangle and point to line (for their substance depends on these and they belong in the account which says what they are) – and also if the things it belongs to themselves belong in the account which makes clear what it is – ex.gr. straight belongs to line and so does curved, and odd and even to number, and prime and composite, and equilateral and oblong; and for all these there belongs in the account which says what they are in the one case line, and in the others number. And similarly in other cases too it is such things that I say belong to something in itself; and what belongs in neither way I call accidental, ex.gr., musical or white to animal".

385 From now on I will use "S" instead of "Subject" and "P" instead of "Predicate" or "Passion".

386 Bronstein, *Aristotle on Knowledge*, part I, chap. 3.

387 For an account of the reception of the *per se* modes in the 13th century and the nuances in their interpretations see A. Corbini, *La teoria della scienza nel XIII secolo. I commenti agli Analitici Secondi*, pp. 102–127. As a token of the breadth of the controversy, see

> ***Per se modus primus*** (= PS1): P belongs to S in PS1 iff (i) P belongs to S, and (ii) P is (part of) the essence of S.[388]

For instance, the statement "Every human being (S) is rational (P)" is in PS1, since P belongs to S by virtue of S's essence (human beings are essentially rational). Hence, "every human being (S) is a living thing (P)" is also a predication in which the predicate belongs to the subject in PS1.[389]

> ***Per se modus secundus*** (= PS2): P belongs to S in PS2 mode iff (i) P belongs to S and (ii) S is part of the essence of P.

This mode requires the introduction of complex definitions for the predicates. This will allow us to fulfil the requirement that P should belong to S in a relation which shows that "belonging to S" is part of P's essence. For instance, if we state that "Reason (Defined P) is the language/thought ordering faculty (Term-2) of the human being (Term-3) for the sake of intellectual virtues (Term-1)", then the predicate P (i.e., reason), defined by the complex definition, will belong to its subject (human being) in PS2: the definition declares that the subject possesses P by virtue of reason's essence through a term (Term-2) which introduces S in P's definition.

Within the 13th-century Latin context, PS2 would be held to be obtained in a proposition in which P is not necessarily the specific difference of S.[390] In that

---

Roger Bacon, *Summulae dialectices*, ed. R. Steele (Oxford: Clarendon, 1940), III, n. 193; Albert the Great, *In libros Posteriorum analyticorum*, I, trac. 2, cap. 9; Thomas Aquinas, *Expositio in libros Posteriorum analyticorum*, I, lec. 10; Giles of Rome, *Super libros duos Posteriorum analyticorum Aristotelis* (Venice: Bonetus Locatellus, 1488), I, p. 46a; Duns Scotus, *Quaestiones in librum Porphyrii Isagoge*, eds R. Andrews, G. Etzkorn, G. Gál, R. Green, T. Noone, R. Wood, vol. 1 (St. Bonaventure NY: The Franciscan Institute Press, 1999), q. 9–11, n. 28; Walter Burley, *Quaestiones super librum Posteriorum analyticorum*, ed. M. C. Sommers (Toronto: Pontifical Institute of Mediaeval Studies, 2000), q. 7.

388 I follow Bronstein's formalisation and account of the *per se* modes 1, 2, and 4. See Bronstein, *Aristotle on Knowledge*, pp. 44–48.

389 Corbini, *La teoria della scienza nel XIII secolo*, pp. 102–106.

390 As a clear token of this interpretation see Thomas Aquinas, *Expositio in libros Posteriorum analyticorum*, I, lec. 10: *Secundus modus dicendi per se est, quando haec praepositio per designat habitudinem causae materialis, prout scilicet id, cui aliquid attribuitur, est propria materia et proprium subiectum ipsius. Oportet autem quod proprium subiectum ponatur in definitione accidentis: quandoque quidem in obliquo, sicut cum accidens in abstracto definitur, ut cum dicimus, quod simitas est curvitas nasi; quandoque vero in recto, ut cum accidens definitur in concreto, ut cum dicimus quod simus est nasus curvus. Cuius quidem ratio est, quia cum esse accidentis dependeat a subiecto, oportet etiam quod definitio eius significans esse ipsius contineat in se subiectum. Unde secundus modus dicendi per se est,*

way, it would be considered that an accident, by virtue of its definition, can be declared to belong to a kind of entity in PS2. The result is that the same accident can be conceptualised as PS4, if the P-S belonging is explained by virtue of the essence of the subject (see below), or PS2, in which case P is explained as belonging to S because of P's definition, despite P being a specific difference or an accident.

> ***Per se modus tertius*** (= PS3): P belongs to S in PS3 mode iff (i) P belongs to S, (ii) S is a particular, (iiia) P and S are the same term, *or* (iiib) P is (part of) the essence of S.

This mode was the one drawn from *Metaphysica* V.[391] There Aristotle defines again the senses of καθ'αὑτό and includes one which cannot be easily assimilated to those introduced in *Analytica posteriora*. We learn that there are some statements showing a P-S καθ'αὑτό relationship which is said to hold some kind of necessity (ἀνάγκη). Two sorts of necessity in this P-S relationship are distinguished:

(1) Trivially, propositions like "Socrates is Socrates" are said to be καθ'αὑτό and necessary.

(2) The second case consists of a statement which predicates the essence (καθ'αὑτὸ τὸ τί ἦν εἶναι ἑκάστῳ) or part of the essence of a particular thing. For instance, the statements "Socrates is a human being" or "Socrates is a living being" are cases of P-S belonging in PS3 mode fulfilling (i), (ii), and (iiib).

Albert the Great noticed other traits of this mode.[392] Firstly, the type of analytic (tautological) belonging presented in "Socrates is Socrates" also holds if the term denotes a non-particular. Although Aristotle only provided examples involving particular entities, the proposition "a human being is a human being"

---

quando subiectum ponitur in definitione praedicati, quod est proprium accidens eius. See Corbini, *La teoria della scienza nel XIII secolo*, pp. 106–114.

391 Aristotle, *Metaphysica*, V, 1022a, l. 26–29: ὥστε καὶ τὸ καθ'αὑτὸ πολλαχῶς ἀνάγκη λέγεσθαι. ἓν μὲν γὰρ καθ' αὑτὸ τὸ τί ἦν εἶναι ἑκάστῳ, οἷον ὁ Καλλίας καθ'αὑτὸν Καλλίας καὶ τὸ τί ἦν εἶναι Καλλίᾳ, ἓν δὲ ὅσα ἐν τῷ τί ἐστιν ὑπάρχει, οἷον ζῷον ὁ Καλλίας καθ'αὑτόν: ἐν γὰρ τῷ λόγῳ ἐνυπάρχει τὸ ζῷον, ζῷον γάρ τι ὁ Καλλίας. Translation in Aristotle, *Complete Works of Aristotle*, vol. 1 (slightly modified): "Therefore 'in virtue of itself' must have several meanings. It applies to the essence of each thing, ex. gr., Callias is in virtue of himself Callias and the essence of Callias; whatever is present in the 'what', ex. gr., Callias is in virtue of himself an animal. For 'animal' is present in the formula that defines him; Callias is a particular animal".

392 These additional observations are clearly expressed, ex. gr. by Albert the Great, *In libros Posteriorum analyticorum*, tract. II, cap. 9 and 10.

is a case of the a = a structure.[393] For Albert, it is also clear that the substitution of the particular for a universal term in PS3 does not hold in the second case. If we take the statement "Socrates is a living thing" and substitute the particular term "Socrates" with Socrates's species, we will obtain not a PS3 but a PS1 proposition: "human beings are living beings". Generally, PS3 was understood as a P-S belonging in which P immediately belongs to S by virtue of S and, thus, as a sort of converse of PS2 (in which P belongs to S because of P). The main difference between PS1 and PS3 is as follows: in PS1, P is always (part of) the definition or essence of S, and P belongs to S because of what S is. In PS3 P also belongs to S because of S but P is not necessarily (part of) S's essence: in the case of "S-Socrates is P-Socrates", P belongs to S because of S, but P does not reveal any essential/definitional feature of S.

As Demange has pointed out, PS3 was commonly held by 13th- and 14th-century authors as unscientific or, at least, as scientifically inferior to other modes.[394] Among other reasons, the fact that PS3 is a mode which involves particular entities made it an unattractive candidate for being taken as a scientific proposition, much less for inclusion in the composition of *demonstratio potissima*.

> ***Per se modus quartus*** (= PS4): P belongs to S in PS4 mode iff (i) P is not part of the essence of S, and (ii) P belongs by necessity to S because of S's essence without being part of S's definition.

The predicate in PS4 can be understood as conveying the ways in which a predicate *per accidens* can be said. There are two: (1) As being κατὰ συμβεβηκότα, or as being (2) καθ'ὑποκειμένου συμβεβηκότα. I will focus on the more scientifically valuable (2): the demonstrable attribute.

The P-S belonging in the PS4 mode obtains when P is a non-essential trait of S which depends on, follows from, and/or is required by an essential feature of S. For instance, following Bronstein's instructive example,[395] suppose that we accept that human beings possess reason, and that speech is one of the effects or signs of rationality. Then, in the statement "human beings possess speech" the P "possessing speech" is a feature which follows from the specific

393 Albert the Great in his *In libros Posteriorum analyticorum*, tract. II, cap. 9: [...] *homo est homo, et asinus est asinus: causa enim quod homo est homo non est alia quam humanitas vel homo.*

394 See D. Demange, *Jean Duns Scotus, La théorie du savoir* (Paris: Sic et Non Librairie Philosophique, 2007), pp. 84–90; see also Corbini, *La teoria della scienza nel XIII secolo*, pp. 114–121.

395 See Bronstein, *Aristotle on Knowledge*, pp. 47–48.

difference of the species functioning as S, namely "human being". Although "possessing reason" and "possessing speech" are not the same feature, the capacity for speech is one of the traits which *ceteris paribus* follows rationality and, the other way around, the genuine capacity for speech attributed to a substance will reveal that such a substance is rational. Thus, although speech is not an essential feature of human beings (the ability to speak could be developmentally truncated or just eliminated from a human being without removing other effects of rationality or rationality itself), "possessing speech" is not just an accident to a human being in the same sense that, ex. gr., "being Athenian" is. The statements "Socrates possesses speech" and "Socrates is Athenian" do not hold the same P-S relationship, although both are considered accidents: whereas speech will *nihilo obstante* naturally be developed as a *per se* accident from Socrates' reason (καθ' ὑποκειμένου συμβεβηκότα), "Athenian-ness" will not follow or be a sign of Socrates' rationality (κατὰ συμβεβηκότα).[396]

2.6.3 Certissimae ergo potissimae: *Piccolomini's Adversaries*

In this section, I will provide some relevant details concerning the authors who influenced Piccolomini's discussions of *demonstratio potissima* and its relationship to mathematical certainty. In the opening section of *De certitudine*, Piccolomini divides his Latin sources into two lists: *Latini antiquiores*, including Albert, Aquinas, Giles of Rome, and Marsilius of Inghen (1340–1396). The *Latini recentiores* are the statesman and mathematician Donato Acciaioli (1428–1478), Zimara, and Nifo (named *Suessanus*).[397] I will deal with both groups.

2.6.3.1 The Question of the Middle-Term

The most prominent formal discussion regarding *demonstratio potissima* in the 13th-century Latin reception of *Analytica posteriora* is that of the middle-term. Since this conceptual problem will be addressed by Piccolomini in *De certitudine*, I will devote here some space to illustrate and discuss it.

---

396 See also H. Granger, "The *Differentia* and the *Per Se* Accident in Aristotle", in *Archiv für Geschichte der Philosophie* 63 (1981), pp. 118–129. For the discussions on PS4 within the Latin context see A. Corbini, *La teoria della scienza nel XIII secolo. I commenti agli Analitici Secondi*, pp. 121–128.

397 Alessandro Piccolomini, *De certitudine*, f. LXXIV: *Quam quidem Averrois authoritatem omnes fere Latini quos ego viderim veluti ex antiquioribus Divus Albertus, Divus Thomas, Marsilius, et Egidius, ex recentioribus vero Zimarra, Suessanus, Acciaiolus, et plerique alii, si quando in eam inciderunt, uno ore quasi alius alium sequens ita interpretati sunt, ut propterea Averroes illud asserat, quia mathematicus ex notioribus et nobis et natura demonstrat, quippe qui vel solus, vel maxime, demonstratione illa, quam potissimam appellant, utatur, qua scilicet simul et quod effectus sit et cur sit liquido innotescit.*

To begin with, in the syllogism

> Every A belongs to B,
> Every B belongs to C,
> Ergo, every A belongs to C

the term B is called by Aristotle μέσον, translated into Latin as *medium*, "middle-term". The middle-term possesses two intertwined functions: (i) a formal one, and (ii) an epistemological one. Its formal function is to allow the deduction of the conclusion by connecting the other two terms A and C. Its epistemic function, according to Aristotle, is to deliver the cause or the explanation (αἰτία) of the conclusion, including (sometimes) the essence of the object under study.[398]

One of the questions at the core of the early Latin debate concerned the traits of the term functioning as the middle-term in the paradigmatically scientific demonstration. The Latin tradition would mainly elaborate on two case-studies delivered by Aristotle: one in mathematics, and another in natural philosophy.

In natural philosophy, the most discussed demonstration was that of the lunar eclipse. It is used by Aristotle in *Analytica posteriora* II, 1 and so forth as a token of "subject *par excellence*" in scientific inquiry (ἁπλῶς τὸ ὑποκείμενον).[399] Some reasons could be argued for this preference: celestial regions and their

---

398 See *Analytica posteriora*, I, 4, 73b, l. 1–2; 5, 74a, l. 30–34; 33, 89a, l. 18; II, 7, 92b, l. 38; *Topica*, VIII, 3, 158b, l. 4; *Metaphysica*, I, 9, 992b; VII, 9, 1034a, l. 30–32; XIII, 4, 1078b, l. 24–25; *De anima*, I, 1, 402b, l. 25–403a, l. 1.

399 *Analytica posteriora*, II, 1, 90a, l. 36–90b, l. 9: ζητοῦμεν δέ, ὅταν μὲν ζητῶμεν τὸ ὅτι ἢ τὸ εἰ ἔστιν ἁπλῶς, ἆρ' ἔστι μέσον αὐτοῦ ἢ οὐκ ἔστιν· ὅταν δὲ γνόντες ἢ τὸ ὅτι ἢ εἰ ἔστιν, ἢ τὸ ἐπὶ μέρους ἢ τὸ ἁπλῶς, πάλιν τὸ διὰ τί ζητῶμεν ἢ τὸ τί ἐστι, τότε ζητοῦμεν τί τὸ μέσον. λέγω δὲ τὸ ὅτι ἔστιν ἐπὶ μέρους καὶ ἁπλῶς, ἐπὶ μέρους μέν, ἆρ' ἐκλείπει ἡ σελήνη ἢ αὔξεται; εἰ γάρ ἐστι τι ἢ μὴ ἔστι τι, ἐν τοῖς τοιούτοις ζητοῦμεν· ἁπλῶς δ', εἰ ἔστιν ἢ μὴ σελήνη ἢ νύξ. συμβαίνει ἄθρα ἐν ἁπάσαις ταῖς ζητήσεσι ζητεῖν ἢ εἰ ἔστι μέσον ἢ τί ἐστι τὸ μέσον. τὸ μὲν γὰρ αἴτιον τὸ μέσον, ἐν ἅπασι δὲ τοῦτο ζητεῖται. ἆρ' ἐκλείπει; ἆρ' ἔστι τι αἴτιον ἢ οὔ; Translation in Aristotle, *Complete Works of Aristotle*, vol. 1 (slightly modified): "We seek, whenever we seek the fact or if it is simpliciter, whether there is or is not a middle term for it; and whenever we become aware of either the fact or if it is – either partially or simpliciter – and again seek the reason why or what it is, then we seek what the middle term is. (I mean by the fact that it is partially and simpliciter – partially: Is the moon eclipsed? or is it increasing? (for in such cases we seek if it is something or is not something); simpliciter: if the moon or night is or is not.) It results, therefore, that in all our searches we seek either if there is a middle term or what the middle-term is. For the middle term is the explanation, and in all cases that is sought. Is it eclipsed? Is there some explanation or not?" For more uses of the eclipse within *Analytica posteriora*, see II, 12, 95a, l. 13–15; II, 16, 98a, l. 34–98b, l. 23.

movements were considered eternal, hence, the truths established about entities in that region were regarded as necessary. Certainly, if *demonstratio potissima* is possible at all in natural philosophy, it should be easier to arrange it in perennial natural substances (like heavenly bodies are) rather than using perishable ones.

The amply discussed mathematical example is the theorem proving that every triangle has internal angles equal to two right angles. This theorem was known in early Greek geometry and was assiduously adopted as a paradigm of scientific knowledge by Aristotle in *Analytica posteriora*.[400] It was afterwards introduced as theorem I.32 in Euclid's *Elementa* – and so commented upon by Proclus.[401]

As mentioned above, when a mathematical demonstration is adduced in this debate, philosophers do not speak of a proof such as those found in Euclid's text. Rather, they refer to the syllogistic arrangement of a given mathematical proof. Thus, in discussing *demonstratio potissima*, Aristotle's syllogistics were adopted as the common meta-language to which mathematical demonstration must be adapted. The tradition likely found the syllogistic formalisation of mathematical proofs both legitimate and natural because of such suggestions by Aristotle and Alexander. In *Analytica posteriora* II, 11, 94a, l. 28–31 Aristotle attempts to clarify the meaning of the term αἰτία by syllogistically formalising the proof of what was to become Euclid's theorem III.31, namely that every angle inscribed in a semicircle is a right angle. On his behalf, Alexander provided, amongst many other quasi-syllogistic examples, the partial formalisation of Euclid's theorem I.5 (that the angles at the base of an isosceles

400 *Analytica posteriora*, I, 1, 71a, l. 17–22: Ἔστι δὲ γνωρίζειν τὰ μὲν πρότερον γνωρίσαντα, τῶν δὲ καὶ ἅμα λαμβάνοντα τὴν γνῶσιν, οἷον ὅσα τυγχάνει ὄντα ὑπὸ τὸ καθόλου οὗ ἔχει τὴν γνῶσιν. ὅτι μὲν γὰρ πᾶν τρίγωνον ἔχει δυσὶν ὀρθαῖς ἴσας, προῄδει· ὅτι δὲ τόδε τὸ ἐν τῷ ἡμικυκλίῳ τρίγωνόν ἐστιν, ἅμα ἐπαγόμενος ἐγνώρισεν. Translation in Aristotle, *Complete Works of Aristotle*, vol. 1 (slightly modified): "But you can become familiar by being familiar earlier with some things but getting knowledge of the others at the very same time – i.e., of whatever happens to be under the universal of which you have knowledge. For that every triangle has angles equal to two right angles was already known; but that there is a triangle in the semicircle here became familiar at the same time as the induction". Aristotle usually returns to this example as a token of paradigmatic scientific knowledge. See *Analytica posteriora*, I, 1, 71b, l. 1–6; 4, 73a, l. 30; 5, 41a, l. 12; 74b, l. 4; 24, 85a, l. 26–85b, l. 15; 24, 86a, l. 24–29; 31, 87b, l. 34–36; II, 3, 90b, l. 6–9; 3, 110b, l. 6–27.

401 For a history of I.32, see Euclid, *The Thirteen Books of the Elements*, ed. T. Heath, vol. 1–3 (New York: Dover Publications, 1956), vol. 1, pp. 316–322. As we shall see, the fact that Proclus commented on proposition I.32 indicating some of its logical flaws meant that his *In Euclidem* served as one of the pathways of this example into the logical debate of the Renaissance.

triangle are equal to each other) in his commentary on the same proposition in *Analytica priora* 41b, l. 14.[402] Alexander's text was widely known only after its 1521 Aldine edition and its 1543 Latin rendering by Giovanni Bernardo Feliciano (*c.*1490–1554) – dedicated to Diego de Hurtado y Mendoza!

2.6.3.2 Parisian Sources

In this section, I will provide some details on Albert's, Aquinas', Giles', and Marsilius of Inghen's opinions regarding the unqualified demonstration.[403] Nonetheless, the reader must be aware that a thorough exposition of their views on *demonstratio potissima* is far beyond the scope of this section. I will present an overview of these figures' theories, with particular emphasis on the details crucial for understanding Piccolomini's perspective on the controversy.

In his *In libros Posteriorum analyticorum* Albert suggested two different versions of *demonstratio potissima*:[404]

(1) *Demonstratio potissima* in mathematics. This demonstrative model adopts the definition of the subject as middle-term.[405] The approximate formula which

---

402 Alexander of Aphrodisias, *In Aristotelis Analyticorum priorum*, pp. 268, l. 10–269, l. 15.

403 I would like to extend my gratitude to Dominic Dold for his invaluable insight and feedback regarding the topics of this section. Some of his views on the topic were part of the presentation "Albert the Great on Definitions in Demonstrations", held at the international conference "Medieval and Renaissance Sciences in Context: Challenges to Aristotelian Methods", Université Paris Cité, 13–14 June 2024, co-organised by Ana María Mora-Márquez and Aurélien Robert.

404 In my treatment of the theories of *demonstratio potissima* in the Latin context I will follow (and try to expand on) the excellent scholarship of Longeway. See Longeway, "Aegidius Romanus and Albertus Magnus vs. Thomas Aquinas on the highest sort of demonstration (*demonstratio potissima*)", in *Documenti e Studi sulla Tradizione Filosofica Medievale* 13 (2002), pp. 373–434; Longeway, *Demonstration and Scientific Knowledge in William of Ockham: A Translation of Summa Logicae III–II: De Syllogismo Demonstrativo, and Selections from the Prologue to the Ordinatio* (South Bend, Indiana: University of Notre Dame, 2007); Longeway, "The Place of *Demonstratio Potissima*". For a general account of the Latin reception of *Analytica posteriora* in the 13th century see Corbini, *La teoria della scienza nel XIII secolo*. See also J. Beumberg-Chaumont, "Découverte, analyse et démonstration chez les premiers commentateurs médiévaux des Seconds Analytiques", in J. Biard (ed.), *Raison et démonstration*, pp. 71–96.

405 Albert the Great, *In libros Posteriorum analyticorum*, II, trac. II, p. 192a–b: *In demonstrationibus autem mathematicis attendendum, quod quamvis mathematicus abstrahat a materia sensibili, et quamvis abstrahat sua subiecta, ut lineam, et superficiem, et corpus a materia sensibili, non tamen passiones subiectorum abstrahit a subiectis, et ideo in diffinitione passionis mathematicae ponitur subiectum, quando dicunt quid et propter quid* […]. *Quod patet in hac passione: habere tres aequos duobus rectis. Haec enim probatur per hanc diffinitionem, figura plana, tribus lineis contenta, habens exteriores angulos extrinsecis sibi oppositis aequales.*

Albert refers to as the paradigm of a *demonstratio potissima* in mathematics is a syllogism with Maior-PS2, minor-PS1, conclusion-PS4.[406] That is to say, a theorem (I.32) is attributed to a mathematical object through the definition of that object (a triangle).

(2) *Demonstratio potissima* in natural philosophy. This form of demonstration uses the definition of the passion as middle-term.[407] As put by Longeway,

406 For the PS2, PS1 premises see Albert the Great, *In libros Posteriorum analyticorum*, I, trac. II, p. 45a–b, s: *Si autem in demonstratione quae concludit passionem de subiecto, sit formalis diffinitio subiecti medium, tunc in maiori propositione est modus secundus, quia praedicatur passio de diffinitione subiecti, et eadem est comparatio passionis ad subiectum et ad suam diffinitionem. In minori autem propositione tunc erit modus primus, quia praedicatur diffinitio de diffinito; cuius exemplum est, omne quod est superficies plana, tribus lineis contenta, habet tres angulos aequos duobus rectis, triangulus est superficies plana, tribus rectis lineis contenta, ergo habet tres angulos aequos duobus rectis*. For the PS2 conclusion see Albert the Great, *In libros Posteriorum analyticorum*, I, trac. II, p. 44b: *Et ad hoc autem intelligendum, notandum quod demonstratio quae simpliciter demonstratio est, aliquando demonstrat passionem de subiecto, et tunc semper secundus modus est in conclusione, tunc enim praedicatur propria passio de subiecto*. As an instance of the influence of the view that *demonstratio potissima* adopts S's definition as the middle-term, see the following testimonies: Simon of Faversham (*ob.*1306), *Questions on the Posterior analytics*, ed. J. Longeway, Ph.D. (Ithaca NY, Cornell University, 2016) q., 31: *Unde in demonstratione potissima in maiore passio attribuitur medio, quod medium est causa efficiens passionis, quia medium in tali demonstratione dicit quid est subiecti et propter quid inhaerentiae praedicati ad subiectum, et illud medium est causa efficiens passionis, et sic est per se quarto modo. In secunda propositione deffinitio subiecti attribuitur subiecto et ita est ibi primus modus. In conclusione passio attribuitur subiecto et ita est ibi secundus modus*; William of Ockham, *Summa logicae*, eds E. A. Moody, Ph. Boehner, G. Gál, S. Brown, A. Gambatese (St. Bonaventure NY: The Franciscan Institute Press, 1999), III, 2, 40: *Et de tali demonstratione potissima, per quam demonstratur passio sive definitio exprimens quid nominis passionis de suo subiecto primo per definitionem subiecti tamquam per medium, intelliguntur multae auctoritates Aristotelis in libro posteriorum. Nam ad talem demonstrationem non sufficit quod syllogismus sit ex necessariis et indemonstrabilibus, quia non omnis syllogismus ex necessariis et indemonstrabilibus est talis demonstratio, quamvis omnis syllogismus ex necessariis et indemonstrabilibus faciens scire conclusionem prius ignotam sit demonstratio magis large sumendo demonstrationem. Tali etiam demonstratione non potest una conclusio nisi unica demonstratione demonstrari, quamvis aliis demonstrationibus possit unica conclusio multis demonstrationibus demonstrari. Et ita tale medium non potest esse nisi unicum respectu unius conclusionis, et extendendo nomen causae, respectu unius effectus non potest esse nisi unica causa*; John Buridan, *Quaestiones in duos libros Aristotelis Posteriorum analyticorum*, ed. H. Hubien, unpublished typescript available at **www.logicmuseum.com**, II, q.7: *Ex istis videtur mihi esse dicendum quod in demonstratione potissima passionis de subiecto definitio subiecti debet esse medium, et maxime definitio causalis, exprimens causas per se illius subiecti*.

407 Albert the Great, *In libros Posteriorum analyticorum*, II, trac. II, p. 189: [...] *si loquimur de potissima et principalissima demonstratione, illa est in qua demonstratur passio de subiecto*

its arrangement would be Maior-PS4, minor-PS2 or PS1, conclusion-PS2.[408] In this model, a demonstrable attribute or *per se* predicate is attributed to a natural entity through the definition of that attribute.

In his examination of the canonical example of the lunar eclipse, Albert finds two major hindrances. Firstly, that the P to be attributed to the subject, the "eclipse", is not always the case. This is so because of the perennial status of change and movement of natural entities: even in heavens, which are certainly eternal, there is local motion and, thus, predicates such as "being eclipsed" are transient.[409] The second observation is that the S to which the P must be attributed, the Moon, is a particular. This prevents us from obtaining a syllogism in which every term is universal and, hence, the figure *Barbara* cannot be arranged.[410]

This difference in the middle-terms of mathematics and physics is justified by the appeal to the metaphysics of their respective objects of study. Albert asserts that the epistemic opacity of the object of study of natural beings determines the formal structure of demonstration in natural philosophy. In other words, Albert sustains that the metaphysics of a given item determines the logical arrangement that must be assumed in producing demonstrations about that item: what an object is like determines how easily and through which means we can get to know it.

Natural objects are subjected to contingency and change, and their essences are far from evident. This compels the natural philosopher to adopt P's definition as the middle-term: unlike mathematicians, natural philosophers do not start from the definitions of the science's object – the first thing that one will find upon opening Euclid's *Elementa*. Rather, natural philosophers attempt to uncover unknown definitions and to get to the essence of natural things through the study of their more apparent attributes. An epistemological

per diffinitionem passionis, quae dicit et quid et propter quid, et in qua eadem ratione diffinitio passionis demonstratur quae dicit quid sicut demonstratur et ipsa passio.

408 See formalisations at Longeway, "Aegidius Romanus and Albertus Magnus vs. Thomas Aquinas on the highest sort of demonstration", pp. 431–434.

409 Albert the Great, *In libros Posteriorum analyticorum*, II, trac. II, pp. 64b–65a: *Eclipsis enim vel alia passio potest comparari ad subiectum tantum, sicut ad lunam, et sic non semper est. Potest etiam comparari ad subiectum cum causa sicut ad lunam existentem in nodo capitis vel caudae latitudine determinata: et sic semper est eclipsis universalis, quae cum eveniat secundum ordinem motus solis et lunae et motum capitis et caudae, semper est necessaria. In se autem haec vel illa considerata, neque necessaria est, neque semper, sive aeterna.*

410 Albert the Great, *In libros Posteriorum analyticorum*, I, trac. II, p. 63b: *Ex quibus quidem propositionibus est talis conclusio non universalis in omnibus et semper, propter quod ex talibus non contigit syllogizare universaliter, sed si syllogizare ex talibus, hoc est, quoniam nunc est, et non simpliciter quia semper est.*

consequence of this, Albert holds, is that mathematical demonstration is more certain (*certior*) than that of natural philosophy. Albert's akribology, probably developed with *Metaphysica* II, 3 in mind, runs as follows:

> And yet, mathematical cognition is nobler and more certain (*certior*) than natural knowledge. This is due to the purity and certainty (*certitudinem*) of the known entities, which it considers through abstraction from motion and matter. Natural knowledge, on the other hand, considers things mixed with matter, and thus, they take on uncertainty, contingency, and mutability.[411]

Aquinas' theory of *demonstratio potissima* can be understood as an answer to Albert's take on it. Aquinas tried to show that, at least in some cases, a demonstration adopting S's definition as the middle-term could be also obtained in natural philosophy.

Aquinas posits that the genuine *demonstratio potissima* is found in mathematics. In it, P is attributed to S through S's definition in logical structure: Maior-PS4, minor-PS1, conclusion-PS2.[412]

Aquinas also rejects the Albertian view that demonstrations in natural philosophy (he refers to the problematic example of the lunar eclipse) necessarily involve non-universal terms.[413] This is achieved by the following logical

411 Albert the Great, *In libros Posteriorum analyticorum*, II, trac. II, p. 193a: *Et tamen nobilior et certior est cognitio mathematica quam naturalis, et hoc est propter puritatem et certitudinem cognitorum, quae in abstractione a motu et materia considerat. Naturalis autem considerat ea permixta ad materiam, et ut accipiunt incertitudinem et contingentiam et mutabilitatem.*

412 Thomas Aquinas, *Expositio in libros Posteriorum analyticorum*, I, lec. 13: *Sciendum autem est quod cum in demonstratione probetur passio de subiecto per medium, quod est definitio, oportet quod prima propositio, cuius praedicatum est passio et subiectum est definitio, quae continet principia passionis, sit per se in quarto modo; secunda autem, cuius subiectum est ipsum subiectum et praedicatum ipsa definitio, in primo modo. Conclusio vero, in qua praedicatur passio de subiecto, est per se in secundo modo.*

413 Thomas Aquinas, *Expositio in libros Posteriorum analyticorum*, I, lec. 16: *Ostendit quomodo eorum, quae sunt ut frequenter, possit esse demonstratio, dicens quod eorum quae saepe fiunt, sunt etiam demonstrationes et scientiae, sicut de defectu lunae, qui tamen non semper est. Non enim luna semper deficit, sed aliquando. Haec autem quae sunt frequenter, secundum quod huiusmodi sunt, idest secundum quod de eis demonstrationes dantur, sunt semper, sed secundum quod non sunt semper, sunt particularia. De particularibus autem non potest esse demonstratio, ut ostensum est, sed solum de universalibus. Unde patet quod huiusmodi, secundum quod de eis est demonstratio, sunt semper. Et sicut est de defectu lunae, ita est de omnibus aliis similibus. Consideranda tamen est differentia inter ea. Quaedam enim*

manoeuvre: according to Aquinas, to achieve a *demonstratio potissima* by following the archetypal structure of mathematics within natural philosophy, predicates that describe states that do not always hold true should nonetheless be treated as universal properties.[414] Therefore, although an eclipse is something that does not always happen to the Moon, we should understand it as universally predicated. Aquinas is here referring to a proposition of the type "whenever p happens, then q necessarily happens". In the present case, it would be a proposition akin to "whenever the Moon and the sun are in the positions *x-y-z* so that the Earth is between them, then an eclipse happens". In other words, the P to be demonstrated of the Moon is the property of "being susceptible to an eclipse", i.e., "eclipsability". Since every S-term in a *demonstratio potissima* must be put universally, the P "eclipsability" must be attributed to the S "every Moon" (although, factually, the Moon is only instantiated once). In this manner, Aquinas evades Albert's objections and obtains a syllogism in *Barbara* in the formalisation of the lunar eclipse. Much like in mathematics, this syllogism uses S's definition as the middle-term.

Regarding the epistemology of natural philosophy vis-à-vis that of mathematics, Aquinas claims that the sort of necessity obtained in our attempts to establish natural science falls short of the *a priori* mathematical necessity because of the metaphysics of natural objects. Demonstration in natural philosophy possesses instead a weaker, *a posteriori* or *ex hypothesi* form of necessity. For instance, *ex hypothesi* an olive seed is expected necessarily to grow into an olive tree, for such is its nature. But the necessity in this statement arises from our accumulated experience with seeds becoming mature olive trees and from the assumption that external, contingent factors, such as disease, would not prevent a particular olive seed from becoming an olive tree.

---

non sunt semper secundum tempus, sunt autem semper per comparationem ad causam, quia nunquam deficit, quin posita tali causa, sequatur effectus; sicut est de defectu lunae. Nunquam enim deficit, quin semper sit lunae eclypsis, quandocunque terra diametraliter interponitur inter solem et lunam. In quibusdam vero contingit quod non semper sunt, etiam per comparationem ad causam, quia videlicet causae impediri possunt. Non enim semper ex semine hominis generatur homo habens duas manus; sed quandoque fit defectus vel propter impedimentum causae agentis vel materiae. In utrisque autem sic ordinandae sunt demonstrationes, ut ex universalibus propositionibus inferatur universalis conclusio, removendo illa, in quibus potest esse defectus vel ex parte temporis tantum vel etiam ex parte causae.

414 Thomas Aquinas, *Expositio in libros Posteriorum analyticorum*, I, lec. 42: *Illud enim est per se causa eclipsis, quod causat universaliter eclipsim. Universale autem non cognoscitur sensu; sed ex pluribus singularibus visis, in quibus multoties consideratis invenitur idem accidere, accipimus universalem cognitionem. Et sic per causam universalem demonstramus aliquid in universali, de quo est scientia.*

In mathematics, though, necessity arises from the definitions of unchangeable objects whose passions always and invariably follow. Therefore, Thomas concludes:[415]

> However, as stated in *Physica* II, necessity is different in natural [conclusions], which are true as a general rule, but may fail in specific cases, and in mathematics, which is always true. In mathematics, there is a priori necessity, while in natural things, it is a posteriori.[416]

Giles of Rome opposed Aquinas in his *Super libros Posteriorum analyticorum*, partially agreeing with Albert: he upheld that it is impossible for the middle-term in a *demonstratio potissima* to be S's definition. Giles' chief argument is that what is sought in doing science is nothing other than S's definition. Hence, if the unqualified demonstration is the arrangement producing scientific knowledge *par excellence*, then S's definition must be its conclusion, not its middle-term. Having S's definition as the middle-term would make a tautology of an unqualified demonstration, for S's definition would be concluded through itself – making the demonstration unnecessary. Giles puts it as follows:

---

415 For a treatment of the types of necessity in Aquinas, see Wallace, "St. Thomas's Conception of Natural Philosophy and its Method", in L. Elders (ed.), *Studi Tomistici, La philosophie de la nature de saint Thomas d'Aquin* (Città del Vaticano: Libreria Editrice Vaticana, 1983), pp. 7–27; Wallace, *Causality and Scientific Explanation*, vol. 1–2 (Ann Arbor: University of Michigan Press, 1972–1974); see vol. 1, pp. 71–80, 102, 104, 143, and vol. 2, p. 247, 250, 293, and 354; Wallace, "Aquinas on the Temporal Relation Between Cause and Effect", in *The Review of Metaphysics* 27/3 (1974), pp. 569–584. For the metaphysical bases of Aquinas' interpretation of *Analytica posteriora* see Longo, "Réflexions ontologiques et renvois à la *Métaphysique* dans le commentaire de Thomas d'Aquin sur les *Seconds Analytiques* d'Aristote", in J. Biard (ed.), *Raison et démonstration*, pp. 97–112.

416 Thomas Aquinas, *Expositio in libros Posteriorum analyticorum*, I, lec. 42: *Necessarium autem, ut dicitur in II Physicorum, aliter est in naturalibus, quae sunt vera ut frequenter, et deficiunt in minori parte; et aliter in disciplinis, id est, in mathematicis, quae sunt semper vera. Nam in disciplinis [i.e., in mathematicis] est necessitas a priori; in naturalibus autem a posteriori.* Wallace has shown that the concept of hypothetical *a posteriori* necessity in natural philosophy is already present in Albert the Great's *In libros Physicorum Aristotelis*, ed. A. Borgnet, vol. 3 (Paris: Ludovicus Vives, 1890), II, 9; and his *De animalibus*, ed. A. Borgnet, vol. 11–12 (Paris: Ludovicus Vives, 1890), XI, I, 2. For an account of the elaboration of this concept by Albert, see Wallace, "Albertus on Suppositional Necessity on the Natural Sciences", in J. A. Weisheipl (ed.), *Albertus Magnus and the Sciences: Commemorative Essays* (Toronto: Pontifical Institute of Mediaeval Studies, 1980), pp. 103–128.

> But just as it is inappropriate for something to be proven by itself, so it is inappropriate for something to be proven about a subject by itself or through the subject itself.[417]

As we will see, this view would not only be adopted by Piccolomini as one of his core arguments but also led to its ultimate consequences regarding the scientificity of mathematics. To solve this hindrance, Giles analyses both canonical cases in the commentary tradition, Euclid's I.32, and the lunar eclipse.[418] He concludes that:

(1) S's definition cannot be the middle-term in either demonstration.
(2) P's definition must be adopted as the middle-term in *demonstratio potissima* both in mathematics and natural philosophy.
(3) Nevertheless, there is a place for S's definition (or part of it) in *demonstratio potissima*. It must be attached to P's, so forming a three-term definition.[419]
(4) *Demonstratio potissima* is possible both in mathematics and natural philosophy.

Giles could be read as proposing the reverse theoretical move to Aquinas: whereas Aquinas took Albert's mathematical *demonstratio potissima* and sought to arrange it in natural philosophy, Giles, after showing that adopting S's definition as middle-term is nonsensical, took Albert's *demonstratio potissima* in natural philosophy, i.e., the one adopting P's definition as the middle-term, and sought to arrange it in mathematics. As we will see, the Albert/Giles model would be partially assumed by Piccolomini: he would propose P's definition as the middle-term and target the use of S's definition as tautological.

Regarding his akribology, Giles strictly follows Averroes' authority and concedes the highest degree of certainty to mathematics:

---

417 Giles of Rome, *Super libros duos Posteriorum analyticorum Aristotelis*, f. 117: *Sed sicut est inconveniens quod aliquid probetur per seipsum, ita est inconveniens quod de subiecto probetur aliquid per seipsum sive per ipsum subiectum*. Transcription by Longeway in "Aegidius Romanus and Albertus Magnus vs. Thomas Aquinas on the highest sort of demonstration", pp. 425–434.

418 See Giles of Rome, *Super libros duos Posteriorum analyticorum Aristotelis* (1488), ff. 117–118. Transcription by Longeway in "Aegidius Romanus and Albertus Magnus vs. Thomas Aquinas on the highest sort of demonstration", pp. 425–434.

419 Giles of Rome, *Super libros duos Posteriorum analyticorum Aristotelis*, f. 119: *Dicendum quod diffinitio subiecti vel aliquid loco diffinitionis subiecti debet ingredi demonstrationem potissimam, non tanquam medium sed tanquam aliquid annexum medio*. Transcription by Longeway in "Aegidius Romanus and Albertus Magnus vs. Thomas Aquinas on the highest sort of demonstration", pp. 425–434.

> Therefore, mathematical demonstrations, which concern quantity, are more known to us and more certain (*certae*). Hence, the Commentator on *Metaphysica* II asserts that such demonstrations are certainly in the first degree of certainty (*in primo gradu certitudinis*). Indeed, in such matters, as he also means in *Physica*, I, the causes are more known to us than the effects; therefore, they are more familiar to both us and by nature.[420]

Giles' argument relies on the nature of mathematical objects as expressed in the *locus communis* of *Metaphysica* II, 3 and Averroes' comments on it: in the examination of quantity the causes (the essence of mathematical objects) are better known, and thus more certain than the effects (theorems). Thus, once it has been established, learning how a theorem necessarily follows from its subject is a straightforward and transparent cognitive activity for us, whereas understanding how a qualitative predicate follows from the nature of a natural subject is a far more uncertain and doubtful process.

Bertagna has shown that the reception of Giles of Rome's teachings on *demonstratio potissima* in the Paduan context by Paul of Venice led to the production of further logical arguments for defending, against Giles, the teaching that the middle-term is S's definition.[421] Nevertheless, it is also to be considered that Paul of Venice was, at the same time, one of the fiercest defenders of *regressus*. Both Randall and Wallace point out that Paul of Venice not only assimilated the demonstrations *quia* and *propter quid* with the parts of *regressus*, but also defended this methodology against the common accusation of circularity in his *Summa philosophiae naturalis*.[422]

---

420 Giles of Rome, *Super libros duos Posteriorum analyticorum Aristotelis*, eds Joanne Jacobo de Puteo, Petro de Nardis (Padua: Maufer, 1478), f. 45: *Demonstrationes ergo mathematicae, quae sunt de quantitate, magis sunt nobis notae et magis certae. Ideo dicit Commentator super II Metaphysicorum quod tales sunt certe in primo gradu certitudinis, immo in talibus, ut idem vult in I Physicorum, sunt magis notae causae quam effectus, itaque eadem sunt plus nota nobis et naturae.*

421 M. Bertagna, "Giles of Rome and Paul of Venice on *Demonstratio Potissima*", in *Documenti e studi sulla Tradizione filosofica medievale* 20 (2009), pp. 375–394.

422 Paul of Venice, *Summa philosophiae naturalis* (Venice: Iodocus Badius Ascensius, 1503), I, IX: *Notitia scientifica causae dependet a notitia effectus, sicut notitia scientifica effectus dependet a notitia causae; patet quoniam prius cognoscimus causam per effectum quam effectum per causam. Et ita habetur principale investigandum, scilicet quod notitia scientifica effectus naturalis exigit priorem notitiam causarum et principiorum. In processu naturali sunt tres cognitiones: prima est effectus sine discursu, et dicitur quia. Secunda causae per notitiam eiusdem effectus, et iterum vocatur quia. Tertia est vero effectus per causam et dicitur propter quid.* See also Wallace, "Circularity and the Paduan *Regressus*"; Randall, *The School of Padua*, pp. 40–41.

The last of the *Latini antiquiores* in Piccolomini's list of adversaries is Marsilius of Inghen.[423] His commentary on *Analytica posteriora* was published in 1516 in a collection of logical writings entitled *Parvorum logicalium liber*. It is therefore very likely that Piccolomini had access to this commentary through the printed text. There Marsilius discusses *demonstratio potissima*, *certitudo*, and mathematics.

In the *quaestio sexta* of his text, Marsilius addresses the question of the middle-term in *demonstratio potissima*. Before giving his opinion, he provides a summary of the previous views, naming Albert and Giles as those who propose P's definition as the middle-term, and Duns Scotus and Grosseteste as those who defend S's definition. Although he does not propose a model of his own, he agrees with the latter group and affirms that the middle-term in the unqualified demonstration must be the formal cause of the S (i.e., S's essence or quiddity) that would work as the efficient cause of the *explanandum* P.[424]

Marsilius also proposes some reflections on the akribology of mathematics and natural philosophy. He swiftly states that mathematics is more certain than all the other sciences.[425] He argues for the superiority of mathematical demonstrations as *potissimae* by emphasizing that they reach their conclusions through a single middle term (something denied by Proclus and Piccolomini, as we shall see).[426] This view is buttressed by appealing to the ontological status of mathematical objects: in demonstrations about more complex entities (such as natural entities), the same P might be potentially proven through each of the four causes. Marsilius argues, invoking Aristotle, that, as it follows from his model of *demonstratio potissima*, the middle-term reveals the formal cause, which is the only form of causation given in mathematics. Therefore, what makes mathematics certain and, at the same time, mathematical demonstrations *potissimae* is that mathematical objects allow us to demonstratively retrace a clear, univocal, and necessary formal cause from the S to the P.

---

423 For an account of Marsilius' logical doctrines regarding *per se* predication, see Corbini, "Jean Buridan et Marsile d'Inghen sur la predication par soi", in J. Biard (ed.), *Raison et démonstration*, pp. 169–184.

424 Marsilius of Inghen, *Quaestiones proutiles in Aristotelis posteriora* in *Parvorum logicalium liber* (Vienna: Singrenius, 1516), f. 234. For Longeway's formalisation see Longeway, "Aegidius Romanus and Albertus Magnus vs. Thomas Aquinas on the highest sort of demonstration", pp. 431–434.

425 Marsilius of Inghen, *Quaestiones in Aristotelis posteriora*, f. 225v.

426 Marsilius of Inghen, *Quaestiones in Aristotelis posteriora*, f. 226: *Scientiae mathematicae non procedunt nisi per unum medium ad unam conclusionem potissime demonstratam. Aliae autem scientiae procedunt per plura media. Ergo scientiae mathematicae sunt aliis firmiores. Patet consequentia et antecedens est notum per commentarium in prologo Philosophi dicentem quod mathematicae non procedunt nisi per unum genus causae, scilicet per formam.*

2.6.3.3 Italian Sources

The Florentine Donato Acciaioli is enumerated amongst Piccolomini's *Latini recentiores* as one of his adversaries. A nobleman of erudition and disciple of the Byzantine émigré John Argyropoulos (*c.*1393–*c.*1487), Acciaioli was renowned for his expertise in classical literature, mathematics, and for his production as a commentator on Aristotle.[427] In his commentary on the *Ethica Nicomachea* (1478, reprinted in 1535), when analysing I, 3 – the section on ἀκρίβεια – he makes two observations on mathematical certainty and the knowledge of mathematical causes and effects:

> Science properly said concerns things that are necessary, perpetual, and always behave in the same way, and its doctrine is demonstrative and most exact, although there are other sciences more certain (*certiores*) than others, and they have degrees of certainty (*gradus certitudinis*) among themselves. For example, mathematical sciences are more certain than natural sciences since they are in the first degree of certainty (*in primo gradu certitudinis*).[428]

And a few lines below he states:

> For some things are known unqualifiedly (*simpliciter*), such as causes and principles, others are known from our part and to us, such as effects, and some are equally known to us and unqualifiedly, as in mathematical sciences.[429]

As indicated by Piccolomini, Acciaioli holds that the most certain science is mathematics because of the perpetual and unchangeable ontological state of its objects. In his second commentary, he explains that, in mathematics,

---

427 E. Garin, "La giovinezza di Donato Acciaiuoli", in *Rinascimento* 1/1 (1950), pp. 43–70; A. D'Addario, "Donato Acciaioli", in *Dizionario biografico degli Italiani*, vol. 1 (Roma: Istituto della enciclopedia italiana, 1960), pp. 80–82. See also G. Steiris, "John Argyropoulos", in *Encyclopedia of Renaissance Philosophy*.

428 Donato Acciaoli, *Ethicorum ad Nicomachum: libri decem, Ioanne Argyropylo Byzantio interprete, ad Graecum exemplar diligentissime recogniti, cum Donati Acciaioli Florentini, viri doctissimi, commentariis* (Lyon: Antonius Vincentius, 1544), 1, 4, p. 4: *Scientia vero proprie dicta est circa ea quae sunt necessaria, perpetua et semper se habent eodem modo, et doctrina illius est demonstrativa et exactissima, licet etiam scientiae sint aliae aliis certiores, et habeant gradus certitudinis inter se, ut mathematicae sunt certiores quam naturales, cum illae sint in primo gradu certitudinis.*

429 Donato Acciaoli, *Ethicorum ad Nicomachum*, 1, 4, p. 5: *Alia enim sunt nota simpliciter, ut sunt causae et principia, alia nostra ex parte et nobis, ut sunt effectus, quaedam vero aeque nobis nota et simpliciter, ut in mathematicis scientiis.*

unlike in other disciplines, we know both the cause and the effect, rendering mathematical objects as "unqualifiedly known". In Piccolomini's reading of Acciaioli this is the same as affirming that mathematical demonstrations are *potissimae*.[430]

Nifo's opinion is presented in the debate regarding the dignity of metaphysics vis-à-vis mathematics and psychology in his commentary on *De anima*. There he pits the certainty of metaphysics against that of mathematics. After discussing some of the opinions of the commentators (such as Averroes and Themistius) he defends his position:

> It is argued so, since the middle-term of mathematical demonstration is known to us and by nature, and this is called "unqualified demonstration" (*simpliciter*), while the middle-term of metaphysics is known only by nature, and the first is called "unqualified demonstration" (*simpliciter*), the second "demonstration of the cause" (*propter quid*) in the commentary [of Averroes] to *De caelo*, II, 35. Therefore, only mathematical demonstration is unqualifiedly so; consequently, it is in the first degree of certainty and to be preferred.[431]

Nifo argues that in mathematical demonstration the middle-term is known both to us and by nature (i.e., it is knowable by itself); this is something that does not take place in metaphysics. Therefore, Nifo concludes that mathematical demonstrations are *potissimae* and have the highest certainty; mathematical proofs possess epistemic pre-eminence over those of metaphysics, which obtain the second place as demonstrations *propter quid*.

Lastly, regarding Marco Antonio Zimara, his view that *demonstratio potissima* is found primarily in mathematics is explicitly targeted by Piccolomini in *De certitudine*.[432] The opinion to which Piccolomini refers was expressed by Zimara in his *Theoremata*:

---

430 Alessandro Piccolomini, *De certitudine*, f. LXXIV: [...] *quia mathematicus ex notioribus et nobis et natura demonstrat, quippe qui vel solus, vel maxime, demonstratione illa, quam potissimam appellant, utatur, qua scilicet simul et quod effectus sit, et cur sit liquido innotescit.*

431 Agostino Nifo, *Super tres libros De anima* (Venice: Alexander Calcedonius, 1503), f. 4: *Arguitur quoniam medium mathematicae demonstrationis est notum nobis et naturae et haec est demonstratio simpliciter; medium metaphysicae est notum naturae tantum et prima dicitur demonstratio simpliciter, secunda propter quid II Caeli com[m]ento, 35. Ergo demonstratio mathematica sola est simpliciter, per consequens in primo gradu [certitudinis] et praeferenda.*

432 Alessandro Piccolomini, *De certitudine*, f. CII: *Fatetur enim [Zimara] demonstrationes mathematicas, saltem in sua puritate, dare causam et effectum, et ex hoc esse potissimas. Nos vero primo istorum huiusmodi interpretationem atque opinionem impugnabimus.*

> Certainly, there is no doubt that mathematical demonstrations are unqualified demonstrations (*simpliciter*), providing both cause and the fact (*esse*), proceeding from things unqualifiedly (*simpliciter*) more known to us and by nature.[433]

Through the examination of the opinions of those seven authors, the background against which Piccolomini designed his own model of *demonstratio potissima* and his criticisms of mathematical causality will be clarified.

As shown, influential authors from the 13th to the 16th century considered that mathematical demonstration were *potissimae* precisely because they were *certissimae*. Piccolomini's target in *De certitudine* is the long-held view *certissimae ergo potissimae*, an opinion that came into being and was spread through Latin philosophy since the 13th century, stemming from Averroes. As seen, the conventional view followed Averroes' conclusion in his comment on *Metaphysica* II, 3 that mathematics is in the first degree of certainty as well as his introduction of a third, superior kind of demonstration in his *Prooemium physicum* – a type of demonstration that the Cordovan attributed mainly to mathematics. The view fusing both of Averroes' tenets holds that (i) the supreme certainty of mathematics, due to the particular ontology of mathematical objects, allows us to know the *quia* and the *propter quid* simultaneously and, hence, (ii) *demonstratio potissimae* is more easily obtained in mathematics.

Piccolomini's akribological argumentation can be read as an attempt to dissolve the traditional link between certainty and demonstrative superiority forged in the Latin tradition: *certissimae* as they surely are because of the ontological status of their objects, Piccolomini would argue, mathematical demonstrations are not, and cannot possibly be, *potissimae*.

This divorce begets a philosophical statement of some epistemic significance, namely that there are contents in our experience and cognition which are certain (and reliable) which, nonetheless, do not fulfil the requirements for being a science. Conversely, it follows from Piccolomini's views that a discourse could be fully scientific according to our standards, but not so certain after all – such as natural philosophy would be in the Sienese's view.

---

433 Marco Antonio Zimara, *Theoremata*, f. 31v: *Simpliciter etiam non est dubium quod demonstrationes mathematicae sunt demonstrationes simpliciter dantes causam et esse, procedentes a notioribus naturae et a nobis simpliciter.*

PART 3

De Certitudine Mathematicarum

Now that the theoretical and historical background underlying *De certitudine* has been provided, we are ready to deliver a detailed account of its philosophy of mathematics.

This third part will address the following arguments: Piccolomini's idiosyncratic model of *demonstratio potissima* (3.1); analysis and mathematical analysis and how they were re-featured and presented in the opusculum using Euclid I.1 as an example (3.2 and 3.3); the epistemic flaws of Euclid I.32 and their origin in Proclus' commentary, the denial of mathematical causation, the "Simplician" reading of Proclus' *In Euclidem*, and Piccolomini's agreements and disagreements with the Paduan master De' Passeri (3.4); Piccolomini's philosophy of mathematics, his theory of *abstractio*, his akribology of mathematics, and the final argument against the scientificity of mathematics (3.5); Piccolomini's recast of his anti-Mertonian views in the opusculum (3.6); and the role of common mathematics in keeping the logical consistency of the *De certitudine* (3.7).

3.1 Piccolomini's *demonstratio potissima*: a Truncated *regressus*

I shall begin by examining the specific model of *demonstratio potissima* portrayed in *De certitudine*. Having understood its roots in the previous tradition, we will better grasp why and how this unqualified sort of demonstration was denied to mathematics.

As we will see, Piccolomini's *demonstratio potissima* stands as a sort of truncated *regressus* consisting just of two intertwined syllogisms: (i) the *quia* syllogism and, mirroring it, (ii) the *propter quid* syllogism. Piccolomini's text only provides the details of the *propter quid* demonstration, leaving the reader with the task of figuring out the precise traits of the first syllogism in the composition. At the end of this section, after having clarified Piccolomini's outline, I will provide a formalised example of his intended arrangement.

3.1.1 *Piccolomini's Requirements*

Three requirements of *demonstratio potissima* are listed by Piccolomini in *De certitudine*.

 | DOI:10.1163/9789004724143_004

(1) *The middle-term is P's nominal definition and it signifies whichever of the four causes.*

Much like Albert and Giles, Piccolomini defends that a case of *demonstratio potissima* will adopt the definition of the P which is attributed to the S as the middle-term:

> Therefore, that the definition of the passion and not that of the subject is the legitimate middle-term of the demonstration could be shown by so many authorities and reasons that it would be almost infinite.[1]

He provides a philosophical reason why it must be the case that P's definition functions as the middle-term:

> If the definition of the subject were the middle-term in the unqualified demonstration (*demonstratio potissima*), it would follow that every demonstration would be superfluous, as it would then be directed towards what is already known. This is absurd, as demonstration should proceed from the known to the unknown (*per nota ad ignotum*).[2]

Piccolomini follows Giles in his argument for defending that S's definition cannot be the middle-term: should that be the case, demonstrations would be either unnecessary or tautological, for what is sought in doing science, namely S's definition, would be already known from the very beginning.

In addition to this, Piccolomini points out that the middle-term for producing a *demonstratio potissima* can adopt whichever of the four causes: "Since there are four causes, any one of them can be demonstrated through the medium of demonstration".[3]

Lastly, Piccolomini states that the definitions (or at least the definitions of Ps) used as the middle-term must be a *definitio nominis*. The core of the argument is the following:

1 Alessandro Piccolomini, *De certitudine*, f. LXXXXI: *Quod igitur diffinitio* [sic] *passionis et non subiecti legiptimum* [sic] *sit medium demonstrationis, tot authoritatibus* [sic] *et rationibus ostendi posset ut paene infinitum esset.*

2 Alessandro Piccolomini, *De certitudine*, ff. LXXXXII–LXXXXIIV: *Si diffinitio subiecti esset medium in demonstratione potissima, sequeretur quod supervacanea esset omnis demonstratio, quia tunc ad nota dirigeretur. Hoc est absurdum, cum demonstratio per nota ad ignotum procedere debeat.*

3 Alessandro Piccolomini, *De certitudine*, f. LXXXVII: *Cum quattor sint causae, quam libet earum per demonstrationis medium ostendi posse.*

> Nor can this foreknowledge be obtained in any other way. For if the true definition of things and subjects and passions were to be foreknown in the sciences, we would in vain seek "the fact" (*quia*) of these very passions.[4]

The Sienese holds that the precise goal of science is obtaining real definitions. Therefore, if the starting point of a science is the real definition either of P or S, it would be pointless to search for the causes from the facts, for they would already be known. Therefore, Piccolomini postulates that in genuine scientific inquiry, we proceed from the nominal definition of the P and the axioms (which are also held to be nominal definitions) towards the real definition of both passions and subjects.[5] Yet again, this view will be crucial in arguing that mathematical demonstrations are not scientific.

(2) *Demonstratio potissima provides the* quia *and the* propter quid *and not all its premises are immediate and indemonstrable.*

> In the second place, I consider it necessary to admit that unqualified demonstrations (*potissimas*), which we particularly discuss in this treatise, and which provide both "the fact" (*quia*) and "the cause" (*propter quid*), are not only from immediate and indemonstrable principles.[6]

Following the Averroistic 13th- to 16th-century Latin tradition, Piccolomini states that *demonstratio potissima* provides the *quia* and the *propter quid*. Preparing the ground for his *regressus* model, he also states that not every premise must fulfil the requirements of indemonstrability and immediacy (see below).

(3) *Maior: PS1-indemonstrable, minor: PS4-demonstrable, conclusion-PS2, Prima figura*

---

4 Alessandro Piccolomini, *De certitudine*, f. LXXXVIv: *Nec alio modo haec praecognitio accipi potest. Nam si vera rei diffinitio et subiectorum et passionum praegnoscenda esset in scientiis, frustra ipsum quia ipsarum passionum inquireremus.* See also Alessandro Piccolomini, *De certitudine*, LXXXVIv: *Qui omnes per quid diffinitionem quid nominis, non autem quid rei, interpretantur. Et hoc non iniuria, nam diffinitiones quae accipiuntur inter dignitates et positiones scientiarum, teste Proclo, Philopono et aliis, sunt diffinitiones quid nominis, non quid rei.*

5 Piccolomini remains silent about how real definitions are reached from nominal ones in scientific inquiry. As it stands, this is something that his model of *demonstratio potissima* cannot achieve.

6 Alessandro Piccolomini, *De certitudine*, ff. LXXXIVv–LXXXV: *Secundo loco fatendum esse duco demonstrationes potissimas, de quibus praecipue loquimur in hoc tractatu, quae scilicet dant quia et propter quid, non solum esse illas quarum ex illis immediatis et indemonstrabilibus sunt.*

> And therefore, in our demonstration, we will have a major premise that is immediate and indemonstrable, wherein the effect, namely the passion, is found. In the minor premise, however, there is no mention of the effect that we seek.[7]

Piccolomini provides some formal insight, although in an obscure way, regarding the P-S belongings of the premises of his *demonstratio potissima*. We are told that the Maior-premise will be an indemonstrable definition where P is included. We know that the middle-term will be P's definition (i.e., the term describing what P is). Thus, both the Maior-premise and the minor-premise should have in common a term which is either P or P's definition. Now, as I understand it, Piccolomini indicates that there is no mention of the sought P in the minor-premise. From this, it follows that the Maior-premise must be composed by P and P's definition. Thus, it is what in the early Latin tradition was named PS1.

Regarding the minor-premise, we have established that P is not mentioned there. Since the middle-term will be P's definition, this must be the term shared by both premises. Therefore, the minor-premise will be a proposition in which P's definition is attributed to S. It cannot be a PS1 statement since a PS1 proposition would include P together P's definition (what we already have in the Maior-premise). It cannot be the definition of S or any of its essential traits, since we are forbidden to use S's definition by requirement (1). Neither can it be a PS2 since the very statement that P belongs to S because of P's definition is what Piccolomini seeks to conclude. Therefore, the minor-premise can only be a PS4 statement: P's definition belongs to the S as a *per se* accident.

Nonetheless, concerning the demonstrability of the minor-premise we also learn that:

> when we say that unqualified demonstrations consist of immediate premises, that is, indemonstrable ones, and that one of them is not prior, we should not understand this of both premises simultaneously, but, for the most part, of those premises which become the major premises in the first figure.[8]

---

7 Alessandro Piccolomini, *De certitudine*, ff. LXXXXIIv–LXXXXIII: *Et ideo, in nostra demonstratione maiorem praemissam immediatam habebimus et indemonstrabilem in qua effectus passio scilicet reperitur. In minori autem de effectu, quem solum quaerimus, non fit mentio.*

8 Alessandro Piccolomini, *De certitudine*, f. LXXXVIIIv: [...] *quod dicimus demonstrationes potissimas constare ex praemissis immediatis, hoc est indemonstrabilibus, et quarum altera non sit prior, non debemus id intelligere de ambabus simul praemissis, sed de praemissis illis ut plurimum quae fiunt maiores in prima figura.*

Piccolomini's insistence that the minor-premise is demonstrable and that one of the premises is not "prior" designedly qualifies two of the Aristotelian requirements of scientific demonstration in *Analytica posteriora* I, 2 (71b, l. 20–22): (iii) from immediate premises ([ἐξ] ἀμέσων, *ex immediatis*), and (v) from prior premises ([ἐξ] προτέρων, *ex prioribus*). This requirement is a paramount clue to understand that Piccolomini is designing a *regressus* structure. In it, the conclusion of each one of both syllogisms must work as the minor-premise in the other syllogism. Therefore, one of the premises will always be demonstrable and, at least in the second syllogism, posterior.

Regarding the conclusion, the *per se* mode of a demonstration in which a P is attributed to S through P's definition can only be a PS2.

In addition, Piccolomini admits that *demonstratio potissima* must be arranged in the *prima figura*, but he does not require it to be a *Barbara* scheme. Therefore, particular affirmative terms as in the *Darii* can be admitted in the most scientific demonstration. By this concession, Piccolomini gets rid of the difficulty raised by Albert that the Moon is a particular without recurring to Aquinas' solution.

3.1.2 Regressus demonstrativus *in* De certitudine

In chapter IV of *De certitudine* (*De demonstratione et eius partibus sive speciebus*) Piccolomini introduces into his discussion of *demonstratio potissima* the theory of *regressus*. The title indicates that the author is going to treat the parts (*partibus*) of demonstration. That said, no mention is made of any premises, modes of belonging, or propositional terms in the section that follows. Only three sorts of demonstration (as was commonly held in his Paduan context) attributed to Aristotle are discussed: *demonstratio quia, propter quid*, and *demonstratio potissima*.

Piccolomini openly declares that he conceives *demonstratio potissima* as a composite of the other two sorts of demonstrations, revealing that what he actually understands by the "parts" of a *demonstratio potissima* are not the propositions in a syllogism, but syllogisms united in the form of a *regressus*:

> Therefore, he who carefully considered book I of *Analytica posteriora*, will find in Aristotle a three-fold demonstration: one he calls "of the fact" (*quia*) or "of existence only" (*esse tantum*), another they call "of the cause" (*proter quid*) or "of the cause only" (*causae tantum*), and the remaining one, as composed of those, that provides both "the fact" and "the cause" simultaneously, which we call the unqualified one (*potissimam*).[9]

9 Alessandro Piccolomini, *De certitudine*, f. LXXXIV: *Triplicem igitur demonstrationem reperiet apud Aristotelem qui dilegenter I librum Posteriorum analyticorum ponderavit, una quam vocat*

Again, in chapter IV, invoking Averroes' authority against those who claim that there is no such thing as *demonstratio potissima*, it is stated that the firmest form of knowledge comes from the conjunction of a *quia* and a *propter quid* demonstrations rather than from any of them taken alone:

> Regarding the two demonstrations, "of the fact" and "of the cause", although their specific distinction is clearly apparent, Averroes states in *Physica* VIII, 90, that when these two demonstrations are simultaneously conjoined (*simul coniunctae*), the science will be firmer than if it were derived from either of them alone.[10]

Piccolomini immediately adds the paraphrase of a regressive reasoning from Averroes to exemplify what he means:

> And therefore, Averroes asserts there that Aristotle, for a firmer science about the eternal first mover, conjoined both processes (*processus*) in *Physica* VIII. After demonstrating the eternal mover in moving from the effect, he then, as if retracing his steps, concluded the eternal moved from the eternal mover.[11]

It is also noteworthy that Piccolomini refers to both accounts as *processus* – the technical term of *regressus* theory for each of the demonstrations *quia* and *propter quid*.

This is clear evidence in favour of a *regressus* interpretation of Piccolomini's theory. Amongst the scholars trying to decipher Piccolomini's *demonstratio potissima*, both Palmieri and Longeway have pointed out the role of the theory of *regressus* in *De certitudine*.[12]

---

quia vel esse tantum, aliam quam dicunt propter quid seu causae tantum, reliquam vero, quasi ex his compositam quae simul et quia et propter quid largitur, quam potissimam nuncupamus.

10 Alessandro Piccolomini, *De certitudine*, f. LXXXIIv: *De duabus vero demonstrationibus, quia et propter quid, quamvis liquido appareat earum specifica distinctio, dicit tamen Averroes, VIII Physicorum, 90, quod quando istae duae demonstrationes simul coniunctae fuerint, erit scientia firmior quam si ex quavis illa tantum haberetur.*

11 Alessandro Piccolomini, *De certitudine*, f. LXXXIIv: *Et ideo asserit ibi Averroes quod Aristoteles, propter firmiorem doctrinam de primo motore aeterno, coniunxit in VIII Physicorum utrosque processus. Postea enim quam ex motu ipso tanquam ex effectu demonstraverat motorem aeternum in movendo, non motum, dehinc quasi regrediens, ex motore aeterno motum aeternum conclusit.*

12 Palmieri, "On *scientia* and *regressus*", see the section entitled 'Regressus and mathematics', pp. 339–343; Longeway, "The Place of *Demonstratio Potissima*".

What then is the exact arrangement of Piccolomini's *demonstratio potissima* as a form of *regressus*? In his account of the formula, we read the following:

> so that they are compelled to posit a certain negotiation (*negotiationem*), as they themselves call it, after the first process, which, pondering it myself, I have never understood what it means. For it will either be a kind of *a priori* process or an *a posteriori* one, and whatever is introduced, we will fall into obscure ambiguities, as is evident to anyone slightly versed in logic.[13]

This rejection of step 2 of *regressus* may mislead us into thinking that Piccolomini was completely ruling out *regressus*.[14] Nevertheless, we possess formal hints and textual evidence to infer that Piccolomini is recreating the structure of *regressus* to make it immune to criticisms of obscurity or lack of rigour. Indeed, Piccolomini's move is arguably a reaction against the role

---

13 Alessandro Piccolomini, *De certitudine*, f. LXXXIII: [...] *ut coacti ponant quandam (ut ipsi dicunt) negotiationem post primum processum, quam ego ponderans nunquam intellexi quod sibi vellet. Aut enim illa erit quidam a priori processus, aut a posteriori, et quidquid detur in obscuras ambages inciderimus, ut cuilibet parum in logica versato manifestum esse potest.*

14 See Palmieri, "On *scientia* and *regressus*", p. 339: "Some, Piccolomini tells us, doubt whether regressus is at all possible in demonstration, raising the objection that, because, in any scientific process, the direction is from the unknown to the known, and because, by *demonstratio quia*, one knows the cause of the effect, it is wholly unclear for what purpose a *demonstratio propter quid* should follow the *demonstratio quia*. Others, Piccolomini continues, in order to defend regressus, admit the existence of an intermediate stage, called "negotiation", between the first demonstration and the second demonstration (i.e., for example, the Zabarellian *mentale examen*). But Piccolomini ironically notes that he has always failed to understand what this so-called negotiation is supposed to be. He claims that it is but an irrational element that vitiates *regressus*. This is a criticism that strikes at the heart of regressus theory. The intermediate stage, as Piccolomini points out, introduces an obscure, ultimately illogical element into the logical-demonstrative process ("*in obscuras ambages inciderimus*"). But Piccolomini does not reject regressus out of hand. He thinks that *regressus* is a conjunction of two types of demonstration, *quia* and *propter quid*, that, when united in a coherent demonstrative process, will give rise to *potissima* demonstration". See also J. Longeway, "The Place of *Demonstratio Potissima*", especially pp. 235–242. Other interpreters have commonly assumed that Piccolomini's *demonstratio potissima* is but a single syllogism or, at least, have not entertained or discussed its relation to *regressus*. See De Pace, *Le matematiche e il mondo*; T. Schöttler, "From Causes to Relations: The Emergence of a Non-aristotelian Concept of Geometrical Proof out of the *Quaestio De Certitudine Mathematicarum*", pp. 29–47; P. Mancosu, *Philosophy of Mathematicsand Mathematical Practice*, pp. 13–35; G. Ferrari, "Dimostrazione Matematiche e Conoscenza Scientifica in Alessandro Piccolomini"; Velilla-Jiménez, "El debate sobre la certeza de las matemáticas".

assigned to the non-demonstrative step 2 in *regressus*. He does so by using the terminology of one of his avowed sources, namely Nifo (*negotiatio*).

However, if the final step of Piccolomini's *regressus* is just a rearrangement of the premises and the conclusion already known as a premise in the first syllogism, how can the Sienese evade the objection that step 2 in *regressus* is circular? Piccolomini offers an epistemic justification for this move:

> I would say, therefore, in response to the argument that, just as the cognition of the definition and the thing defined is the same, except that the former is more concealed, the latter more open, so the cognition in the regress from the second process, although it is the same as that from the first one, is nevertheless more open and thus clearer.[15]

I interpret Piccolomini's argument as follows: in *demonstratio quia* we obtain insight regarding a fact leading to its cause, whereas in the *demonstratio propter quid* we learn the same content, but in such a way that it reflects the ontological underpinning of the explored terms, which thus expands our insight. Piccolomini compares it with the fact of knowing a cluster of properties of an entity to be defined and knowing the very definition of that entity. In formal terms, this means that we conclude that a P which appeared to be a mere accident in the first syllogism is revealed as an "accident *per se*" in the conclusion of the second syllogism. Thus, we discover that the accident under scrutiny actually follows from S's essence (see example below). In other words, what was known as a PS4 in the first syllogism is concluded as a PS2 in the second one. Therefore, Piccolomini argues, the resulting *demonstratio propter quid* is a deeper epistemic moment in which already familiar terms are known more conspicuously.

As made explicit in the previous section, Piccolomini's formula for *demonstratio potissima* must include: (1) that the middle-term is P's nominal definition, (2) it must provide the *quia* and the *propter quid* (i.e., it consists of two syllogisms), and (3) the *per se* modes must be Maior-PS1, minor-PS4, conclusion-PS2.

Requirement (3) raises a further question: can this arrangement be fulfilled by both syllogisms, the *quia* and the *propter quid*, in the *regressus* arrangement? The answer is no.

---

15 Alessandro Piccolomini, *De certitudine*, f. LXXXIII: *Dicerem igitur ad argumentum quod, quemadmodum eadem est re diffinitionis et deffiniti notitia, praeterquam quod haec magis est involuta, illa expansior, sic notitia in regressu ex secundo processu, quamvis sit eadem quae ex primo, sit expansior tamen et ita clarior.*

Let us suppose that one of the syllogisms possesses the prescribed structure. Then, it must be:

> *Maior*: P is X (PS1, the proposition providing the definition of P, where P is a predicate, and X is P's nominal definition)
> *minor*: X belongs to S (PS4, where X is P's definition and S is the subject to which X belongs as a fact)
> *conclusion*: P belongs to S (PS2, using X [= P's def.] as middle-term, P is attributed to S)

Given the former arrangement, the only way in which the same terms/propositions can be rearranged for obtaining a *regressus* is:

> *Maior*: P is X (PS1, the proposition providing the definition of P, where X is P's nominal definition)
> *minor*: P belongs to S (it must be a PS2, as being the conclusion of the previous syllogism)
> *conclusion*: X belongs to S (it must be PS4 since X [P's def.] is attributed to S through P, not through P's definition, which is required for being a PS2).

From this, it is clear that Piccolomini's pattern cannot be applied to both syllogisms in his *regressus* structure. Given the prescribed disposition in either syllogism, it is impossible to arrange the second syllogism in the same way, because (a) we must concede that the PS2 conclusion of the first syllogism is a PS4 when functioning as a premise in the second syllogism. This makes no sense if the point of *regressus demonstrativus* is to gain insight into the same propositions by a combination of demonstrations. Also, (b) even if we admit that the minor-premise is a PS4, the middle-term will necessarily be P, not P's definition, thus violating requirement (1) and generating a conclusion in PS4.

Therefore, requirements (1) and (3) cannot be simultaneously satisfied in both syllogisms. And yet, Piccolomini insists that this is the right arrangement for *demonstratio potissima*.

I find that this can be easily explained by assuming that, much like Nifo and Zabarella after him, Piccolomini held that the *demonstratio propter quid* in the *regressus* is the real *demonstratio potissima*. Since the requirements (1) and (3) are met by one of the two intertwined syllogisms, a quite plausible conjecture is that, much like other *regressus* theorists, Piccolomini held that the traits of the unqualified demonstration are those of the *demonstratio propter quid* in his *regressus*.

3.1.3 *Piccolomini's* demonstratio potissima*: an Example*

To arrange an instance of Piccolomini's *demonstratio potissima* (something that he did not) we require a P that must be defined in PS1 as the Maior-premise; then we need the same P to be attributed to an S in PS4 as the minor-premise; finally, we conclude that P's definition belongs to S through P as a PS4.

Let us suppose that we notice that the Moon (S) is sometimes eclipsed (P). After some research into the nature of eclipses, we produce a complex-truncated (nominal) definition of "eclipsability:"

> Eclipsability (P) is being susceptible to a failure of light (Term-2) due to the interposition of the Earth or of any other celestial body (Term-1).[16]

Thus, we can produce the following *demonstratio quia*

> Step 1: *processus, demonstratio quia*
> Being susceptible to a failure of light (Term-2) due to the interposition of the Earth or of any other celestial body (Term-1) is eclipsability (P) (PS1)
> To the Moon (S) belongs eclipsability (P) (PS4)
> Therefore, to the Moon (S) belongs being susceptible to a failure of light (Term-2) due to the interposition of the Earth or of any other celestial body (Term-1) (PS4)

Through this demonstration, we learn the fact that the nominal definition of eclipsability (P) belongs to the Moon (S) in PS4. Now an epistemically more insightful syllogism can be produced whereby we return from the explanation attained in the prior demonstration to the same fact that was our starting point. In other words, we are now entitled to demonstrate the minor-PS4 in the former syllogism as a PS2-conclusion. This is tantamount to showing that P belongs to S not just as a matter of fact, but by virtue of P's very essence/definition. In the *demonstratio propter quid* the Maior-PS1 will remain the same as in the former demonstration (i.e., P's definition), and the minor-premise will be the PS4-conclusion of the former syllogism. Thus, we obtain:

16 This is a nominal definition according to Aristotle's *Metaphysica*, 97b, l. 5–26. There we are told that nominal definitions include through disjunction many parallel characteristics of the definiendum. In this case it is accomplished by including the disjunction "or by any celestial body", for, let us assume, we may be unsure whether a Moon eclipse could be caused by celestial bodies other than Earth. On Aristotle's notion of nominal definition see D. Modrak, "Nominal Definition in Aristotle", in D. Charles (ed.), *Definition in Greek Philosophy* (Oxford: Oxford University Press, 2010), pp. 252–285; M. Deslauriers, *Aristotle on Definition*, chap. 2, section III.

Step 2 *regressus, demonstratio propter quid*
Being susceptible to a failure of light (Term-2) due to the interposition of the earth or of any other celestial body (Term-1) is an eclipse (P) (PS1)
To the Moon (S) belongs being susceptible to a failure of light (Term-2) due to the interposition of the Earth or of any other celestial body (Term-1) (PS4)
Therefore, to the Moon (S) belongs eclipsability (P) (now known as PS2)

Through this demonstration, we have learned in PS2 that eclipsability (P) belongs to the Moon (S) through its cause "being susceptible to a failure of light due to the interposition of the Earth or any other celestial body" using P's nominal definition as the middle-term. This is a token of Piccolomini's *demonstratio potissima* in natural philosophy.

As said, the clarification and expansion of our understanding indicated by Piccolomini (*ita clarior et expansior*) occurs when we re-learn that the minor-PS4 we departed from in the first syllogism can be put as a conclusion-PS2 in the regressive syllogism: we learn the same fact again, but now as an effect proved from its cause.

This form of *regressus* is the combination of two syllogisms in the first figure *Darii*, as postulated by Piccolomini.

Requirement (1) or using P's nominal definition as middle-term is met by the *demonstratio propter quid.*

Requirement (2) is satisfied by the conjunction of both syllogisms, giving thereby the *quia* and *propter quid* accounts.

Requirement (3) or being Maior-PS1-indemonstrable, minor-PS4-demonstrable, conclusion-PS2 is fulfilled by the *demonstratio propter quid.*

The demonstrability and posteriority of the minor-premise constitutes a critical piece of evidence in favour of the *regressus* reading. Piccolomini's goal in postulating such a thing fittingly corresponds to the formal requirements of a truncated *regressus*. If the minor-premises in either of the syllogisms were required to be indemonstrable, this sort of *regressus* would be impossible.[17]

It is noteworthy that (a) by allowing *demonstratio potissima* in *Darii*, Piccolomini resolves the main problem worsening demonstration in natural

17 Should he like to remain strictly faithful to Aristotle, Piccolomini's qualification of the epistemic status of the minor-premises in his *demonstratio potissima* would also require the retreatment of two of the Aristotelian intrinsic criteria (ii) primary ([ἐξ] πρώτων, *ex primis*), and (iii) immediate ([ἐξ] ἀμέσων, *ex immediatis*), and of all the three extrinsic criteria (iv) better known than ([ἐξ] γνωριμωτέρων, *ex notioribus*), (v) prior to ([ἐξ] προτέρων, *ex prioribus*), and (vi) explanatory of the conclusion ([ἐξ] αἰτιῶν τοῦ συμπεράσματος, *ex causis conclusionis*). There is no explicit treatment of this problem in *De certitudine*.

philosophy in the Latin tradition of commentary on *Analytica posteriora*: demonstrating about a particular (the Moon).

Secondly, (b) the fact that Piccolomini only admits that the *quia* and the *propter quid* can be provided simultaneously in the *regressus* structure suppresses the traditional view that it is the superior certainty of mathematical items that allows arranging demonstrations where the *quia* and the *propter quid* are simultaneously given with great easiness. This last point is one of the cruxes of *De certitudine*'s argumentative strategy: by making the double explanative account which defines *demonstratio potissima* dependent upon the logical arrangement and not on the ontological status of the object of study, Piccolomini accomplishes the divorce of certainty and demonstrative superiority.

In other words, the certainty with which an object's essence and properties are grasped by us does not tell anything about the scientific rank of the demonstrations concerning such an object. Conversely, a science concerning itself with epistemically opaque and uncertain objects, provided that it can produce the right logical arrangement, would be able to produce unqualified demonstrations.

3.2 Mathematical Analysis and *demonstratio potissima*

How did Piccolomini interpret Proclus' references to analysis and synthesis? And how did Proclus' account of analysis influence Piccolomini's logical views in *De certitudine*?

Piccolomini's elaborations on mathematical analysis predate the publication of Pappus' *Mathematicae collectiones*, made widely available as late as 1589 in a Latin translation by Federico Commandino (1509–1575). Piccolomini's understanding of analysis in mathematics is based on four Greek authorities: some lines from Aristotle's *Analytica posteriora*, Themistius' and Philoponus' commentaries on it, and, the most quoted source, Proclus' *In Euclidem*.[18]

---

18 Alessandro Piccolomini, *De certitudine*, f. LXXIX: *De qua resolutione loquitur Aristoteles in I Posteriorum, et praesertim Cap. 9 [12], quando dicit resolutionem faciliorem esse in mathematicis quam in dialecticis. Et ibi Themistius, cap. 26 hanc resolutionem sibique oppositam compositionem declarat, necnon Philoponus super eodem loco multis et claris verbis geometrarum huiusmodi resolutionem ad vivum resecat, de qua inferius sufficienter pertractabimus.* Piccolomini refers to *Analytica posteriora*, I, 12, 78a, l. 10: ἀντιστρέφει δὲ μᾶλλον τὰ ἐν τοῖς μαθήμασιν, ὅτι οὐδὲν συμβεβηκὸς λαμβάνουσιν (ἀλλὰ καὶ τούτῳ διαφέρουσι τῶν ἐν τοῖς διαλόγοις) ἀλλ' ὁρισμούς. Translation by Barnes in Aristotle, *Complete Works of Aristotle*, vol. 1: "In mathematics, things convert more because they assume nothing accidental – and

Secondly, it is also important to take into consideration the expressions used to translate the terms ἀνάλυσις and σύνθεσις from the Greek to the Latin in the reception of Proclus' text. To do so, I shall compare the Latin versions of the passage f. 8, 4–6 in Friedlein's edition. Table 8 in the Appendix contains the details of this search.

The comparison of the Latin translations (from Valla's 1501 partial translation to Mariner's 1638 translation) shows that ἀνάλυσις and σύνθεσις were rendered as the Latin *resolutio* and *compositio* (the only exception is Mariner, who straightforwardly introduced the Greek *analyses*). This terminological choice is consistently applied within the same sources.

This wording is not casual or surprising: an earlier Latin tradition was already established regarding the rather vague concepts of *resolutio* and *compositio*. They came to the Latin readership from (i) the translations of some passages of Aristotle in which ἀνάλυσις and σύνθεσις are used as a paradigmatic form of explanation and thought (*Ethica Nicomachea* 1112b, l. 15–27, *Analytica priora* 49a, l. 11–20, *Analytica posteriora*, I, 13, 78a, l. 22), (ii) Gerard of Cremona's Latin translation of Anaritus' *Euclidis commentarii* (mostly book II), and (iii) the Galenic logical tradition of the *Ars parva* operating from signs/symptoms to causes and vice versa.

As studied by Sweeney,[19] other Latin sources transmitted the concepts of ἀνάλυσις and σύνθεσις to the Medieval Latin *litterati*, such as (iv) Calcidius' (fl. 4th century) commentary on the *Timaeus*,[20] through which a notion of analysis per decomposition into elements reached Latin authors including those of the so-called school of Chartres and even Albert and Aquinas, and (v) some passages in the works of Eriugena, who shows a distinctively Neoplatonic understanding of this logical operation as the finding of superior genera.[21] As

---

in this too they differ from argumentations – but only definitions". For more mentions of Proclus by Piccolomini regarding this point, see Alessandro Piccolomini, *De certitudine*, f. Cv. See also section 3.2.3 below.

19 E. Sweeney, "Three Notions of Analysis (*Resolutio*) and the Structure of Reasoning in Aquinas", in *The Thomist* 58/2 (1994), pp. 197–243.

20 Calcidius, *Timaeus a Calcidio translatus commentarioque instructus*, eds W. J. Verdenius, J. H. Waszink, vol. 4 (London Leiden: Warburg Institute, Brill, 1962), sec. 303, p. 305: *Si ergo has qualitates et quantitates, etiam formas figurasque volemus ratione animi separare, tum demum deliberare, quid sit illud, quod haec omnia inseparabiliter adhaerens complexumque contineat, inveniemus nihil aliud esse quam id quod quaerimus, silvam. Inventa igitur est origo silvestris. Et hoc quidem est unum duarum probationum genus, quod resolutio dicitur*. Quoted by Sweeney, "Three Notions of Analysis", pp. 206–207.

21 John Scot Eriugena, *Expositiones super hierarchiam caelestem S. Dionysii*, ed. J. Barbet (Turnhout: Brepols, 1975), VII, l. 578–587: *Duae quippe partes sunt dialecticae disciplinae, quarum una διαιρετική, altera ἀναλυτική nuncupatur. Et διαιρετική quidem divisionis*

Sweeney establishes from these witnesses, the 13th-century Latin West inherited a three-fold sense of resolution, present in authors such as Albert the Great and Thomas Aquinas: a mathematical resolution from the mathematical tradition (mostly Anaritus), the Calcidian notion of decomposition into elements, and a (Neo-)Platonically inspired notion of resolution into superior genera.

Coming back to the 16th-century reception of Proclus' *In Euclidem*, the evidence indicates that scholars from northern Italy were likely to assimilate Proclus' mentions of ἀνάλυσις and σύνθεσις as parts of the already deeply elaborated theories of *resolutio* and *compositio* as a syllogistic arrangement in the form of a *regressus*.

A token of this interpretation of mathematical analysis is provided by Niccolò Leoniceno (1428–1524) in his *De tribus doctrinis ordinatis secundum Galeni sententiam liber* (1508).[22] In this work, Leoniceno indexes the various existing uses of the term *resolutio* to ultimately elucidate what its right meaning may be and how it relates to demonstration.[23] Amongst the many cases of resolution he provides there is the *resolutio geometrica*. The definition provided is as follows:

> Resolution is the discovery of a demonstration. Those who say that they are resolving start from the conclusion. If they find that they have suitably

vim possidet; dividit namque maximorum generum unitatem a summo usque deorsum, donec ad individuas species perveniat, inque eis divisionis terminum ponat; ἀναλυτική vero ex adverso sibi positae partis divisiones ab individuis sursum versus incipiens, perque eosdem gradus quibus illa descendit, ascendens cumvolvit et colligit, eosdemque in unitatem maximorum generum reducit, ideoque reductive dicitur et reditiva. Quoted by E. Sweeney, "Three Notions of Analysis (*Resolutio*) and the Structure of Reasoning in Aquinas", pp. 217–218.

22 For some details about Leoniceno and the medical school of Ferrara see L. Premuda, "Un discepolo di Leoniceno tra filologia ed empirismo: G. Manardo e il «libero esame» dei classici della medicina in funzione di più spregiudicati orientamenti metodologici", in *Atti del Convegno internazionale per la celebrazione del v Centenario della nascita di Giovanni Manardo, 1462–1536. Ferrara, 8–9 dicembre 1962* (Ferrara: Università degli studi di Ferrara, 1963), pp. 43–56; E. C. Streeter, "Leoniceno and the School of Ferrara", in *Bulletin of the Society of Medical History of Chicago* 1 (1916), pp. 18–22. See also A. Touwaide, "Galen and Galenism", in *Encyclopedia of Renaissance Philosophy*.

23 Niccolò Leoniceno, *De tribus doctrinis ordinatis secundum Galeni sententiam liber*, ed. A. Leenius, *Opuscula* (Basel: Andrea Cratandrus et Ioannes Bebelius, 1532), ff. 62r–83r, f. 63v: *Quid igitur de resolutorii vocabulo dicemus? Videtur quidem impositum ad significandum progressum ab his quae ex principiis ortum habent ad ipsa principia, quod et ipsum resolutionis nomen ostendit. Est enim resolutio effectuum posteriorum in priora ac superiora ex quibus constant, videturque per translationem sumptum ab his qui post peregrinationem ad proprios lares revertuntur.*

resolved it, later, by putting together what they found through resolution, they make a demonstration.[24]

Leoniceno describes the mathematical procedure of analysing a given proposition into its presuppositions for discovering a demonstration of it. He attributes this view to Geminus. His likely source is either Hero of Alexandria's quotation in the second book of An-Nayrīzī's commentary on Euclid, Proclus' *In Euclidem*, or both.[25] That Leoniceno made use of Proclus' commentary is confirmed by the subsequent passages, where Proclus' first prologue is invoked in establishing the relevance of *resolutio* and *compositio* amongst the logical procedures of mathematical disciplines.[26]

Another of the cases of resolution and composition proposed by Leoniceno, this time in natural philosophy, features one of the examples of *Analytica posteriora*: it is the traditional example of the lunar eclipse.[27] In the manner of the *regressus* theorists, Leoniceno adopts now the terms *resolutio* and

24 Niccolò Leoniceno, *De tribus doctrinis*, f. 62v: *Resolutio est inventio demonstrationis: a conclusione enim incipiunt qui resolvere dicunt. Quod si invenerint se opportune resolvisse, postmodum componentes quae per resolutionem invenerint, faciunt demonstrationem.*

25 See An-Nayrīzī, *The Commentary of al-Nayrizi on Book II–IV of Euclid's Elements of Geometry*, ed. A. Lo Bello (Leiden Boston: Brill, 2009), p. 22.

26 Niccolò Leoniceno, *De tribus doctrinis*, f. 66: *Nec minus Proclus, philosophus eximius, qui praecipue Platonis dogmata excoluit in* ***I libri Elementorum Euclidis*** *expositionis praefatione ita de quattuor dialectices facultatibus scribit. Sed dialectica philosophiae purissima pars super disciplinas proxime extenditur, omnemque illarum continet involutionem atque earundem doctrinis varias ex ipsa tribuit facultates, quae perfectivae sunt, indicativae atque intellectuales. Resolutivam dico divisivam, definitivam, ac demonstrativam quibus adiunctae atque perfectae hae disciplinae quae a Graecis mathematicae vocantur, nonnulla quidem per resolutionem inveniunt, nonnulla per compositionem, et alia quidem per definitionem, alia per modum divisionis ponunt, nonnulla autem quaesita per desmonstrationes absolvunt, hasce facultates suis adoptantes subiectis, singulisque ad mediorum utentes contemplationem.* See also Niccolò Leoniceno, *De tribus doctrinis*, f. 67: *Atqui Proclus, et Euclidis expositor et omnium Graecorum iudicio mathematicus excellentissimus, testatur mathematicos non minus resolutiva quam reliquis aliis tribus dialectices facultatibus uti, quemadmodum eiusdem auctoris verba, quae supra de quattuor doctrinis recitavimus, apertissime ostendunt; ac quemadmodum etiam sine aliquo Procli testimonio, mathematicorum libri pleni resolutionum ac compositionum, ut diximus, manifestant.*

27 Niccolò Leoniceno, *De tribus doctrinis*, f. 64: *Quod enim Luna interiectu terrae deficiat, ex posterioribus ad priora ascendentes agnoscimus, ac postmodum inventa causa demonstrationem componimus. Nam Lunam quidem deficere sensu comprehendimus; huius deinde causam intellectus invenies, de eadem scientiam scribit hoc pacto: Luna quidem deficit, quod deficit aliquod habet quod obicitur. Hic ab effectibus ad causas progressus, resolutio appellatur, post* **quem sequitur demonstratio** *descendens a causis ad effectus hoc modo: Lunae aliquid obiicitur, cuicumque aliquid obiicitur deficit.*

compositio to express the two reciprocal effect-to-cause and cause-to-effect accounts explaining the eclipse.

Lastly, after examining all the opinions of the ancients, Leoniceno proposes his own judgement on the confusing question:

> Although indeed some of what has been explained above from the viewpoint of ancient philosophers seems to support the idea of the mode of resolution in which we proceed from compounds to simpler things, or from posterior things to prior ones, or from effects to causes, it seems to confirm the opinion of those who say that the resolutive doctrine is demonstration, which is now called by more recent philosophers "demonstration of the fact" (*quia*). However, if someone carefully and subtly examines not only the writings but also the inner thoughts of these same ancient philosophers, they will find that almost all of them suggest that there is no true demonstration unless we proceed from the prior things which have been discovered through resolution to the posterior ones by composing them so that the saying of Geminus is true: resolution is nothing other than the discovery of demonstration.[28]

Firstly, Leoniceno singles out the view that he would oppose: that *resolutio* is a form of demonstration. He then corrects this opinion by establishing the difference between heuristic and demonstrative procedures: he tells us that resolution is the methodology through which the contents of demonstrations are found, just as it is done in mathematics. Once the researcher has resolved the proposed object of study into its causes, the process can be formalised by composition in the form of a syllogism: this is the real demonstration. Leoniceno's opinion stands in some opposition to the Galenic tradition: whereas the *via resolutiva* or *demonstratio quia* are not real demonstrations, the *via compositiva* is proposed as the genuine demonstration. In order words, Leoniceno assimilates the so-called "demonstration of the fact" to analytic heuristics and

28 Niccolò Leoniceno, *De tribus doctrinis*, f. 66v: *Etsi vero nonnulla quae supra ex veterum philosophorum sententia sunt a nobis exposita de eo resolutionis modo, quo a compositis ad simpliciora, aut a posterioribus ad priora, sive ab effectibus ad causas accedimus, probare videantur illorum opinionem qui aiunt doctrinam resolutivam esse demonstrationem, quae a recentioribus philosophis "demonstratio quia" nuncupatur, si quis tamen eorundem veterum philosophorum non modo scripta, sed etiam intimos sensus diligenter subtiliterque scrutetur, inveniet id omnes fere innuere* ***non esse veram demonstrationem****, nisi cum a prioribus quae per resolutionem inventa sunt, ad posteriora componendo procedimus, adeo ut vera comprobetur Gemini sententia,* ***resolutionem nihil aliud esse, nisi demonstrationis inventionem****.*

deprives it of any demonstrative value: we do not discover through demonstration but demonstrate from discovered things.

Nevertheless, the opinion that both Galenic *viae* were wholly equivalent to the Aristotelian demonstrative procedures *quia* and *propter quid* was destined to endure. Leoniceno's work notwithstanding, various approaches to interpreting the resolutive structure as the Aristotelian "demonstration of the fact" would remain. For instance, one generation later, Zabarella would still adopt the syntagm *methodus resolutiva* as a synonym to *demonstratio quia* and consider it an apodictic step.[29]

All this reveals a highly problematic state of the question regarding analysis in the mid-16th century. Divergent puzzles existed: how analysis precisely works; whether it is the same logical procedure as demonstration, or a part of it, or perhaps a research strategy; or whether it is practised in the same manner in every discipline, all these were common questions.

Given this convoluted state of the art, mathematical analysis, as conveyed by Proclus, could be read in a variety of ways, including the Galenic fashion. This was a danger that Piccolomini had to address.

3.2.1 *Piccolomini's Five-Fold Classification of* resolutio

In this section, I will present evidence that Piccolomini sustained the anti-Galenic opinions that (i) *resolutio* and *demonstratio quia* are distinct logical procedures and that (ii) resolution has no demonstrative or heuristic value. Following this, I will explain why establishing this distinction is a crucial argumentative step in *De certitudine*.

The theme of *resolutio-compositio* is addressed at different points throughout *De certitudine*.[30] The first time such concepts are mentioned is in Piccolomini's explanation of his intentions in the presentation of his opusculum. There he refers to some of the items he should explore, amongst which he counts resolution and composition.[31] Already in this first mention Piccolomini specifically

---

29 See Jacob Zabarella, *De methodis*, lib. III, q. 18, in *Opera logica* (Venice: Paulus Meietus, 1578), pp. 178–179. See N. W. Gilbert, *Renaissance Concepts of Method*, p. 171.

30 Piccolomini considered this topic of such importance as to include it in the complete title of his opusculum, namely *Commentarium de certitudine mathematicarum disciplinarum in quo de* ***resolutione****, diffinitione et demonstratione necnon de materia et fine logicae facultatis quam plura continentur ad rem ipsam, tum mathematicam, tum logicam maxime pertinentia.*

31 Alessandro Piccolomini, *De certitudine*, f. LXXIII: *Haec igitur, quae dixi, est nostra intentio in hoc tractatu. Sed ut commodius haec fieri possint, brevissime primum explicandum erit, quae vis, quod opus, quae materia sit mathematicarum scientiarum, quae problemata, quae theoremata, et horum partes, necnon quid in illis resolvere, quid componere, et alia id genus.*

refers to *resolutio-compositio* in relation to mathematics. Nevertheless, it would soon be revealed that resolutive procedures do not only belong to mathematics, but they are generally related to logic.[32]

Piccolomini establishes that, just as the logical division is the tool through which definitions are obtained, so resolutive procedures are somehow related, and subservient, to demonstration. Therefore, from the outset, the Sienese establishes that resolution is not to be identified with any kind of demonstration. But what is the precise link between *resolutio* and *demonstratio*? This is tackled in chapter III of *De certitudine*. Here, Piccolomini reveals the following:

> For no one will be able to form a judgment about anything, either discussing or demonstrating, unless he has first discovered the middle-terms of the arguments by which the judgment is sustained. But, unless he does not ignore to judge the discovered arguments with the help of resolution, he will easily fall into the snares of deception.[33]

Following Cicero, Piccolomini proposes that logic contains two moments other than demonstration itself: (i) *logica inveniendi* and (ii) *logica disserendi* or *iudicandi*.[34] *Logica inveniendi* is the procedure through which middle-terms (i.e.,

---

32 Alessandro Piccolomini, *De certitudine*, f. LXXIIIv: *Quod enim resolutio et compositio demonstrationis gratia sint, divisio vero simul cum illis ad diffinitionem dirigatur, non est quis neget.*

33 Alessandro Piccolomini, *De certitudine*, f. LXXIX–LXXIXv: *Nullus enim de re quapiam poterit vel disserendo vel demonstrando iudicium ferre, nisi inventis prius argumentorum mediis quibus iudicatio fulciatur. Rursus nisi resolutione auxiliante iudicare argumenta iam inventa non ignoraverit in deceptionis laqueos facillime paratus erit.*

34 Alessandro Piccolomini, *De certitudine*, f. LXXIX. Piccolomini attributes the distinction between *logica inveniendi* and *disserendi* to Cicero, *De finibus bonorum et malorum*, ed. H. Hackham (Cambridge MA: Harvard University Press, 2015), IV. Most likely, Piccolomini refers to IV, 10: *Cumque duae sint artes, quibus perfecte ratio et oratio compleatur, una inveniendi, altera disserendi, hanc posteriorem et Stoici et Peripatetici, priorem autem illi egregie tradiderunt, hi omnino ne attigerunt quidem. Nam e quibus locis quasi thesauris argumenta depromerentur, vestri ne suspicati quidem sunt, superiores autem artificio et via tradiderunt. Quae quidem res efficit, ne necesse sit isdem de rebus semper quasi dictata decantare neque a commentariolis suis discedere. Nam qui sciet ubi quidque positum sit quaque eo veniat, is, etiamsi quid obrutum erit, poterit eruere semperque esse in disputando suus. Quod etsi ingeniis magnis praediti quidam dicendi copiam sine ratione consequuntur, ars tamen est dux certior quam natura. Aliud est enim poetarum more verba fundere, aliud ea, quae dicas, ratione et arte distinguere.* Other sources invoked by Piccolomini for his distinction are Alexander of Aphrodisias' commentary on *De sophistis elenchis*, Boethius (no work is referred), and Rodolphus Agricola's (1443–1485) *De inventione*. See Rudolf Agricola, *De inventione dialectica libri omnes et integri et recogniti* (Amsterdam: Alardus, 1539), chap 2, pp. 6–9, where *ars disserandi* or the way to confirm that opinions and arguments are true

the causes or explanations) required to arrange a demonstration are obtained; *logica iudicandi* or *resolutio* is the process whereby the truth and valid logical connexions in the obtained demonstration are checked, lest deductive mistakes, imprecisions, or other fallacies deceive the inquirer. Therefore, Piccolomini holds that analysis is a logical operation to be applied as a checking-tool for the validity of the argument once the middle-term has been found. In other words, against the Galenic view, resolution is portrayed as part of the justificatory moment, not the heuristic one, let alone as equivalent to demonstration.

Piccolomini proceeds then to distinguish five different types of analysis. The first two types of resolution are introduced as follows:

> It must be said that since resolution stands oppositely to composition, logical resolution itself (setting aside all natural resolutions from which logical resolution is metaphorically called) is indeed manifold. Those who break down compound syllogisms into simpler ones and those who resolve simple [syllogisms] into their propositions, from which they are composed, are said to resolve logically. Also, those who make imperfect ones into perfect ones are said to resolve. Furthermore, those who arrange and reduce each proposed syllogism into its proper figures [...].[35]

Piccolomini firstly remarks that for every kind of resolution he is about to address, there exists an opposed operation of composition, completing it and sharing the same traits. Hence, for every resolution-type-x Piccolomini should be dealing with there will be an analogue composition-type-x completing it.

The first type of resolutive process presented is the so-called *resolutio naturalis*. Piccolomini gives no details of what this sort of resolution might consist of. He declares that this is the operation in analogy to which other resolutive procedures receive their name. Details on the traits of *resolutio naturalis* shall be explored below (3.2.2).

---

is expounded on. The distinction between the inventive and judicative aspects of logic was maintained in Piccolomini's later vernacular writings and was also embraced by figures close to him, such as the *infiammato* Varchi. See M. Sgarbi, *The Italian Mind*, p. 105 and p. 194.

35 Alessandro Piccolomini, *De certitudine*, f. LXXVIIIv: [...] *dicendum erit quod, cum resolutio opposito modo se habeat ad compositionem, resolutio ipsa logica (omissis omnibus resolutionibus naturalibus a quibus logica resolutio metaphorice forte dicitur) multiplex quidem est. Resolvere enim logice dicuntur ii, qui syllogismos compositos in simpliciores redigunt, et qui simplices in suas propositiones, ex quibus confecti sunt; necnon qui imperfectos revocant ad perfectos. Quin etiam et ii qui propositos quosque syllogismos in propias eorum figuras disponunt ac reducunt* [...].

The second sort of resolution is the *resolutio logica*, which is part of the *logica iudicandi* proper. The *resolutio logica* would be divided into four species, amounting to a total of five different ways in which resolution is spoken of in Piccolomini's account. Since the Sienese does not provide proper names for the species of logical resolution, I will coin a label for each of them for the sake of clarity.

The first species of *resolutio logica* is what I shall call *resolutio formalis*. This sort of resolution relates to the formal structure of a syllogistic argument: it consists of reducing complex syllogistic structures (such as *regressus*) into their parts, reducing a syllogism into its propositions, or rearranging a syllogism so that it is rendered into an epistemically superior, more sound form.[36]

Piccolomini then introduces the second species of logical resolution, which I label *resolutio ad universalius*. He does so as follows:

> There are also other logical resolutions, as Eustratius says, such as when from less universal to more universal, as if ascending from more compound to simpler, we aim towards the first genera.[37]

Although, as mentioned above, Sweeney has shown that this type of procedure was spread into the Latin West through the works of Eriugena,[38] Piccolomini's source here will be revealed to be Eustratius' treatment of ἀνάλυσις in his commentary on book II of *Analytica posteriora*. This species of resolution stands as the converse of the logical operation of division. Through it, a given genus (or relative species) is related to a more universal genus and then to the first genera. It is crucial to remark that, whereas in other (Neo-)Platonically inspired intellectual frameworks the prior genera in which inferior genera are resolved stand for extra-mental metaphysical realities, Piccolomini's *resolutio ad universalius* must be understood as a purely conceptual operation in which superior genera, *qua* genera, have no extra-mental counterpart of any sort.

The third species of logical resolution is what I have called *resolutio quia*. It is presented in this way:

---

36 Aristotle, *Analytica priora*, 24b, l. 23–25. See also H. Weidemann, "Aristotle on the Reducibility of All Valid Syllogistic Moods to the Two Universal Moods of the First Figure (Apr A7, 29b1–25)", in *History and Philosophy of Logic* 25 (2004), pp. 73–78.

37 Alessandro Piccolomini, *De certitudine*, f. LXXVIIIv: *Aliae etiam sunt resolutiones logicae, ut dicit Eustratius, veluti si a minus universalibus ad universaliora, quasi a compositioribus ad simpliciora ascendentes, ad prima genera tendimus.*

38 E. Sweeney, "Three Notions of Analysis (*Resolutio*) and the Structure of Reasoning in Aquinas".

> Moreover, those also resolve who proceed by reasoning from effects to causes, as attested by Eustratius. Aristotle deals with this resolution, which is called by some people "demonstration of the fact" in *Analytica posteriora*, [II], 13.[39]

Invoking again Eustratius' authority, Piccolomini describes a resolutive process which retraces effects to the causes in a syllogism to check the logical validity of the causal links established by middle terms in an argument. Piccolomini says that it is also called *demonstratio quia* by some people (*a quibusdam*), arguably meaning theorists of Galenic persuasion such as Jacob of Forlì, Hugh Benzi, or Nifo. He refers to the exposition of this species of resolution to *Analytica posteriora* II, 13 where the ὅτι account is introduced.[40]

Finally, the fourth kind of logical resolution in Piccolomini's classification is what he, importantly, calls *resolutio mathematica*. It is described as follows:

> Finally, there is another logical resolution, which, because it is of great use in mathematics, is also called mathematical. With this resolution, we break down the ultimate conclusion into its premises and the premises of the premises until we arrive at some granted first principles.[41]

The species of logical resolution called "mathematical resolution" consists of reducing a proven mathematical proposition (either demonstration or construction) to the premises required by its proof. The obtained premises may, on their behalves, be either axioms or other propositions. In the latter case, those propositions are also reduced to their presuppositions, until the initial proposition is retraced back through the demonstrative chain to the set of axioms it ultimately depends upon. As we will see below (section 3.2.3) the process of

---

39 Alessandro Piccolomini, *De certitudine*, f. LXXVIIIv: *Praeterea illi resolvunt etiam, teste eodem Eustratio, qui ex effectibus ad causas ratiocinando procedunt, de qua resolutione, quae a quibusdam demonstratio quia nuncupatur, pertractat Aristoteles Posteriorum analyticorum capitulo* X<*III*>. Piccolomini wrongly wrote "chapter X"; for chapter X examines the objects/terms of demonstrative science, not logical procedures of any kind. Piccolomini refers to chapter XIII in book I of *Analytica posteriora* as his subsequent indication reveals: it is the locus where Aristotle treats the demonstration of the fact or effect (*quia*).

40 *Analytica posteriora*, II, 13, 78a, l. 22-b, l. 3.

41 Alessandro Piccolomini, *De certitudine*, f. LXXIX: *Alia tandem est logica resolutio quae, quia mathematicis magno usui est, mathematica etiam dicitur, qua conclusionem ultimam in suas praemissas et praemissarum praemissas eo usque resolvimus, donec ad quaedam prima concessa deventum sit.*

mathematical resolution, understood as a logical validity checking-tool, will require formalising a mathematical proof as syllogisms to ensure its logical soundness.

What is the difference between this kind of resolution and resolution *quia*? Arguably, Piccolomini holds that, whereas the causes formalised in the analytic checking of a natural proposition and its syllogism *quia* describe real causal chains, mathematical resolution is a mere logical exercise for checking the formal validity of arguments relating signs instead of causes.[42]

Piccolomini's classification of resolution can synthesised in the following table:

TABLE 4 Piccolomini's account of *resolutio*

Type	Divisions	Description
Resolutio naturalis		The kind of *resolutio* from which the other *resolutiones* received their name by homonymy.
Resolutio logica	*Resolutio formalis*	Reducing a syllogism into a simpler form or dividing a syllogism into its simpler propositions or rearranging a syllogism so that it is rendered epistemically better.
	Resolutio ad universalius	Finding the more universal genus to which a given genus belongs.
	Resolutio quia	Checking that the logic and causal links from effect to causes in a *demonstratio quia* are correct.
	Resolutio mathematica	Reducing the premises of a mathematical proposition into other equivalent or presupposed propositions and repeatedly looking for the premises of the premises until the first indemonstrable principles are reached.

3.2.2 *Eustratius' Four-Fold Analysis and Piccolomini's Downgrading of* resolutio

The main source for Piccolomini's classification of analysis is the (twice called upon) commentary on *Analytica posteriora* of Eustratius of Nicaea (*c.*1055–*c.*1120), a work which is not mentioned in Leoniceno's examination of

42 See section 3.3.3 for an account of Piccolomini's theory of the causal idleness of mathematics.

resolution.[43] In his commentary, Eustratius provides an exhaustive list of all the types of analysis:

> As definition, demonstration, and division are said in many ways, so also analysis. (1) There is the analysis which is the way from the composites to the simple things out of which they are composed from ordered principles, the so-called elements. (2) There is the analysis of syllogisms into figures as principles of some kind or into technical and scientific arrangements, and again, into the propositions or the definitions of which [syllogisms] are composed. (3) There is an analysis which is like an ascent from the more particular things to the more universal, such as from particulars to most specific [species], and from these to the genera above them. Which [analysis] Aristotle also uses in the preceding book and, through it, he proceeds to find the definitions, so that the common reason found in each thing falling under the same species is the definition of the species, and the common reason found in the species under the same genus stands as the definition of the genus. Which [analysis] [Aristotle] complemented with division, so that it may easily and more firmly find definitions. (4) There is also the analysis of some known thing towards its principles and causes, from which it derives its being and its knowability. According to this mode, it is also called "demonstrative analysis". Because principles are found by us mostly through this sort of analysis, going from the things which are prior to us towards those which are prior according to nature, i.e., from the caused towards the causes, so, it is called "analytic" and the syllogism according to it "analytic syllogism".[44]

---

43 Eustratius' commentary is also extensively referenced in Chapter 1 of *De Certitudine*, where Piccolomini discusses the goal of logic.

44 Eustratius, *In Analyticorum posteriorum librum secundum commentarium*, ed. M. Hayduck (Berlin: Academia Litterarum Regiae Borrusicae, 1907), p. 3, l. 4–18. In the original Greek: ὥσπερ δὲ ὅ τε ὁρισμὸς ἡ ἀπόδειξις ἥ τε διαίρεσις πολλαχῶς λέγεται, οὕτω δὴ καὶ ἡ ἀνάλυσις. (1) ἔστι μὲν γὰρ ἀνάλυσις ἡ ἀπὸ τῶν συνθέτων ὁδὸς εἰς τὰ ἁπλᾶ, ἐξ ὧν συνετέθησαν ὡς ἐκ συντεταγμένων ἀρχῶν, ταὐτὸν δὲ εἰπεῖν στοιχείων. (2) ἔστι δὲ τῶν συλλογισμῶν ἀνάλυσις εἰς τὰ σχήματα ὡς εἰς ἀρχάς τινας ἤτοι κανόνας τεχνικοὺς καὶ ἐπιστημονικούς. Καὶ ἑτέρα αὖθις ὡς στοιχεῖα τὰς προτάσεις ἢ τοὺς ὅρους, ἐξ ὧν πρώτων συντέθειντααι. (3) ἔστιν ἀνάλυσις καὶ ἡ ἀπὸ τῶν μερικωτέρων ἄνοδος ἐπὶ τὰ καθολικώτερα, οἷον ἀπὸ τῶν καθ'ἕκαστα ἐπὶ τὰ εἰδικώτατα καὶ ἀπὸ τούτων ἐπὶ τὰ ὑπὲρ αὐτὰ γένη, ᾗ καὶ χρῆται ὁ Ἀριστοτέλης ἐν τῷ προκειμένῳ βιβλίῳ, δι'αὐτῆς εὑρίσκειν τοὺς ὁρισμοὺς προτρεπόμενος, ὡς τὸν κοινῶς εὑρισκόμενον λόγον τοῖς ὁμοειδέσι τοῖς καθ'ἕκαστα ὁρισμὸν εἶναι τοῦ εἴδους, καὶ τὸν κοινῶς εὑρισκόμενον λόγον τοῖς ὁμογένεσιν εἴδεσιν ὁρισμὸν ὑπάρχειν τοῦ γένους. ἣν καὶ τῆς διαιρετικῆς ὑπερτίθησιν ὡς ῥᾷον ἢ ἐκείνη καὶ ἀσφαλέστερον τοὺς ὁρισμοὺς θηρᾶν δυναμένην. (4) ἔστι καὶ γνωστοῦ τινος ἀνάλυσις εἰς τὰς ἀρχὰς καὶ τὰ αἴτια, ὅθεν τὸ εἶναι ἔχει καὶ τὸ γινώσκεσθαι, καθ'ὃν δὴ τρόπον καὶ ἡ ἀποδεικτικὴ ἀναλυτικὴ

Eustratius begins by declaring, in an Aristotelian manner, that analysis is said in many ways (πολλαχῶς λέγεται). Then, he proceeds to offer a four-fold classification of the meanings of analysis.

Firstly, (1) we are told that analysis is the way or the manner of reducing a composite into its simple parts (ἡ ἀπὸ τῶν συνθέτων ὁδὸς εἰς τὰ ἁπλᾶ). The simplest parts forming a composite are called principles or elements (ἀρχῶν, στοιχείων). Given that the other three members of Eustratius' classification match Piccolomini's list in their order and specific traits, it is sensible to assume that this rather loose description of analysis was taken by Piccolomini as *resolutio naturalis*. From the semantics of the terms adopted (simple parts, elements) it can be easily concluded that Eustratius is here referring to the resolution or decomposition of a natural body into its fundamental elements. Hence, taking this from Eustratius, Piccolomini affirms that (a) this is the most common meaning of *resolutio* in resemblance to which the logical procedures receive their name, and (b) it is placed under the "natural" label, for the items dissolved or analysed are supposed to be natural bodies instead of logical statements.

Secondly, (2) Eustratius addresses the analysis of syllogisms (ἀνάλυσις τῶν συλλογισμῶν). His description was clearly paraphrased in Latin by Piccolomini: this sort of analysis consists of reducing a syllogism into its principles (i.e., its propositional parts) or reshaping a syllogism so that it is rendered into an epistemically superior (more scientific) arrangement.

Thirdly, (3) Eustratius elucidates the use of analysis into the more universal genera (ἀνάλυσις ἐπὶ τὰ καθολικώτερα). Yet again, Piccolomini's account is a shortened paraphrase of Eustratius' exposition: this sort of analysis consists in finding the more universal genera to which a given species belongs. Nevertheless, Eustratius stresses an aspect that Piccolomini omitted, namely that this model of analysis also includes finding the common species for different particulars and, thereby, it is related to the logical operation of definition and division.

Lastly, (4) Eustratius dwells on how analysis reduces a proposition into principles (ἀνάλυσις εἰς τὰς ἀρχὰς), also called "analytic syllogism" (συλλογισμὸς ἀναλυτικός). This is the procedure of going from that which is first to us, namely the effects, to that which is first by nature or the causes. Eustratius equates the analytic procedure of moving from the caused into its prior cause with the

---

ὀνομάζεται. ἐπεὶ γὰρ ὡς ἐπὶ πᾶν ἐξ ἀναλύσεως ἡμῖν αἱ ἀρχαὶ ταύτης εὑρίσκονται ἀπὸ τῶν ἡμῖν προτέρων ἀνιοῦσιν ἐπὶ τὰ τῇ φύσει πρότερα, τουτέστι τῶν αἰτιατῶν ἐπὶ τὰ αἴτια, διὰ τοῦτο αὐτή τε ἀναλυτικὴ ὀνομάζεται καὶ ὁ κατ' αὐτὴν συλλογισμὸς ἀναλυτικός.

syllogism ὅτι, or analytic syllogism. The analytic nature of the ὅτι account is stressed in later sections of Eustratius' text, where analysis is depicted as the logical process preceding the arrangement of a demonstration, i.e., of a διότι syllogism or synthetic demonstration.[45]

This evinces that almost the whole section on *resolutio* in chapter III of *De certitudine* is a paraphrase of Eustratius' text. The parallelisms and discrepancies between Eustratius' and Piccolomini's views are offered in the following synoptic table:

TABLE 5 Eustratius and Piccolomini on analysis

Eustratius' account of ἀνάλυσις		**Equivalent in Piccolomini's list**
Type	**Description**	
ἀνάλυσις εἰς στοιχεῖα	Dividing a composite entity into its simpler parts or elements.	*Resolutio naturalis*
ἀνάλυσις τῶν συλλογισμῶν	Reducing a syllogism into its principles (i.e., its propositional parts) or rearranging a syllogism so that it is rendered into an epistemically superior form.	*Resolutio logica*
ἀνάλυσις ἐπὶ τὰ καθολικώτερα	Finding the more universal species or genus to which a given genus or collection of items belongs.	*Resolutio ad universalius*
ἀνάλυσις εἰς τὰς ἀρχὰς or συλλογισμὸς ἀναλυτικός	Finding the causes (that what is first according to nature) by scrutinising the caused things (the effects; that which is first to us).	*Resolutio quia*
No equivalent in Eustratius' list		*Resolutio mathematica*

As is clear, there is no place for a specific mathematical analysis in Eustratius' list. This is an addition made by Piccolomini. Moreover, Eustratius adopts mathematical examples as instances of genuine science: ex. gr., following the Aristotelian tradition, he considers the demonstration of Euclid's I.32 to be a

45 Eustratius, *In analyticorum posteriorum*, p. 4, l. 2 and further.

paradigm of scientific explanation.[46] Thus, in Eustratius' account, mathematical proof would be subsumed as a case of analytic syllogism. In other words, to the question "Where does mathematical demonstration belong in this list of analytic processes?" Eustratius would have answered "It is a case of ἀνάλυσις εἰς τὰς ἀρχάς, going from that which is first to us (the given mathematical proposition) to that which is first according to nature (the axioms)". Thus, in Eustratius' account, the *demonstratio quia* and *resolutio* (including *resolutio mathematica*) are held to be the same procedure.

Much like the Galenic interpretation, this position is deleterious to Piccolomini's epistemological purposes. The reason is the following:

If resolution and composition are exact equivalents to *demonstratio quia* and *propter quid*, the inevitable consequence would be that, since, as established by Proclus, ἀνάλυσις-σύνθεσις is the most distinctive mathematical procedure, then intertwinements of syllogisms in the form of *regressus* could be obtained in mathematics. In other words, Proclus' mentions of analysis, if interpreted as the followers of Galen and Eustratius do, could be legitimately read as describing a *demonstratio potissima* as it was held by the *regressus* theorists.

The fact that Piccolomini denies that analysis is a demonstrative or even heuristic process and that, in addition to this, he introduces a new kind of resolution specifically for mathematics is a telling one. By declaring that resolution is a part of the justificatory step in science (*logica iudicandi*) and setting mathematical analysis apart from analysis, Piccolomini deliberately opens a methodological chasm between mathematics and natural philosophy. He also shows that he understands analysis *only* as a syllogistic formalisation of an already given mathematical proof, hence greatly downplaying the heuristic aspects attributed to analysis by the Galenic tradition and Proclus himself. This chasm is arguably intended to avert any reading that would assimilate the syllogistic arrangements produced in the use of mathematical analysis (see next section) to *regressus*: analytical structures resembling a *regressus* in mathematics ultimately belong to a different order of things. They are merely the outcome of the validity-checking method of analysis when applied to mathematics and do not meet the stringent criteria of a *demonstratio potissima* (see next section).

In this way, amongst the themes from Proclus' *In Euclidem* which made their way into *De certitudine*, mathematical analysis is perhaps the one which underwent the most severe adjustments. From being cherished by the Diadochus as the method which better resembles the procession of being and as an egregious heuristic tool in the fields of mathematics, physics, and metaphysics, analysis became in Piccolomini's hands a sort of syllogistic safety-check to apply once the science is done.

---

46 Eustratius, *In analyticorum posteriorum*, p. 24, l. 26–28.

3.2.3 Resolutio mathematica: *Euclid's I.1 as Case-Study*

What are the specific formal features of Piccolomini's *resolutio mathematica* and what sets it formally apart from *demonstratio potissima*? After its introduction in chapter III of *De certitudine*, Piccolomini provides a token of a resolutive process in mathematics later, in chapter X.

The chapter opens by introducing Proclus' *In Euclidem* twice as the main authority regarding mathematical resolution:

> Euclid frequently proceeds in this manner. I say "frequently" because sometimes he uses deduction to the impossible, employing the second mode of hypothetical syllogisms, as Proclus and Philoponus say. These deductions can also be resolved, as is evident in logic, and when I was expounding Euclid, I also used to resolve them. However, when we want to judge this kind of compositive process (*processu compositivo*), then, coming back from the final conclusion through the premises and the premises of the premises, finally ascending to the first principles, we are said to resolve. By this resolution, we ascertain whether the conclusion was composed through true and proper premises. Proclus mentions this in several places, in books I and II of [his commentary on] the *Elementa*.[47]

Firstly, we are told that analysis is the most frequent method in Euclid's *Elementa*, except for the theorems proved *ad impossibile*, which, strictly speaking, do not count as cases of analysis, as they do not go back to first principles. Rather, they are formalised as hypothetical syllogisms.

---

47 Alessandro Piccolomini, *De certitudine*, f. Cv: *Et hac via procedit frequenter Euclides, dico frequenter, quia quandoque deductione utitur ad impossibile, per secundum scilicet modum hypotheticorum syllogismorum, ut dicit Proclus et Philoponus. Quae deductiones etiam resolvi possunt, ut patet in logica, et ego Euclidem exponens, etiam resolvebam. Cum vero iudicium facere volumus de huiusmodi compositivo processu, tunc regredientes ab ultima conclusione, per praemissas et praemissarum praemissas, tandem ascendentes ad prima principia resolvere dicimur. Qua resolutione dignoscimus an per veras et proprias praemissas, composita prius fuerit illa conclusio. Hoc dicit Proclus pluribus in locis, in I et II libro Super I Elementorum.* Piccolomini refers to the *loci* in the prologues found in Proclus, *In Euclidem*, p. 8, l. 4–6; p. 18, l. 17–28; p. 43, l. 1–2. Outside the prologues, see p. 211, l. 18–23. Regarding Philoponus, Piccolomini probably refers to the commentary on *Analytica posteriora*, 78a, l. 6–7, where Philoponus compares scientific to dialectical syllogisms when dealing with analysis in geometry. See Philoponus, *In Aristotelis Analytica posteriora commentaria*, ed. M. Wallies (Berlin: De Gruyter, 1962), pp. 163, l. 33–164, l. 5. For an account of Piccolomini's analysis of I.1 in the wider context of the conflict between syllogistic and mathematical decisions see Mancosu, Mugnai, *Syllogistic Logic*, chapter 3. See also Freguglia, *Ars analytica*; M. Pedrazzi, "Sul tentativo di Alessandro Piccolomini di ridurre a sillogismo la I dimostrazione degli Elementi di Euclide", *Cultura e Scuola* 13 (1974), pp. 221–230.

Secondly, after stating yet again that this sort of analysis resolves a mathematical conclusion into its premises with the only goal of checking that the deduction is correct (*per veras et proprias praemissas*), Piccolomini identifies the type of analysis he is going to convey here with the one about which Proclus teaches in the first and second books (i.e., the two prologues) of *In Euclidem*. It is also noteworthy that the methodology of mathematical analysis, as transmitted by Proclus, is rephrased in terms of the *regressus* tradition. Following this, Piccolomini labels Euclid's proof as "compositive process".

We should bear in mind that, in Piccolomini's view, *resolutio quia* and *resolutio mathematica* are two species of the same genus, namely of *resolutio logica*. By this, Piccolomini means that they share some feature by which they belong to the same genus. In this case, both procedures operate through formalisation into deductive (syllogistic) structures. Thus, to resolve Euclid I.1 into its axioms and indemonstrable principles to show how this justificatory moment of the inquiry is supposed to work, Piccolomini syllogistically formalises the demonstrative step of the proposition.[48] The section reads as follows:

> Indeed, to make this clearer, because it is of great importance, we will explain an example of the same first problem of Euclid. The figure for this problem is as follows:

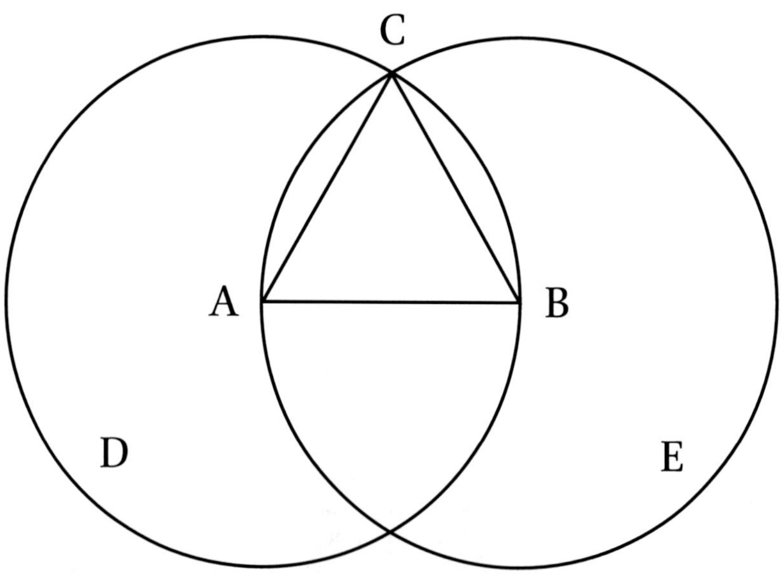

FIGURE 2 Diagram of Proposition I.1

48 Proclus, *In Euclidem*, p. 203, l. 1–15; Alessandro Piccolomini, *De certitudine*, f. CII.

Euclid, therefore, in that part of the problem which (as I said) is called "ostension", composes syllogisms in this way after making the construction. For, having made the construction, he immediately begins to demonstrate:

Because A is the centre of the circle CDB, AB is equal to AC.

Again, because B is the centre of the circle ACE, AB is equal to BC.

Now, it has been shown that AC is equal to AB.

Therefore, both BC and AC are equal to AB.

And things that are equal to the same thing are equal to each other.

Therefore, AC is equal to BC.

So, the three lines AB, BC, and AC are all equal.

This concludes the demonstration.

In this demonstration, there are virtually four syllogisms, which we will recognize by resolving. However, when we want to judge the aforementioned process by resolving, we will proceed in reverse order. Therefore, the last syllogism in the compositive way (*via compositiva*) will be completed first in the resolutive way (*resolutiva*), and it is as follows:

Every figure contained by three equal lines is an equilateral triangle.

Figure ABC is contained by three equal lines, AB, BC, and AC.

Therefore, figure ABC is an equilateral triangle.

The major premise is evident from the definition of an equilateral triangle, and the minor is proven as follows, completing the second syllogism:

Things that are equal to the same third thing are equal to each other.

But AC and BC are equal to the third line AB.

Therefore, AC and BC are equal to each other.

The major is evident from the first conception of the mind. The minor is proven on both sides and thus two syllogisms arise. First, regarding the first part, namely that BC is equal to AB, the third syllogism will be completed in order:

All lines from the centre of a circle to the circumference are equal.

But BC and AB are from the centre of the circle ACE to the circumference.

Therefore, BC and AC are equal.

The major premise is evident from the 14th definition. The minor is proven from the construction previously made. The other part of that minor in the preceding syllogism, namely that AC is equal to AB, is proven as follows, completing the fourth syllogism in order:

All lines from the centre of a circle to the circumference are equal.

But AC and AB are from the centre of the circle BCD to the circumference.

Therefore, AC is equal to AB.

> The major is evident from the same 14th definition, and the minor from the same construction previously made. This completes the resolution because when we have reached immediate and indemonstrable [principles], there is nothing more to be resolved.[49]

Euclid's I.1 shows why, even if the syllogistic arrangement produced by the resolution of a mathematical proof is taken as the *demonstratio quia* together with its inverse regress working as the *demonstratio propter quid*, the resulting *regressus* would not instantiate a *demonstratio potissima* according to Piccolomini's model. The formal reasons are the following:

1) The demonstration does not use P's definition as the middle-term.

To begin with, the middle-term of the first syllogism "*Omnis figura tribus rectis aequalibus contenta*" is the definition of S. Piccolomini stresses this fact (*Maior patet ex diffinitione trianguli aequilateri*). This arrangement violates Piccolomini's requirement that the middle-term must be P's definition. Thus, in

49 Alessandro Piccolomini, *De certitudine*, ff. CV–CII.: *Nos vero, ut melius haec res appareat, quia magni ponderis est, circa idem primum problema Euclidis huius rei exemplum explicabimus. Cuius problematis haec est figura: / Euclides igitur in illa problematis parte quae (ut dixi) ostensio nuncupatur, sic componendo syllogismos componit. Facta enim constructione, sic statim incipit demonstrare: / Quoniam A signum centrum est circuli CDB, aequalis est AB ipsi AC. Rursus, quia B signum centrum est circuli ACE, aequalis est AB ipsi BC. Ostensum autem est quod aequalis AC ipsi AB. Utraque igitur ipsarum BC et AC aequalis ipsi AB. Quae autem sunt aequalia uni eidem, inter se sunt aequalia. Aequalis ergo erit AC ipsi BC. Tres igitur lineae AB, BC, AC sibi invicem sunt aequales. Hactenus ostensio. In hac ostensione quattuor in virtute sunt syllogismi, quos resolvendo conoscemus. Cum vero iudicare praedictum processum resolvendo volemus, retrogrado ordine procedemus. Ultimus igitur syllogismus in via compositiva, erit primus in resolutiva completus, et est talis: Omnis figura tribus rectis aequalibus contenta est triangulus aequilaterus. Figura ABC continetur a tribus lineis aequalibus AB, BC, AC. Ergo figura ABC est triangulus equilaterus. Maior patet ex diffinitione trianguli aequilateri, minor probatur sic, et erit secundus syllogismus completus: Quae sunt aequalia uni tertio sunt aequalia inter se. Sed AC et BC sunt aequales tertiae lineae AB. Ergo AC et BC sunt inter se aequales. Maior manifesta est ex primo conceptu mentis. Minor quod ad utramque partem probatur, et sic insurgent duo syllogismi. Et primo quoad primam quod scilicet BC sit aequalis ipsi AB, et erit tertius syllogismus completus in ordine: Omnes lineae a centro circuli ad circumferentiam sunt aequales. Sed BC et AB sunt a centro circuli ACE ad circumferentiam. Ergo BC et AC sunt aequales. Maior patet ex 14 diffinitione. Minor vero ex constructione prius facta. Altera autem pars illius minoris praecedentis syllogismi. Quod scilicet AC sit aequalis ipsi AB sic probatur, et est quartus completus syllogismus in ordine: Omnes lineae a centro circuli ad circumferentiam sunt aequales. Sed AC et AB sunt a centro circuli BCD ad circumferentiam. Ergo AC est aequalis AB. Maior patet ex eadem diffinitione 14, minor vero ex eadem contructione prius facta. Et est completa resolutio, quia cum ad immediata et indemonstrabilia deventum sit, nihil restat amplius resolvendum.*

the resulting *regressus*, both the *demonstratio quia* and the *propter quid* would adopt the definition of equilateral triangle as the middle-term. Moreover, the *per se* modes required in the Sienese's model[50] fit in the *quia*, but they cannot be arranged in the *propter quid* as the middle-term is S's definition.

2) The *via resolutiva* and the *via compositiva* consist of four syllogisms each. Piccolomini's *demonstratio potissima* consists of solely two intertwined syllogisms. On its part, the formalised analytic process of proposition I.1 requires no fewer than four syllogisms, which means that the complete would-be *regressus* would be composed of a total of eight syllogisms.

This reveals another paramount detail regarding axiomatized systems vis-à-vis the Aristotelian scientific requirements as proposed in *Analytica posteriora* I, 2 (71b, l. 20–22). In an axiomatic system, the more advanced a proposition is, the farther it stands from the axioms. A syllogistic analysis of an advanced theorem would involve lengthy chains of syllogism going back to all its indemonstrable assumptions. This entails a larger number of demonstrable propositions serving as premises in such syllogistic structures. Therefore, there is no guarantee, as Aristotle postulates, that a theorem solely dependent on indemonstrable premises will be found among the more advanced proofs within the axiomatic system. Instead, advanced theorems are typically dependent upon propositions that have already been demonstrated.

Generally speaking, from the Aristotelian tenet that the superior scientific rank of an unqualified demonstration is signalled by the immediacy of the premises (requirement (iii): [ἐξ] ἀμέσων, *ex immediatis*), it follows that, the further we proceed in an axiomatic system, the less scientific are the results it produces.

Piccolomini was evidently aware of this formal limitation of axiomatical systems and utilized it as part of his argumentation. He declares that Euclid's proofs do not (ever) depend upon two indemonstrable principles, and (by invoking Proclus) that not every proposition in Euclid depends upon the axioms: "no one [demonstration] will be found which, by its nature, has both of the premises undemonstrable" and "not all theorems or problems of Euclid actually depend on the axioms, but only a few".[51] The clash between

50 Maior-PS1 and indemonstrable, minor-PS4 and demonstrable, conclusion-PS4 in the *demonstratio quia*; Maior-PS1 and indemonstrable, minor-PS4 and demonstrable, conclusion-PS2 in the *propter quid*.

51 Alessandro Piccolomini, *De certitudine*, f. LXXXIX: *Ponderentur quotquot demonstrationes in Euclide sunt: nulla invenietur quae ambas praemissas sua natura indemonstrabiles habeat, nisi forte in constructionibus, sed in ostensionibus minime*. Alessandro Piccolomini, *De certitudine*, f. LXXXVI: *Amplius dicit Proclus in I Elementorum, libro III, non omnia theoremata vel problemata Euclidis ex dignitatibus actu constare, sed pauca quaedam.*

the Aristotelian requirements of scientific demonstration and the factual and practical mathematical proof is a topic with which early modern apologists of mathematical vis-à-vis syllogistic proof were bound to wrestle.[52]

It is to be noted that, notwithstanding this formal obstacle, Piccolomini managed to preserve in each of the four syllogisms the Maior-premise indemonstrable in his analysis of Euclid's I.1, whereas the minor-premise is always the one further resolved – as indicated by his model of *demonstratio potissima*. This is achieved by admitting that visually checking the diagram in this case is a permissible move whose factual statement counts as an indemonstrable principle (*ex constructione prius facta*).

However, if the analysis of Euclid's first proposition requires four syllogisms and uses S's definition as the middle term – a troublesome feature that will be revealed as common to all mathematical demonstrations – then there is no hope that more advanced propositions dependent on it could be reduced to axioms in fewer than four syllogisms or avoid the adoption of S's definition as the middle-term at some point.

3) The *progressus* and the *regressus* use accidental features.

This is the most severe obstacle revealing the discrepancy between geometrical analysis and Piccolomini's model of *demonstratio potissima*: the resolution of I.1 involves demonstrating through accidental features. For instance, the proposition *sed BC et AB sunt a centro circuli ACE ad circumferentiam* indicates an accidental property of the lines BC and AB: it is not necessary or essential to a line (it does not belong to its definition) to be the radius of a circle. Overall, whenever the diagram is invoked to introduce indemonstrable propositions *ex constructione*, accidental features of a particular entity as observed in the image are introduced (*Figura ABC* [...], *lineae AB, BC, CA*).[53] This formal hindrance would be explicitly voiced by Piccolomini (see 3.3.1).

Piccolomini refers to Proclus' final comments on proposition X of Euclid's first book, where he values as epistemically superior proofs directly dependent on the axioms. See Proclus, *In Euclidem*, p. 280, l. 9–11: πολλῷ δὴ οὖν κρείττων ἢ τοῦ στοιχειώδους ἀπόδειξις καὶ ἁπλουστέρα καὶ ἀπὸ τῶν ἀρχῶν. In Morrow's translation: "The proof given by the author of the *Elements* is therefore much better since it is simpler and proceeds from the principles".

52 For an excellent history of this topic see Mancosu, Mugnai, *Syllogistic Logic*. See also Pietro Catena, *Oratio pro idea methodi ad nobilem Federicum Sforciam* (Padua: Gratiosus Perchacinus, 1563), ff. 5v–7v. See also Giacobbe, "La riflessione metamatematica di Pietro Catena".

53 See also Alessandro Piccolomini, *De certitudine*, f. LXXXIXv: *Isti solent negare hanc consequentiam, nam in geometria, verbi gratia, minores illae erunt tandem ostensae ex constructione, quae est quarta pars theorematis vel problematis* (*dividuntur enim in sex partes, ut dicit Proclus*).

In sum, Piccolomini's formalisation of Euclid's problem I.1 as a token of mathematical analysis provides a syllogistic arrangement that fulfils only two of the postulated requirements of his model of *demonstratio potissima*, namely that (a) in every syllogism the Maior-premise is indemonstrable (at the cost of introducing accidental features), and that (b) producing the syllogistic composition after the resolution does not require any non-demonstrative intermediary step 2, but just a rearrangement of the syllogistic order.

Regarding the place of Piccolomini's position in the history of logic, two points are to be stressed:

Firstly, Piccolomini is not the first author to oppose the possibility of a *regressus* in mathematics. Ex. gr., before Piccolomini, Nifo had denied that this logical structure could be achieved in mathematics.[54]

Secondly, some of Piccolomini's comments concerning resolution in mathematics were echoed by later authors, showing that such themes became *loci communes* in the northern Italian intellectual milieu. An example of a *regressus* theorist who held views akin to those of Piccolomini is Zabarella. He admitted that *demonstratio potissima* may be instantiated by a syllogism with a demonstrable minor-premise.[55] Nonetheless, unlike Piccolomini, Zabarella considers this a lesser case of *demonstratio potissima* – something never entertained by Piccolomini. Zabarella sustained that, nevertheless, the scientific knowledge produced by this lesser *demonstratio potissima* is not inferior to that of the perfect one.

Regarding mathematical resolution, much like Piccolomini, Zabarella did not identify *resolutio mathematica* with the *methodus resolutiva*.[56] Zabarella's

---

54 Randall, *The School of Padua*, p. 45; Agostino Nifo, *Expositio super octo Aristotelis Stagirite libros De physico auditu*, p. 13b: *Magna fieri digraduatio solet de regressu in naturalibus demonstrationibus; dico in naturalibus demonstrationibus quia in mathematicis non fit regressus. Nota bene*: For Nifo a *demonstratio potissima* does not require a *regressus*. He upheld that *demonstratio potissima* is obtained in mathematics where, on the other hand, a real *regressus* cannot be arranged.

55 Jacob Zabarella, *De medio demonstrationis*, in *Opera logica*, p. 381: *Dicimus igitur necessariam esse potissimae demonstrationis condicionem ut propositio maior sit immediata; sed non esse necessarium ut sit immediata minor, quum etiam ex minore mediata fiat potissima demonstratio, quae hac ratione erit ignobilior atque imperfectior demonstratio et ab illa pendens, sed tamen non imperfectiorem scientiam pariens.*

56 Jacob Zabarella, *De methodis*, pp. 178–179: *Nam resolutio illa mathematica qua, post factas omnes demonstrationes, retrocedimus et posteriora theoremata in priora et haec denique in prima principia resolvimus,* ***est potius quaedam eruditorum exercitatio quam methodus resolutiva****, de qua in praesentia loquimur. Est enim processus ab ignotioribus ad notiora, qui cuilibet rudi ad eam scientiam capessendam accedenti esset prorsus inutilis, quia nullam cognitionem pareret; nos autem de illa resolutiva methodo sermonem facimus, quae rerum*

argument in his *De methodis* aligns with Piccolomini's view that *resolutio* in mathematics does not produce new knowledge: mathematical resolution is depicted as an erudite exercise in logic rather than an actual part of the research. Nonetheless, Zabarella closes with a claim which opposes Piccolomini's views on analysis and aligns with the Galenic views: that there is such a thing as a resolutive method of inquiry – primarily found in natural science.

3.3 Euclid's I.32: Formal Flaws of Mathematical Demonstration

In this section, I shall analyse Piccolomini's treatment of the much-debated theorem I.32 of Euclid's *Elementa*.[57] This theorem demonstrates that the predicate "having internal angles equal to two right angles" belongs to "every triangle". It was adopted by Aristotle in explaining what he meant by "unqualified science" (ἐπιστήμη ἁπλῶς) in *Analytica posteriora* I, 1 and 4. Hence, it became one of the most debated mathematical propositions (if not the most debated) in the Latin tradition of commentary. The use of this theorem as a case of genuine scientific knowledge is rather unfortunate and undermines Aristotle's point. Philosophers such as Proclus, Giles of Rome, and Piccolomini would point out that this proof is logically flawed by the Stagirite's own standards.

According to Aristotle, as repeatedly stated in the first book of *Metaphysica* and famously summarised in the Latin motto *verum scire est per causas scire*, knowing something is knowing through causes.[58] In logical terms, the cause is provided by the middle-term in the syllogism.[59] From this, Piccolomini shall conclude that mathematical demonstrations cannot instantiate a *demonstratio potissima* not only regarding their pure logical form but also from a metaphysical point of view. By using the paradigmatic theorem I.32 as a case-study, Piccolomini buttressed his view that the middle-term of a syllogistically formalised mathematical proof does not express any of the four Aristotelian causes. Therefore, mathematical demonstrations cannot be held to be *real* explanations, where "real" refers to non-phantastic entities bound by strict causal chains.

---

ignotarum ex notioribus cognitionem parit, et in aliis scientiis locum habet, praesertim in scientia naturali.

57 See Euclid, Heath (ed.), *The Thirteen Books of the Elements*, vol. 1, book 1, p. 316 ff.

58 See, for instance, sections 980a–983a.

59 See *Analytica posteriora*, I, 4, 73b, l. 1–2; 5, 74a, l. 30–34; 33, 89a, l. 18; II, 7, 92b, l. 38; *Topica*, VIII, 3, 158b, l. 4; *Metaphysica*, I, 9, 992b; VII, 9, 1034a, l. 30–32; XIII, 4, 1078b, l. 24–25; *De anima*, I, 1, 402b, l. 25–403a, l. 1.

For his criticisms of theorem I.32, Piccolomini drew arguments from Proclus' *In Euclidem*, widely using the former text as a source of mathematical insight and as an authority. In this section, I will provide the *loci* of *In Euclidem* from which Piccolomini constructed his argumentation and address his understanding of them.

The criticisms of theorem I.32 which Piccolomini produces are:

(1) Theorem I.32 is proved through an accident.
(2) Theorem I.32 can be demonstrated through more than one proof (and through more than one middle-term).
(3) Theorem I.32, although provable of "every triangle", is a case of a more general theorem which also belongs to "every rectilinear figure", i.e., to the genus to which triangles belong.

3.3.1 *First Flaw: Demonstration through an Accident*

Piccolomini begins his criticisms in the following way:

> For example, if theorem 32, in book I of the *Elementa*, discussed a thousand times, is examined, it will appear that the external angle, which is put as the middle-term to conclude the passion, that is, to have three [internal angles each equal to two right angles] of a triangle, is not a definition, neither of a triangle (as it is evident), nor of the passion. For a triangle does not need to have an external angle as part of its definition to have three [internal angles each equal to two right angles], and even in its absence [of the external angle], it is still a triangle and has three [internal angles each equal to two right angles]. The same will be evident in almost all other theorems and problems of Euclid.[60]

Piccolomini, aware of the Latin tradition on the topic, introduces his case-study by declaring that it is a "theorem a thousand times proposed". He claims that this demonstration is accidental: the middle-term through which P is attributed to S is neither part of S's definition nor of P's definition. Instead, an accidental feature of the triangle is taken as the middle-term, namely that

60 Alessandro Piccolomini, *De certitudine*, f. CIIII: *Exempli gratia, si theorema millies allegatum 32, I Elementorum perpendatur, cognoscetur angulus extrinsecus, qui ponitur ibi medium ad declarandam passionem, quod est habere tres* [*angulos internos duobus rectis aequales*] *de triangulo non est diffinitio, neque trianguli* (*ut patet*), *nec passionis. Tam enim triangulus habere tres* [*angulos internos duobus rectis aequales*] *non indiget in sui diffinitione angulo extrinseco, quo non existente, etiam est triangulus, et habet tres* [*angulos internos duobus rectis aequales*]. *Idem patebit in omnibus fere aliis Euclidis theorematibus et problematibus.*

by extending one of its sides an external angle is produced. As Piccolomini immediately notes, having an extrinsic angle is not part of the definition of a triangle. Moreover, the theorem remains true even in the absence of the external, accidental angle. Therefore, the theorem is not attributed to the triangle by virtue of any of its essential features.

Controversially enough (and without further proof for it), Piccolomini makes a general statement after his scrutiny of this flaw: that almost every demonstration of Euclid is of this kind, namely a proof through accidental, supervened features.

Piccolomini's criticism follows Proclus' comments on the deficiencies of the same proof:

> For example, when from the fact that the exterior angle of a triangle is equal to the two opposite interior angles it is shown that the sum of the interior angles of a triangle is equal to two right angles, how can this be called a demonstration based on the cause (ἀπ'αἰτίας ἡ ἀπόδειξις)? Is not the middle term used here only a sign? For even though there would be no exterior angle, the interior angles are equal to two right angles; for it is a triangle even if its side is not extended.[61]

Proclus declares in this passage that theorem I.32 is not demonstrated through a cause (ἀπ'αἰτίας ἡ ἀπόδειξις), but that the middle-term is a "sign" or "hint" (πῶς δὲ ουχὶ τεκμήριόν ἐστι τὸ μέσον). By this, Proclus refers to a property supervening the triangle through which the conclusion is reached but which does not essentially belong to the triangle, i.e., an accident. This is rendered clear by his final remark (paraphrased by Piccolomini) that even if the triangle's side is not elongated (to produce an external angle), it remains a triangle.[62]

---

61 Morrow's translation. Original Greek in Proclus, *In Euclidem*, p. 379, l. 19–26: ὅταν γὰρ διὰ τοῦ τὴν ἐκτὸς τοῦ τριγώνου γωνίαν ἴσην εἶναι δύο ταῖς ἐντὸς καὶ ἀπεναντίας δεικνύηται τὸ τρίγωνον ἴσας ἔχον τὰς ἐντὸς τρεῖς γωνίας δυσὶν ὀρθαῖς, πῶς ἀπ'αἰτίας ἡ ἀπόδειξις αὕτη, πῶς δὲ ουχὶ τεκμήριόν ἐστι τὸ μέσον; καὶ γὰρ μήπω τῆς ἐκτὸς οὔσης γωνίας αἱ ἐντὸς οὖσαι δυσὶν ὀρθαῖς ἴσαι εἰσίν. ἔστι γὰρ τὸ τρίγωνον καὶ τῆς πλευρᾶς μὴ ἐκβεβλημένης.

62 The same observation present in both Proclus and Piccolomini, namely that the proof of theorem I.32 requires the introduction into the syllogism of an accidental feature or a demonstrative-accident, was made by Giles of Rome in his commentary on *Analytica posteriora*. See Bertagna, "Giles of Rome and Paul of Venice on *Demonstratio Potissima*"; see f. 119d in Longeway, "Aegidius Romanus and Albertus Magnus vs. Thomas Aquinas on the highest sort of demonstration".

3.3.2 *Second Flaw: Middle-Term Multiplicity*

Piccolomini cites Proclus once more concerning the same theorem to further highlight the shortcomings of mathematical demonstrations:

> And we see that this is true because mathematicians demonstrate the same passions of the same subjects through various assumed middle-terms. For example, Theon or Euclid demonstrate one way that a triangle has three [internal angles each equal to two right angles], Campanus demonstrates it in another way, and Proclus also demonstrates it differently, as he himself observed what I am saying. And the demonstration that Proclus provides regarding having three [internal angles each equal to two right angles] seems to be from a more intrinsic middle-term than Theon's demonstration. For he does not take anything external but demonstrates the proposed [theorem] from an internal perpendicular.[63]

Piccolomini refers to the two alternative ways of demonstrating I.32 given in *In Euclidem*, one by Proclus himself, the other attributed to the Pythagoreans by Eudemus.[64] Piccolomini also adds Campanus of Novara's demonstration of the same proposition, making a total of four different ways of proving the same proposition. This leads to the second criticism above, namely (2) that I.32 can be demonstrated through more than one proof and by adopting different middle-terms.[65]

As we will see below (see 3.4), one of Piccolomini's central arguments against mathematical causality hinges on the fact that the same theorem can be demonstrated through various, potentially infinite, middle terms.

---

63 Alessandro Piccolomini, *De certitudine*, ff. CIIIIV–CV: *Et quod hoc sit verum videmus, quod mathematici easdem passiones de eisdem subiectis variis assumptis mediis demonstrant. Aliter enim triangulum habere tres [angulos internos duobus rectis aequales] ostendit Theon vel Euclides, aliter Campanus, et aliter etiam Proclus, qui hoc ipsum quod ego dico animadvertit. Et demonstratio quam facit Proclus de habere tres [angulos internos duobus rectis aequales], videtur esse ex intimiori medio quam demonstratio Theonis. Non enim accipit quiddam extrinsecum, sed ex perpendiculari intrinseca ostendit propositum.*

64 Proclus, *In Euclidem*, p. 379, l. 2–15; see also Euclid, *The Thirteen Books of the Elements*, vol. 1, p. 320.

65 See also Alessandro Piccolomini, *De certitudine*, f. CV: *Quamvis tamen diversimode (ut dixi) possint demonstrari passiones mathematicae, videtur tamen, sicut innuit Proclus super 32, I, dicendum quod illa ex omnibus demonstrationibus eiusdem passionis erit laudabilior quae breviori tramite et per pauciores syllogismos concluserit intentum.*

3.3.3 *Third Flaw: Demonstration of the Genus*

The third flaw of I.32 is related to the concepts of genus and species and completes the other two criticisms. Following Proclus, Piccolomini concludes that the only way in which a mathematical demonstration could be considered superior to another one is because of its concision. Nonetheless, regardless of the many alternative proofs of the theorem, none of them uses the specific difference of the triangle as the middle-term in attributing the theorem to the figure:

> Although it [the diversity in their conciseness] indicates to be some difference amongst them [i.e., amongst the various passions used as middle-terms in the alternative proofs] since it is not a specific difference, there is nothing to prevent, in terms of necessity, that the middle-terms be considered as equal, that is, as all concluding from the last necessity.[66]

It follows that all the middle-terms are equally flawed. Factually, the alternative proof offered by Proclus also requires the use of an accidental angle to demonstrate the theorem (the angle is produced within the triangle in the case of Proclus' proof).[67] Thus, all such demonstrations conclude from the last necessity, i.e., from the mere formal necessity due to the very structure of the proof.

Following Proclus, Piccolomini criticizes the demonstrations of I.32 by pointing out that the conclusion does not pertain exclusively to triangles, as it can be proven for the entire genus of rectilinear figures to which triangles belong. This means that, by similar procedures (i.e., using an accident), any other rectilinear figure will show an analogous passion regarding the sum of its interior angles. This is shown by Proclus, who introduced the general theorem in the form of a corollary to this proposition:

> Therefore, the number which is double the number of the constituent triangles will give the number of right angles to which the angles of a many-sided figure are equal. Hence every four-sided figure has angles equal to four right angles, for it is composed of two triangles; and every five-sided figure has six right angles; and similarly for the rest.[68]

---

66 Alessandro Piccolomini, *De certitudine*, f. CV: *Quod quamvis aliquam differentiam arguat illarum, tamen cum non sit specifica differentia, nil prohibet quin, quoad necessitatem, ipsius medii aequales dici possint, hoc est, omnes ex necessitate ultima concludentes.*

67 Proclus, *In Euclidem*, pp. 382, l. 15–385, l. 31.

68 Proclus, *In Euclidem*, p. 382, l. 15–21: διπλάσιος ἄρα καὶ τῶν γωνιῶν ἀριθμὸς αὐτῶν τῶν συντεθέντων τριγώνων γενόμενος παρέξεται τὸ πλῆθος τῶν ὀρθῶν, ὅσαις ἕκαστον πολύγωνον ἴσας ἔχει

Proclus' observation can be formalised more concisely by stating that "the sum of the interior angles of a convex rectilinear figure is equal to twice as many right angles as the figure has sides, less four".[69] This theorem can be proven of the genus "every rectilinear figure", thereby making theorem I.32 just a case of a more general truth as it is instantiated in triangles. Thus, I.32 does not belong to the triangle *qua* triangle and, hence, it violates Aristotle's requirement in *Analytica posteriora*, I, 4, 73b, l. 27 that, in scientific demonstration, P must be predicated of S "as such" (ᾗ αὐτό). Therefore, Piccolomini concludes, this theorem cannot be a case of *demonstratio potissima.*

3.4 *Quasi* παράδοξον: Piccolomini's Denial of Mathematical Causality

In his *Praefatio* to *De Certitudine*, Piccolomini expresses his fear that his view may be perceived as a paradox (*quasi* παράδοξον *a me dictum videretur*).[70] Indeed, the claim that mathematical demonstrations are scientifically flawed seems still paradoxical to us – and perhaps even more perplexing for us than it was for Piccolomini's contemporaries.

One aspect of this potentially paradoxical doctrine is articulated by Piccolomini in Aristotelian terms. Specifically, he posits that none of the four causes can be identified in mathematical entities. Therefore, real causation is inexpressible as the middle-terms of mathematical demonstrations. Although Piccolomini's argument employs Aristotelian causes and draws on Averroes for its metaphysical foundations, I will argue that Priscian of Lydia's philosophy of mathematics, as discussed in section 2.1.5, is behind this critique.

Piccolomini's denial can be considered as an expansion on the reasons of the second flaw above (3.3.2.), sc., that in mathematical demonstrations the same P can be proven with equal validity through a variety of demonstrations. The following question arises: What distinguishes the causes expressed in mathematical middle-terms from those in the demonstrations of natural philosophy? In other words, what is it in mathematics that allows one to conclude the same truth through a plurality of middle-terms, something that, according to Piccolomini, does not happen in natural philosophy?

According to Piccolomini, real causation is defined by what he calls "active principles" (*principia activa*) belonging to substances. The flowing (*fluxus*) of

---

γωνίας. διὸ πᾶν μὲν τετράπλευρον τέτρασιν ὀρθαῖς ἴσας ἔχει γωνίας. ἐκ δυεῖν γὰρ συνέκειτο τριγώνων. πᾶν δὲ πεντάπλευρον ἕξ, καὶ τοῦτο ἑξῆς ὁμοίως.

69 Euclid, *The Thirteen Books of the Elements*, vol. 1, p. 322.

70 Alessandro Piccolomini, *De certitudine*, f. LXXII.

the active principles of substances secures a natural order according to which passions are produced within a tight structure of priority and anteriority and thus, every effect is preceded and produced by only one cause. In Piccolomini's own words:

> Mathematical passions are found in subjects without any order of priority. The true order amongst them arises from the fact that they flow from the subject and its form. Because, when such a flow is given, the order of priority of nature is also given, because one, insofar as it is one, can only immediately produce one, as is commonly stated in Aristotle's works. However, mathematical properties cannot have such an order or process or flow from a subject, because quantity is not amongst the active principles (*de principiis activis*), as stated by Averroes in his commentary on Aristotle's *Physica*, book IV, comment 84. And we see that this is true, as mathematicians demonstrate the same properties of the same subjects through various assumed middle-terms.[71]

Piccolomini states that mathematical passions, i.e., theorems, do not follow an order of priority (*prioritatis ordo*). Such order is secured by the active principles flowing from the subject and its form. Moreover, according to Piccolomini, active principles also ensure that this order of causation is unique; specifically, for a given P in S, there is only one middle term that can be correctly invoked to immediately explain why P belongs to S. In accounting for this, he mentions the Avicennian principle stating that "one *qua* one can immediately produce but one" and attributes this view to Aristotle.

Nevertheless, such a causal order does not hold in mathematics. As shown above, theorem I.32 can be attributed to its subject through various middle-terms. Even worse, calling upon Proclus' authority, Piccolomini reminds the reader that some of Euclid's theorems are reciprocal:

> Proclus, in his commentary on book I of the *Elementa*, page 21, says that several elements, meaning Euclid's propositions, could be elements of

71 Alessandro Piccolomini, *De certitudine*, ff. CIIIIv–CV: *Passiones mathematicae nullo prioritatis ordine in subiectis reperiuntur. Verus enim illarum ordo ex hoc procedit, quod fluunt ex subiecto et eius forma. Quia cum primum datur huiusmodi fluxus, datur etiam ordo prioritatis naturae, quia unum in quantum unum non potest immediate producere nisi unum, ut passim habetur apud Aristotelem. Sed passiones mathematicae non possunt habere talem ordinem vel processum sive fluxum a subiecto, quia quantitas non est de principiis activis, teste Averroe* IV *Physicorum, commentario 84. Et quod hoc sit verum videmus, quod mathematici easdem passiones de eisdem subiectis variis assumptis mediis demonstrant.*

> each other, which indeed would not be the case if they were demonstrated through true causes.[72]

Piccolomini refers to Proclus' elaboration on the term στοιχεῖα in the second prologue, where the notion of reciprocal demonstrability is introduced:

> For what proves is called an element of what is proved by it; thus, in Euclid, the first theorem is an element of the second, and the fourth of the fifth. In this sense, many propositions can be called elements of one another, when they can be established reciprocally.[73]

This means that through theorem A, theorem B can be demonstrated, and, conversely, through theorem B, theorem A can also be proved. This implies that the entire axiomatic system of geometry can be derived by beginning with different sets of theorems and problems, leading to the discovery of what was previously a foundational theorem in one system as a derivative consequence in another version of the system.[74] According to Piccolomini this fact contravenes the view that there is real causation in mathematics: mathematical explanation does not seem to follow the unidirectionality of *real* causality; it lacks strict order and uniqueness in the immediate causes adduced.

Piccolomini proceeds to explain this puzzling fact by appealing to the metaphysics of the mathematical object of study: quantity. He claims that, unlike real substances, quantity is not related to active principles (*quia quantitas non est de principiis activis*). This means that, unlike a natural substance, a triangle is an idle item: it does nothing, it produces nothing by itself, it does not act. Whereas from the form of a human being there immediately follows (or flows) rationality, from rationality flows the language-capacity, and thus there cannot be a human being without the flowing of rationality, a triangle does not act

72 Alessandro Piccolomini, *De certitudine*, f. CVII: *Proclus enim Super I Elementorum, pagina 21, dicit plura elementa, id est, plures propositiones Euclidis, posse esse sibi invicem elementa, quod profecto non esset si per veras causas demonstrarent.*

73 Morrow's translation. Proclus, *In Euclidem*, pp. 72, l. 24–73, l. 1: καὶ γὰρ τὸ κατασκευάζον ἐστὶ τοῦ κατασκευαζομένου στοιχεῖον, ὡς τὸ πρῶτον παρ' Εὐκλείδῃ τοῦ δευτέρου, καὶ τοῦ πέμπτου τὸ τέταρτον. Οὕτω δὲ καὶ ἀλλήλων εἶναι πολλὰ στοιχεῖα ῥηθήσεται. κατασκευάζεται γὰρ ἐξ ἀλλήλων.

74 See Alessandro Piccolomini, *De certitudine*, f. CVII–CVIIv: *Addit etiam non omnia Euclidis theoremata quasi per catenam sibi ipsis semper ordine quodam inservire, quia nimis arduum esset, ne dicam impossibile, illum ordinem invenire. Et ideo (dicit ipse) diversi diversimode demonstrant, non alius alio melius, nisi quia forte brevius.*

in any way to produce a theorem, nor does theorem A flow from the triangle neither necessarily before, nor after theorem B.

Piccolomini's portrayal of genuine causation precludes a given phenomenon from being causally explained by more than one middle term. This is so because the middle-term in a *propter quid* demonstration must reveal the proximate cause of the *explanandum* phenomenon and the immediate cause in the case of *demonstratio potissima*. And such causes are held to be unique.[75]

The Sienese's explanation of the causal power and order of real substances, in contrast to the causal idleness and arbitrariness of mathematics, follows Averroes' commentary on book IV of Aristotle's *Physica*, as the Sienese himself acknowledges.[76] There Averroes attempts to explain how the augmentation of a given quality, warmth in his example, occurs. The Cordovan targets the view that the cause of something becoming warmer is to be found in the category of quantity. It may seem to be so, since, without deep consideration, it might be postulated that the cause of S becoming "more" P is a change in the quantity of P and, thus, that the qualitative intensification can be explained through quantity. Averroes swiftly points out that this is a mistake, because quantity is not something which may act, and thus, it is not something that may cause any qualitative change. It is here that the Latin translation of Averroes' text introduces the syntagm referring to active potencies: [...] *quia non fit per aliam quantitatem secundum viam generationis cum non sit quantitas de potentiis activis*. Therefore, lacking the active potency of real substances, quantity is not explanative of qualitative intensification and remission (a common topic in the northern Italian anti-Mertonian controversy, as shown in section 2.5.2 above). Quantity cannot add or generate anything in quality, nor can a qualitative change be explained through quantity.

Piccolomini's argument employs this view. Nonetheless, he takes Averroes' point on active potencies in its most general and extreme version: quantity does not act, *ergo* quantity cannot be causally explanative in any sense, even within the realm of mathematics. Therefore, demonstrations in sciences concerned with quantity are not demonstrations *sensu stricto*. This train of

75 Alessandro Piccolomini, *De certitudine*, f. CIIIIv: *Passionis in subiecto unum tantum immediatum et verum debet esse medium ex quo conficiatur demonstratio potissima, sed passiones mathematicae non habent talia unica immediata media* [...].

76 Averroes, *Commentaria in libros De physico auditu Aristotelis*, IV, 3, f. 79v: *Et differentia est inter illa cum caliditas sit ab alia caliditate in actu, quod non invenitur in quantitate, scilicet quia non fit per aliam quantitatem secundum viam generationis cum non sit quantitas de potentiis activis, sed sequitur potentias activas et operationes earum, scilicet calidum et frigidum*. For an illuminating account of Averroes views' on quantity and his notion of active potency as well as its reception in Paduan Averroism see Higashi, *Penser les mathématiques*, pp. 48–76.

thought ultimately leads Piccolomini to one of his most momentous conclusions in *De certitudine*: in chapter XI he attempts to show that, in their lack of action as pseudo-substances, mathematical entities are not endowed with any of the four Aristotelian causes.

He begins by rejecting efficient and final causes:

> First, it must be held that mathematicians cannot demonstrate through the efficient and final cause, which are two of the four types of causes. Regarding the efficient cause, there is no doubt since mathematicians do not consider motion except metaphorically. However, we should not demonstrate through metaphors, as Aristotle and Averroes argue in *Analytica posteriora* and the books *De caelo* and *De anima*. Regarding the end, some have attempted, with great effort, to show that good and, therefore, the end is found in mathematics. The end, according to Aristotle in book I of the *Ethica Nicomachea* is interchangeable with the good. However, all their labour is in vain as they are deceived in this point because they believe that achieving good in some science and demonstrating it through the final cause are the same. They [both things] greatly differ, as good in mathematics comes from the outside.[77]

Against efficient causes, in consonance with his observation that mathematical objects lack active potencies or principles, Piccolomini declares that there is no movement in mathematics; when mathematicians speak of movement, they do so metaphorically. That is to say, mathematicians speak about "motion" by phantasising, say, a point to be here, and then there, supposing a sort of continuum,[78] although real movement remains completely alien to mathematical objects.[79]

---

77 Alessandro Piccolomini, *De certitudine*, f. CIIv: *Primum igitur tenendum est ex quattuor generibus causarum, per causam efficientem et finalem mathematicum demonstrare non posse. De efficiente nullus dubitat cum mathematicus non consyderet motum nisi metaphoricum. Demonstrare autem per metaphoras non debemus, teste Aristotele et Averroe in Posteriorum et in libro De caelo, et De anima. De fine vero aliqui sunt conati magno labore ostendere in mathematicis inveniri bonum, et ideo finem. Finis enim cum bono convertitur, teste Aristotele I Ethicorum. Sed vanus est omnis eorum labor quia decipiuntur in hoc, quia credunt quod bonum ad aliquam scientiam consequi et per causam finalem demonstrare idem sint. Sed valde differunt, quoniam bonum in mathematicis de foris advenit.*

78 Piccolomini will explicitly refer to phantastic continuum as the way in which movement is represented in phantasy in chapter XII. I will deal with this topic in section 3.6.

79 For an analysis of Proclus' theory of phantastic motion and a detailed analysis of its reception in early modern mathematics, see A. Axworthy, *Motion and Genetic Definitions in the Sixteenth-Century Euclidean Tradition* (Birkhäuser Cham: Springer, 2021).

Final causes are then denied. Piccolomini argues that, although in their usefulness (i.e., their goodness, *bonum*) mathematical disciplines may be applied to an end (such as in mechanics), this finality does not proceed from mathematical items themselves, but it is accidentally added by human beings from the outside. By the same token the wood of the tree is used to fabricate a chair, but becoming a chair does not belong to the substance of a tree; it does not flow from its form. Final causes and accidental purposes given by human action are quite different things. And so it happens that those maintaining that final causes can be found in mathematics through such arguments are deceived (*decipiuntur*). Therefore, much like motion, finality is a human addition, and no genuine, intrinsic final cause can be found in mathematics.

The sources for the arguments which deny these two causes are Aristotle himself and Alexander of Aphrodisias' comments on the first aporia in *Metaphysica* III, 996a, l. 21, where the question of whether all causes are found in all kinds of beings is discussed. Piccolomini openly attributes the view that there are no final causes in mathematics to Aristotle and Alexander in this commentary, mentioning the appropriate *loci* in both texts.[80] Elaborating on Aristotle's claim, Alexander proposes that neither the efficient cause (αἴτιον ποιητικόν) nor the final cause (αἴτιον τελικόν) holds in mathematics:

> He [Aristotle] includes the mathematicals too among the unmoved, for the source of the principle of motion, that is, the efficient cause, is not a cause in their case either; for how could an efficient and moving (αἴτιον ποιητικὸν καὶ κινητικόν) pertain to things that are ingenerated and unmoved? But neither does the nature of the good, this is the end and that for the sake of which (ἐστι τὸ τέλος καὶ τὸ οὗ ἕνεκα), pertain to these things. For that which is a cause qua good is that for the sake of which (τὸ οὗ χάριν) other things are, and which is itself not for the sake of anything, and such a thing is an end. But the end, and the doing of things preliminary to the end for the sake of something, pertain to actions, and to the sciences of production and action. But where there is production and action, there is motion. Hence neither could [be] this cause, that is, the final cause (ἡ αἰτία ἡ τελική).[81]

80 Alessandro Piccolomini, *De certitudine*, f. CIII: [...] *cum Aristoteles expresse dicat in III Metaphysicorum 3 non esse finem in mathematicis, et idem asserunt etiam Alexender et Commentator nec aliter dicere possunt*. See Aristotle, *Metaphysica*, III, 2, 996a, l. 25–996b, l. 1: [...] τὰς δὲ μαθηματικὰς οὐθένα ποιεῖσθαι λόγον περὶ ἀγαθῶν καὶ κακῶν.

81 Translation by W. E. Dooley and Arthur Madigan (slightly modified). Alexander of Aphrodisias, *On Aristotle Metaphysics 2 and 3* (Bloomsbury Academic: London, 1992), p. 104. Original at Alexander of Aphrodisias, *Commentary on Aristotle, Metaphysics*

And thereafter:

> To prove that this is the case, Aristotle cites the mathematicals. Neither the productive cause nor the final cause pertains to them. Nor is there in geometry any demonstration of anything that uses the argument "because it is better that way". For example, [there is no demonstration] that the angles of a triangle are equal to two right angles because it is better that way (in the way that that for the sake of which figures in the case of things that come to be by nature and in the case of things that are moved according to art), nor does anyone who does geometry attend to the final cause at all.[82]

Alexander's reasoning is plain: Aristotle locates mathematics amongst the unmoved beings, therefore implying (in Alexander's reading) that there is no efficient cause in it. Secondly, Aristotle explicitly denies that geometers reason by "the better" in producing their proofs. Conveniently for Piccolomini's goals, Alexander offers as an example Euclid's I.32 to show that the final cause is alien to mathematical reasoning. He thus plainly concludes that "in those [mathematics] there is neither efficient nor final cause" (ἐν γὰρ τούτοις οὐκ ἔστιν οὔτε ποιητικὴ οὔτε ἡ ὡς τέλος αἰτία [...]). It is, therefore, evident how Piccolomini's remarks build upon Alexander's exposition.

It is also possible that Piccolomini's view that mathematical movement is a metaphor could be an interpretation of Proclus' comments on postulates I–III. In this passage, Proclus, while accounting for the generation of the line,

---

(*Books I–III*), *Commentaria in Aristotelem Graeca et Byzantina*, ed. P. Golitsis (Berlin: De Gruyter, 2022), pp. 181, l. 34–182, l. 4: ἐν δὲ τοῖς ἀκινήτοις [ὁ Ἀριστοτέλης] τίθησι καὶ τὰ μαθηματικά· ἐν γὰρ τούτοις οὔτε τὸ ὅθεν ἡ ἀρχὴ τῆς κινήσεώς ἐστιν αἴτιον, τοῦτο δέ ἐστι τὸ ποιητικόν· πῶς γὰρ ἂν εἴη ἐν τοῖς ἀγενήτοις καὶ ἀκινήτοις αἴτιον ποιητικὸν καὶ κινητικόν; ἀλλ' οὐδὲ ἡ τοῦ ἀγαθοῦ φύσις (αὕτη δέ ἐστι τὸ τέλος καὶ τὸ οὗ ἕνεκα) ἐν τούτοις ἐστίν· τὸ γὰρ ὡς ἀγαθὸν αἴτιόν ἐστι τὸ οὗ χάριν τὰ ἄλλα, αὐτὸ δὲ οὐδενός, τὸ δὲ τοιοῦτον τέλος. ἀλλὰ μὴν τὸ τέλος καὶ τὸ τινὸς χάριν τὰ πρὸ τοῦ τέλους ποιεῖν ἐν πράξεσίν ἐστι καὶ ἐν ταῖς ποιητικαῖς καὶ πρακτικαῖς τῶν ἐπιστημῶν· ἐν οἷς δὲ ποίησις καὶ πρᾶξις, καὶ κίνησις ἐν τούτοις, ὥστε ἐν τοῖς ἀκινήτοις οὐκ ἂν εἴη οὐδὲ αὐτὴ ἡ αἰτία ἡ τελικὴ [...].

82 Translation by Dooley and Madigan (slightly modified) in Alexander of Aphrodisias, *On Aristotle Metaphysics 2 and 3*, p. 104. Original at Alexander of Aphrodisias, *Commentary on Aristotle, Metaphysics* (*Books I–III*), p. 182, l. 22–28: ὅτι δὲ τοῦτο οὕτως ἔχει, παρέθετο τὰ μαθηματικά· ἐν γὰρ τούτοις οὐκ ἔστιν οὔτε ποιητικὴ οὔτε ἡ ὡς τέλος αἰτία, οὐδὲ ἔστιν ἀπόδειξίς τις ἐν γεωμετρίᾳ τινὸς προσχρωμένη τῷ βέλτιον γὰρ οὕτως, οἷον τοῦ τὰς τριγώνου γωνίας δυσὶν ὀρθαῖς ἴσας εἶναι, διότι τοῦτο ἄμεινον, ὥσπερ ἐπὶ τῶν φύσει γιγνομένων καὶ τῶν κατὰ τέχνην κινουμένων τὸ οὗ ἕνεκα· οὐδὲ τὴν ἀρχὴν μέμνηταί τις γεωμετρῶν τοῦ ὡς τέλους αἰτίου.

speculates on the incorporeal kind of movement present in phantasy.[83] He distinguishes corporeal movement from phantastic movement, identifying the latter as a unique type of κίνησις, wherein (as absurd as it may seem) even partless entities, such as points, can "move". By this, Proclus arguably suggests that a point is successively envisioned in phantasy in two (or more) different coordinates: first in (3, 4), then in (7, 1). The movement of the point travelling from (3, 4) to (7, 1) is then hypothesised, even though it has not taken place in a literal sense. It would not be difficult for Piccolomini to interpret Proclus' concept of phantastic movement as metaphorical.

The next cause which is denied is the material cause:

> Regarding matter, Aristotle in *Analytica posteriora*, II seems to provide a mathematical example of this type of cause, and others, such as Giles and the Lincolnian [i.e., Grosseteste], as reported by Zimara, also give different examples where demonstrations are assigned from intelligible, not from sensible, matter. Intelligible matter is itself placed in phantasy, as we discussed earlier when dealing with mathematical matter.[84]

Piccolomini refers to Zimara's influential work *Theoremata* and singles out the opinions of Giles of Rome and Grosseteste there expounded by Zimara: all of them, including Zimara himself, had sustained that mathematics explains through material causes.[85] To refute this, Piccolomini would expand on the notion of *materia intellegibilis* adopted by Zimara in his justification and

---

83 Proclus, *In Euclidem*, p. 186, l. 9–14. τὴν δὲ κίνησιν μὴ τοῦ σωματικὴν ἀλλὰ φανταστικὴν νοήσωμεν καὶ τὰ ἀμερῆ τὰς μὲν σωματικὰς κινήσεις κινεῖσθαι μὴ συγχωρῶμεν, τὰς δὲ αὖ φανταστικὰς διεξόδους ὑπομένειν. καὶ γὰρ ὁ νοῦς ἀμερὴς ὢν κινεῖται καὶ οὐ τοπικῶς καὶ ἡ φαντασία κατὰ τὸ ἑαυτῆς ἀμερὲς ἔχει κίνησιν ἰδίαν. In Morrow's translation (slightly modified): "But let us think of this motion not as bodily, but as phantastic, and admit not that things without parts move with bodily motions, but rather that they are subject to the ways of the imagination. For intellect, though partless, is moved, but not spatially; and phantasy has its own kind of motion corresponding to its own partlessness".

84 Alessandro Piccolomini, *De certitudine*, f. CIII: *De materia autem videtur Aristoteles II Posteriorum ponere exemplum mathematicum in hoc genere causae, et alii etiam, ut Egidius, et Linconiensis, referente Zimara, alia exempla ponunt ubi ex materia intelligibili tamen, non sensibili, assignatae sunt demonstrationes. Materia autem intelligibilis, quantitas ipsa est in phantasia collocata, ut declaravimus superius de materia mathematica pertractantes.*

85 Marco Antonio Zimara, *Theoremata*, f. 31v (F, l. 12): *Sed ista opinio stare non potest, nam scientia mathematica in suo genere demonstrat, quandoque per medium quod est causa materialis in illo genere, ut in geometria demonstratur quod angulus est rectus ex eo quod est medietas duorum rectorum.* This passage and its relation to Piccolomini's philosophy of mathematics is commented on in Higashi, *Penser les mathématiques*, chapter 2, sections 2 and 3, especially pp. 67–82, and chap. 3, sections 2 and 3, pp. 135–168.

use it against him. Indeed, similar to Zimara's position in his *Theoremata*, Piccolomini argues that the object of mathematical sciences is not a natural substance regarded as quantity; rather, it is intelligible matter.

Piccolomini's ontology of mathematical objects contravenes the Aristotelian tenet declared in *Physica* II, 2 that geometry studies the very natural objects but only *qua* quantity, leading to the explicit view that geometry somehow deals with natural entities. This is an inconvenient view of Aristotle. As we saw, 15th- and early 16th-century anti-Mertonian reactions in the Italian context emphasised the separateness of mathematics from real objects. Both Zimara and Piccolomini assumed, each in their own way, that intelligible matter is not to be identified with sensible matter, let alone with matter in motion.[86]

The proceeding question, then, is: at which point does Piccolomini set himself apart from Zimara? How did he dismiss the material cause in mathematics? Piccolomini echoes an argument that we already encountered in Boccadiferro's controversy against Aquinas,[87] namely that the name *materia intellegibilis* is an improper denomination.[88] Phantasy is called "matter" as a metaphor for describing the sensible-like and quantitative traits of mathematical entities, shared to some degree by sensible matter. Nevertheless, although *phantasia* is equivocally named as "matter", this does not mean that it is a real substratum in which substances could be enacted. Factually, Piccolomini continues, there is no difference between intelligible matter and "quantity itself located in phantasy".[89] "Intelligible matter" is, thus, an improper term applied to phantastic quantities due to their capacity – similar to sensible matter – to be stretched, compressed, divided, multiplied, added, or reduced.[90]

Therefore, unlike Zimara, the Sienese does not admit that *materia intellegibilis* can be deemed as a genuine material substratum which is the cause of phantastic quantity. Rather, Piccolomini's view is that, just as mathematical

86 For an account of Zimara's doctrine see Higashi, *Penser les mathématiques*, p. 51. For the exposition of this doctrine by Zimara see Marco Antonio Zimara, *Theoremata*, prop. 81.

87 Ludovico Boccadiferro, *Lectiones super tres libros De anima*, lec. 34, pp. 57–58: *Secundo errat quia abutitur nomine materiae intelligibilis Divus Thomas quia Metaphysicorum II, 37 et XV, 8. Non ergo materia intelligibilis est materia prima, ut dicit vir iste Divus Thomas, quare et cet.*

88 Alessandro Piccolomini, *De certitudine*, f. LXXXXVIIv: *At id a plerisque, quamvis non satis propie, materia intellegibilis nuncupatur.*

89 Alessandro Piccolomini, *De certitudine*, f. CIII: *Materia autem intelligibilis, quantitas ipsa est in phantasia collocata, ut declaravimus superius de materia mathematica pertractantes.*

90 I will elaborate on Piccolomini's theory of intelligible matter in sections 3.5.2 and 3.5.3 below.

objects are pseudo-substances, the stuff they are actually made of is nothing but "pseudo-matter" i.e., quantity itself.

Lastly, Piccolomini denies that middle-terms in mathematical demonstrations express a formal cause. Although he divides his argument into three parts, all of them boil down to the following line of thought:

> From the form, since we have already excluded matter, how will the passions depend, if there is no action in quantity, nor a reason for action, as we will explain below? Therefore, no one can say how in the definition and form of a triangle is this, namely "the external angle is greater than any opposite interior angle", which is indeed proven by Euclid as a passion in Proposition 17 in book I.[91]

Piccolomini stresses, yet again, that there is no action in quantity and no reason to suppose action from it. This means that, as has been explored above, mathematical objects do not produce their passions themselves from an inner flow (*fluxus*) stemming from substantial forms, for they lack substantial forms. Piccolomini presents proposition I.17 of the *Elementa* as an instance of this. There, the passion "having the sum of any two angles less than two right angles" is attributed to a triangle by using an accident, namely (yet again) an exterior angle always greater than the sum of two right angles produced by elongating the side. Piccolomini points out that these passions follow neither from the form of the triangle nor from any internal action proper to the form of the triangle. As in the case of I.32, I.17 is also demonstrated through an accident.

A consequence of this formal idleness is that the same mathematical proposition may be proved through a variety of middle-terms (i.e., pseudo-causes): no real cause or strict causal order of the properties prevents an arbitrary attribution of mathematical passions through different middle-terms to mathematical objects.

In conclusion, mathematical objects, as portrayed in *De certitudine*, are inert entities waiting for the mathematician to make their properties manifest by phantastically acting upon them. In their inactivity and insubstantiality, mathematical objects offer no opposition to the adoption of different middle-terms,

91 Alessandro Piccolomini, *De certitudine*, f. CIIIIv: *Ex forma vero (materiam enim iam exclusimus), quomodo dependebunt* [*passiones*], *si in quantitate non est actio, nec actionis ratio, ut inferius declarabimus? Et ideo nullus est qui dicere possit quomodo in ratione et forma trianguli sit hoc quod angulus extrinsecus sit maior quolibet opposito interiore, quod quidem tanquam passio probatur ab Euclide in propositione 17, I libri.*

or a variety of ways of accounting and justification.[92] The structure of the mathematical realm is a game bounded by rules far looser than the strict laws of the natural realm. Hence, greater demonstrative freedom is given to those playing the game of mathematics at the big cost of the unreality of their conclusions and the pseudo-substantial ontological status of their items.

3.4.1 *Behind the Denial: Genua De' Passeri and the Risks of "Simplician" Averroism*

Was the view that there is no causation in mathematics and that mathematical disciplines are not real an unheard-of one?

In accounting for Piccolomini's intellectual background, I have introduced the Paduan master Marco Antonio Genua De' Passeri, prince of the 16th-century "Simplician" Averroists as, arguably, one of the main sources, together with Priscian, of Piccolomini's point of departure.

In his commentary on Aristotle's *De anima*, III, 4, Genua engages, as it was the *locus communis*, with some themes of the philosophy of mathematics. There, following Pseudo-Simplicius as his main guide, he proposes the following observations:

> For the intellect connected to sensation knows matter and all accidents, of which accidents the intellect knows the composed substances according to itself, as its proper reasons. However, as Simplicius rightly notes, here the question does not concern intellect in itself, but intellect connected to sensation or phantasy, and, thus, he said that it cannot understand without phantasy and sensation. Therefore, as it has been said about the practical intellect that it cannot understand without phantasy and sense because, as intellect has progressed (*intellectus progressus est*), it is united to sensation and phantasy, in the same way it must be said about the understanding of mathematical things.[93]

---

92 This view may have been suggested to Piccolomini by the passage in Pseudo-Simplicius' *In De Anima* which he references in chapter XII (comm. in 402b, l. 11). See Simplicius, *On Aristotle's On the Soul* 1.1–2.4, p. 29: "Scientific knowledge is knowledge of activities from essence (ἐκ τῆς οὐσίας τῶν ἐνεργειῶν), but the reverse is clearer to us (πρὸς ἡμᾶς δὲ σαφεστέρα)". See section 2.3.3.

93 Marco Antonio (Genua) De' Passeri, *In De anima*, p. 174v: *Nam intellectus sensui connexus materialia novit, atque accidentia omnia, quorum accidentium intellectus secundum se substantias compositas veluti proprias rationes cognoscit. At hic non agitur, ut recte Simplicius, de intellectu secundum se, verum ut connexo sensui vel phantasiae, ideo dixit non posse sine phantasia intelligere. Quare, ut de practico dictum est, sine phantasia, et sensu intelligere*

And a few lines later in the same column:

> It is to be noted, as can be gathered from Simplicius, that sometimes this power [progressed intellect] operates together with that which is of the same order (such as the senses with that which is sensed), sometimes by itself alone. Our reason operates in the same manner when it knows natural things: sometimes it acts together with the sense and is joined with it; sometimes without the sense, knowing the substances of natural forms and the substances of their compositions, which are determined according to themselves, when it judges the affections and accidents of these from substances as from causes. And, thus, the natural [philosopher] differs from the mathematician, because the latter deals only with accidents, while the former contemplates both causes and substances. Therefore, natural philosophy is called in this way [i.e., "philosophy"] because it knows both causes and substances. However, mathematical expertise (*mathematica peritia*) should by no means be called "philosophy", as it deals only with accidents.[94]

Genua's elaboration follows Priscian's doctrine of the insubstantiality, causal idleness, and, thus, the low epistemic status of mathematics as expounded in section 2.1.5. Before dealing with it, some notes on Genua's general noetics are in order.

As illustrated by Spruit in his analysis of Genua's doctrine, he sought, against Nifo, an Averroistic interpretation of *De anima* III, 4 and 5 through Pseudo-Simplicius, able to explain how the intellect was at the same time one and many.[95] He held as his key concepts that of an *intellectus manens* (identified with the *intellectus agens*), and the progression of such intellect in a second moment of the intellect named *intellectus progressus*, identified with

---

non potest, quia, ut sic intellectus progressus est et unitus sensui et phantasiae, sic est dicendum de intellectione mathematicarum.

94 Marco Antonio (Genua) De' Passeri, *In De anima*, p. 174v: *Notandum, ut ex Simplicio colligitur, quod vis illa potest interdum cum ea, quae eiusdem ordinis est, ut sensus scilicet cum sensatis, interdum ex se ipsa sola. Sic ratio nostra cognoscens naturalia, quandoque una cum sensu agit, et cum eo coniuncta est; quandoque sine sensu substantias formarum naturalium atque substantias eorum compositorum, quae secundum ipsas determinata sunt, cognoscens ubi ex substantiis tanquam ex causis, affectiones et accidentia illorum diiudicat, et sic naturalis a mathematico differt, quia is duntaxat accidentia, at naturalis ex formis, veluti ex causis, sive ex compositis substantiis et figuras et reliqua quae substantiis adveniunt, contemplatur. Naturalis itaque philosophia est, quia et causas et substantias cognoscit. At mathematica peritia nequaquam philosophia dicenda est, cum solum de accidentibus sit.*

95 Spruit, *Species Intelligibilis*, pp. 164–173.

Averroes' infamous, unique لعقل الهيولاني (*al-ʿaql al-hiyūlānī*) or *intellectus materialis*. According to De' Passeri's view, a unique intellect progresses as a cascade as it departs from itself, thereby giving birth to a multi-level variety of operations as it conjoins itself to particular powers. Hence, in Averroes' manner, one and the same intellect – if it is considered in itself (*intellectus secundum se*) – can be understood as a spectrum of functions and individuated cognitive acts, depending on the powers and/or entities with which it unifies to produce some kind of intellect-like insight.

This cascade of intellect-related powers is introduced by Genua in the first quotation above. We are told that the intellect knows sensible things as its own reasons (*veluti proprias rationes cognoscit* – note the Neoplatonic phrasing in referring to the λόγοι) as it is considered as conjoined to senses or phantasy. In this way, Genua explains one of the most heatedly discussed phrases of Aristotle's noetics, namely that intellection does not occur without phantasy (*De anima* III, 431a, l. 9–15). This, of course, is also applied to the exercise of mathematical disciplines, which is one of the possible operations of the *intellectus progressus* when it joins to sense and phantasy.[96]

After this, Genua introduces his interpretation of Priscian's *digressio mathematica* in commenting on 431b, l. 12. He points out that the proper place for such an argument, i.e., the denial of mathematical substantiality and causation, is the commentary on book III of the *Metaphysica* or *Physica* II.[97] Genua's argument intends to show that mathematics, having a causally idle accident as its object of study, is not exercised by the *intellectus speculativus* – whose function is knowing. Rather, mathematics is enacted by the *intellectus practicus* – which ranks in both Priscian's and Genua's classifications as the lowest manifestation of the intellect.[98] In other words, in Genua's philosophy of science, mathematics, much like carpentry or tailoring, is an art dealing with accidental properties of things, a practical affair. And when it is exercised in the seeking of mathematical truth, it does not "contemplate" in a strict sense, devoid of substance that it could use as causative principles.

Genua provocatively introduces his view by comparing mathematics to natural philosophy and refusing even to call mathematics "knowledge" in his Latin phrasing. Instead, he refers to it as a *peritia*, i.e., a "skill" or "technical expertise".

96 Genua arguably thinks that the mere action of numbering sensible items is a way in which *intellectus progressus* joins with sense perception in a direct, mathematical use.

97 Marco Antonio (Genua) De' Passeri, *In De anima*, p. 174v: […] *cum negotium sit vel metaphysici in III Metaphysicorum vel philosophi naturalis in II Physicorum.*

98 Simplicius, *In De anima*, comm. in III, 407b, l. 2, p. 48, l. 32–34: εἴ γε κρείττων καὶ θεωρητικὴ φιλοσοφία τῆς πρακτικῆς, καὶ τῆς θεωρητικῆς καὶ περὶ τὰ χωριστὰ τῆς φυσικῆς, καὶ ἡ μηδὲν σώματι χρωμένη τῆς χρωμένης, καὶ τῆς ἔξω φερομένης ἡ εἰς αὑτὴν συμπτυσσομένη.

His point here is that, whereas in doing metaphysics and natural philosophy the *intellectus progressus* unifies with *phantasia* in an activity which renders it closer to its own nature and function, i.e., the contemplation of substances and causes, intellect does not do so when it conjoins to *phantasia* to exercise mathematics.

The importance of Genua's extremely anti-mathematical interpretation of Pseudo-Simplicius and his long university teaching at Padua for reading *De certitudine* cannot be overstated. It evinces that a kind of philosophy of mathematics denying substantiality, causality, and the rank of theoretical philosophy to mathematics was held by the *Simpliciani* readers of Priscian's *De anima* at Padua. Of this Simplician philosophy, Piccolomini's *De certitudine* can be taken as a rather special, well-argued, and enriched expansion. Therefore, Piccolomini's παράδοξον, the opinion he feared to voice before encountering Proclus' *In Euclidem*, namely that *demonstratio potissima* is impossible in mathematics, most likely came to his mind as he became familiar with Genua, Priscian, and the *Simpliciani* at Padua.

3.4.2 *Piccolomini's* speculatio operativa

But a question still lurks.

Did Piccolomini wholly assume Genua's view on the philosophy of mathematics, point by point, in all its consequences? Or does he differ in any aspect? Although the influence of Genua De' Passeri and Priscian on the Sienese's approach to the ontology of mathematics is evident, there is a crucial aspect in which Piccolomini's view is stridently dissonant with that of his professor. For, as we read in the *Praefatio* to the paraphrase in the *Quaestiones mechanicae*, Piccolomini declared that mathematical disciplines (and also mechanics) are, without any doubt (*nullum est dubium*), part of contemplative philosophy (see 2.5.1). This claim is even more puzzling given the fact that Piccolomini wrote the *Praefatio* to the Latin paraphrase after completing *De certitudine* and proposed its contents as a natural continuation of the opusculum. How is this possible?

Resuming, Piccolomini's outlook holds that:

(a) Mathematics is not endowed with (a.1) substantiality, (a.2) causation, and (a.3) real demonstrations.

And, at the same time, he says that:

(b) Mathematics (and mechanics) is (b.1) contemplative (b.2) philosophy/science.

It cannot be doubted that Piccolomini embraces the two "Simplician" (a.1), (a.2) tenets, and his view (a.3). Therefore, the answer must lie in the meaning of either (b.1) or (b.2) or both.

I believe that the solution to this enigma is found in the very first lines of the *Praefatio* to the *Quaestiones mechanicae*. There Piccolomini addresses and cautiously defines what he means by (b.1) "contemplative/speculative" and by "operative (b.2) philosophy:"

> Following Ammonius, Philoponus, and Simplicius, we divide philosophy only into two parts, namely one operative, and the other contemplative. We are therefore allowed to philosophise by speculating both while operating and while contemplating. But because operative speculation (*operativa speculatio*) reaches its end in work, and there it concludes, while contemplative speculation is perfected and remains in itself, for that reason we designate the former by the name of "operation" and the latter by the name of "contemplation".[99]

This represents a subversive twist on Piccolomini's behalf, arguably introduced as he has foreseen the unfavourable implications of his views as stated in *De certitudine*. Attributing this opinion to Simplicius (and others), Piccolomini distinguishes between *contemplatio* and *speculatio*. He subsumes contemplation under speculation as one of its species. Speculation, for its part, is identified with philosophy (*philosophari*). Hence contemplation is classified as one of the possible cases of *intellectus speculativus*, i.e., one of the modes of philosophising, but not as the only one. Indeed, Piccolomini adds that *operatio* is another of the species of *speculatio*, admitting that *intellectus speculativus* is also enacted in operation. What does this mean?

As I interpret it, Piccolomini thinks of the very science he is about to introduce in the *Praefatio*: mechanics. As we saw when we addressed his *Praefatio*, Piccolomini stated the following about the principles of mechanics: "Although these principles are directed towards practical work (*opus*), the contemplative faculty (*facultas contemplatrix*) of these principles will not be disregarded for that reason".[100] Therefore, as explained (see 2.5.3), Piccolomini upheld that

99 Alessandro Piccolomini, *In mechanicas quaestiones*, ff. I–IV: [...] *Ammonium, Philoponum, ac Simplicium secuti, philosophiam in duas tantummodo partimur partes, alteram scilicet operativam, contemplatricem reliquam. Licet enim tam operantes quam contemplantes speculando philosophemur, quia tamen operativa speculatio ad opus exit, ibique desinit, contemplativa vero in se ipsa perficitur ac conquiescit, idcirco hanc contemplationis, illa vero operationis nomine nuncupamus.*

100 Alessandro Piccolomini, *In mechanicas quaestiones*, ff. VV–VI: *Mechanica enim ars illa tantum existimanda est qua excogitantes quamplures machinas, constructiones ceteris artibus sellularibus invenimus, quam philosophiae contemplatricis partibus annumerandam esse nemini dubium esse debet, quippe quae, geometriae subiecta, multarum artium principia*

applied mathematics has two moments: one theoretical, *qua mechanicus* (i.e., *qua* mathematician), and a second practical, *qua* technician. This means that, as he puts it, the specific difference setting apart applied mechanics from, say, metaphysics, is just the moment of application. Both disciplines engage in speculation of their own kind, attempting to discover universal principles as a token of speculative intellect. Nonetheless, whereas metaphysics remains always in the field of speculation as belonging to the "speculative-contemplative" species, mechanical principles are intended to be applied to some practical ends – but the prior speculative moments remain as such. When the application happens, mechanical principles are *also* a token of practical intellect, as oriented to a particular goal or end and concerned with a particular sensible object (this balance). Thus, the fact that mechanics is susceptible to being put to work while it remains a speculative philosophy means that it falls under the species of "speculative-operative" philosophy. Beautifully, Piccolomini even considers that discovering a machine or designing a building while using phantastic principles is a sort of philosophical contemplation.

Piccolomini is therefore establishing that speculation and philosophy do not require substantiality, causation, or real demonstrations. In other words, he, perhaps in his keenness for mathematical disciplines and his familiarity with mathematically minded technicians, architects, and astronomers (as he himself was), refused to wholly accept Genua's conclusion that finding mathematical truths is not a case of contemplation. On the contrary, he held that grasping the universal truth of quantity as such is a form of speculation, regardless of whether this speculation concerns a substantial item or whether it is directed to practical ends.

Secondly, regarding (b.2), it is highly likely that Piccolomini took mathematics as a sort of homonymous science. Regardless of his total denial of mathematical causality, substantiality, and the scientificity of its demonstrations, nevertheless, Piccolomini labels mathematical disciplines (all of them) as *scientia* and *philosophia*.[101]

---

excogitata speculatur. Quae quidem principia, etsi dirigantur ad opus, non ob id tamen horum principiorum facultas contemplatrix non existimabitur.

101 See, for instance, the *Index* of *De certitudine* in Alessandro Piccolomini, *In mechanicas quaestiones*. It is noteworthy, and perhaps part of Piccolomini's rhetoric, that, whereas in the names of chapter VII and VIII in the *Index* mathematics appears labelled as *scientia*, in the titles inside the very opusculum it appears as *disciplina*. As we saw in Biringucci's testimony in section 2.5, Piccolomini held that mechanical demonstrations possess some cognitive value, i.e., they were able to reveal a "reason" (*ragione*, adopting the Vitruvian language). However, I will refrain from using Biringucci's testimony to elucidate Piccolomini's philosophy of mathematics in *De certitudine* and the *Praefatio*, as Biringucci's vernacular translation and his introduction to it (where the memoir about

If this is so, mathematics appears as a discourse which resembles a science with an object of study of its own about which it is able to produce demonstrations of some sort. It demonstrates through epistemically lesser *signa*, working as justificatory reasons, not through real *causae*. Inversely to natural philosophy, mathematical speculation starts from the easily knowable definitions of its objects, trying hence to deduce their passions (see 3.5.3.3 below). This notwithstanding, Piccolomini is generous in his classification: it is sufficient for him that a discipline engages at some moment in speculation (i.e., the obtention of universal truths with no end beyond themselves) to be labelled as *philosophia* or, equivocally, as *scientia*.[102] Thus, according to Piccolomini's standards, even the most modest carpenter in the arsenal, who abstractly considers mechanical principles while designing a new model of a catapult, is engaging in the lofty activity of *intellectus speculativus*.

Regarding his possible sources, it could be argued that Piccolomini's appeal to Simplicius' authority in support of his argument in the opening of his *Praefatio* stemmed from a specific passage in Priscian's *In De anima*, whose interpretation he seems to dispute:

> Directly after the practical [intellect] comes in the ascent the theoretical [intellect] that is concerned with mathematical objects. Mathematical objects are themselves enmattered, but they do not exist in the matter that is generated and in different states at different times [i.e., sensible matter]. They are enmattered as they exist under all circumstances in extension and some division, and for that reason, they are also themselves phantasisable beings. The intellect that considers these objects imposes boundaries and order upon them and both adds what is lacking and removes what is superfluous and investigates them as objects in phantastic items (φαντάσματα), [whereas] the practical [intellect] knows the phantastic items in reference to sense-objects. In this respect, the mathematical [intellect] surpasses the practical.[103]

Piccolomini is contained) were produced approximately 30 years after the publication of *De certitudine*.

102 Alessandro Piccolomini, *In mechanicas quaestiones*, p. 111v: *Mechanicus tamen artifex, ea ratione qua mechanicus est, solummodo causas et principia earum consyderans, in contemplatione ipsa quiescit et sistit. Nota bene*: Piccolomini should be using the term *causae* to mean "reason" or "explanation", rather than "causal explanation", as he has just argued that causation in mathematics is impossible. If he were to use it otherwise, he would risk being accused of inconsistency.

103 Translation by Steel (slightly modified) in 'Simplicius', *On Aristotle On the Soul* 3.6–13, p. 76. See Simplicius, *In De anima*, comm. in III, 431b, l. 10, l. 27–35, p. 276, l. 3–10: προσεχέστατα δὲ τοῦ πρακτικοῦ ἐπαναβέβηκεν καὶ τῶν μαθηματικῶν πραγμάτων θεωρητικός, ὃ

Thus, Piccolomini's hermeneutics of "Simplicius" against those of Genua and his like hold that the speculative intellect engages in mathematics, albeit working upon in-substantial and causally idle phantastic items. Nonetheless, this activity remains a contemplative one. To this Piccolomini would add that, even when mathematics has practical goals, as far as the speculation deals with principles and universal truths, it remains a form of speculation.

The "Simplician" philosophy of mathematics of Genua De' Passeri casts new light on Piccolomini's views. Surprisingly, Alessandro Piccolomini, the young man whose audacious opusculum questioning the status of mathematical demonstration ignited a continent-wide controversy, actually held more moderate and supportive views on mathematics than many of his contemporaries in Padua – and certainly more favourable than many of his epigones. After all, if compared with De' Passeri's views, it turns out that Piccolomini was not attacking, but, in a certain way, defending mathematics.

3.4.3 *The "Simplician" Proclus and Piccolomini's Hermeneutics*

What is Proclus' precise role in all this?

As we saw in section 3.3, Piccolomini's scrutiny of theorem I.32 in the use of Proclus' commentary is intended to conclude a general statement, namely that "every *potissima* demonstration has a middle-term which is the immediate cause of the effect itself, that is, the passions. However, no mathematical demonstration is found to be such".[104]

This is a controversial logical leap. Indeed, from the fact that theorem I.32 in Euclid's *Elementa* is flawed, it does not follow that every mathematical demonstration in every mathematical discipline is defective.[105] Nevertheless, the generality intended by Piccolomini could be supported by the fact that (i) by using I.32 as his paradigm, he has pointed out pervasive features of mathematical demonstration and ontology, such as the lack of causality allowing the possible adoption of a multitude of middle-terms. Hence it would follow

καὶ αὐτὰ ἔνυλα μέν, οὐκ ἐν τῇ γενητῇ δὲ ὕλῃ καὶ ἄλλοτε ἄλλως ἔχουσα, ἀλλ'ὅμως ἔνυλα ὡς ἐν διαστάσει πάντως καὶ μερισμῷ τινὶ ὑφεστῶτα καὶ διὰ τοῦτο καὶ αὐτὰ φανταστὰ ὄντα, τοῦ θεωροῦντος αὐτὰ νοῦ ὅρον αὐτοῖς καὶ τάξιν ἐπιτιθέντος καὶ τό τε ἐνδέον προστιθέντος καὶ τὸ περιττὸν ἀφαιροῦντος καὶ σκοποῦντος αὐτὰ ὡς φανταστά, τοῦ πρακτικοῦ καὶ τὰ φανταστὰ κατὰ τὴν πρὸς τὰ αἰσθητὰ ἀναφορᾶς γινώσκοντος, καὶ ταύτῃ τοῦ μαθηματικοῦ τὸν πρακτικὸν ὑπερβάλλοντος.

104 Alessandro Piccolomini, *De certitudine*, f. CIIII: *Omnis demonstratio potissima medium habet, quod est causa immediata ipsius effectus, id est passionis. Sed nulla demonstratio mathematica reperitur talis.*

105 This logical leap, namely that from "some mathematical demonstrations are not *potissimae*" Piccolomini concludes "no mathematical demonstration is *potissima*" was already pointed out by Barozzi. See Francesco Barozzi, *Opusculum in quo una oratio, et duae quaestiones, altera de certitudine, et altera de medietate mathematicarum continentur* (Padua: Grazioso Percacino, 1560), pp. 29–30.

that (ii) *demonstratio potissima* is impossible in mathematics. Additionally, (iii) targeting this very theorem makes Piccolomini's argument much more persuasive: as said, I.32 was the theorem proposed by Aristotle as an example of *scire simpliciter* and commented upon by the Latin tradition on *Analytica posteriora*.[106]

Now we know that Piccolomini's position is a consequence of the more general claim made by Priscian and sustained by the Paduan master Genua. Nevertheless, since the primary source adduced by Piccolomini in his argumentation is Proclus, we can ask ourselves in which precise way was *In Euclidem* read for writing *De certitudine*. I will attempt to show that *De certitudine*'s Proclus can be fittingly described as a "Simplician" Proclus, i.e., one who denies mathematical causality and substantiality and who, as written in the first pages of the opusculum, predated Piccolomini's conclusion that mathematical demonstrations are not *potissimae*.

Let us examine some of the explicit uses of Proclus' commentary in Piccolomini's opusculum:

In chapter XI of *De certitudine* (*An certitudo mathematica ex vi demonstrationum potissimarum oriri dicenda sit*) Piccolomini invokes Proclus' commentary to reinforce his main position, namely that mathematical certainty is neither a symptom nor a consequence of the superior status of mathematical demonstrations. So it happens that Proclus is invoked three times in the same sentence:[107]

> On page 57, indeed, Proclus states that the reasons of geometers have force and necessity from their subject-matter, but not from the nature of their demonstration. What could be clearer? Moreover, on page 10 he says that not all mathematical demonstrations have the same certainty. Furthermore, on page 56, he writes these very clear and explicit words: "Geometry seemed to many not to contemplate the cause and the 'why' (τήν γε μὲν αἰτίαν καὶ τὸ διὰ τί). This is Amphinomus' opinion, originated by Aristotle". What more do we seek than this view?[108]

---

106 Potential objections questioning the universality of the conclusions reached in *De certitudine* will be countered by Piccolomini's inclusion in chapter VII of a *communis mathematica scientia* guaranteeing the general scope of his theory (see section 3.7 below).

107 In this quotation, Piccolomini erred twice in referring to the folia from Grynaeus' 1533 edition. I offer the right pagination between brackets. The Greek terms between brackets indicate the words present in Friedlein's edition omitted in this quotation.

108 Alessandro Piccolomini, *De certitudine*, f. CVII–CVIIV: *Pagina vero 47* [*57*] *inquit Proclus geometrarum rationes habere vim et necessitatem ex subiecta materia, non autem ex natura demonstrationis. Quid clarius? Pagina autem 10 dicit non omnes demonstrationes*

Firstly, Piccolomini paraphrases the passage in f. 206, l. 16–18,[109] where Proclus, using theorem I.32 as an example, establishes (i) that not every mathematical proof embodies the demonstrative perfection (ἀποδείξεως τελειότης) for sometimes it demonstrates by signs or symptoms (ἐκ τεκμηρίων); (ii) that the necessity of mathematics stems from its object of study, but does not come from its demonstrative methods in every case (διὰ τῶν ἀποδεικτικῶν μεθόδων); and, (iii) regarding I.32 particularly, that this theorem does not demonstrate from causes, but from signs (πῶς ἀπ'αἰτίας ἡ ἀπόδειξις αὕτη).

Regarding (i), it is clear that Piccolomini understood ἀποδείξεως τελειότης as referring to *demonstratio potissima*. Piccolomini was familiar with a demonstration ἐκ τεκμηρίων not only through Aristotle, but arguably also through the elaborations of Philoponus and Simplicius as a form of imperfect demonstration able to establish *de facto* correlations between phenomena without providing causes. Common examples of this type of reasoning come from medicine.[110]

Regarding (ii), Piccolomini interpreted Proclus' words as saying that the necessity (τὸ ἀναγκαῖον, interpreted as *certitudo*, see 2.3.4.1) in mathematical demonstration comes in every case from its object. This interpretation departs from Proclus' text which entertains that, although mathematical demonstrations are not perfect in every case, some of them can be deemed as such.

---

mathematicas habere eandem certitudinem. Praeterea pagina 96 [56] ponit haec verba apertissima, et manifesta: τήν γε μὲν αἰτίαν καὶ τὸ διὰ τί πολλοῖς ἔδοξεν ἡ γεωμετρία μὴ θεωρεῖν. [καὶ] ταύτης γάρ ἐστι καὶ ὁ Ἀμφίνομος τῆς δόξης Ἀριστοτέλους κατάρξαντος. *Quid amplius quaerimus pro hac sententia?*

109 Proclus, *In Euclidem*, p. 206, l. 12–23: τὴν δὲ λεγομένην ἀπόδειξιν ὅτε μὲν καὶ τὰ ἴδια τῆς ἀποδείξεως ἔχουσαν εὑρήσομεν ἀπὸ τῶν ὁρισμῶν μέσων τὸ ζητούμενον δεικνύουσαν – αὕτη γὰρ ἀποδείξεως τελειότης – ὅτε δὲ ἐκ τεκμηρίων ἐπιχειροῦσαν. καὶ δεῖ μὴ λανθάνειν. πανταχοῦ μὲν γὰρ τὸ ἀναγκαῖον ἔχουσιν οἱ γεωμετρικοὶ λόγοι διὰ τὴν ὑποκειμένην ὕλην, οὐ πανταχοῦ δὲ περαίνονται διὰ τῶν ἀποδεικτικῶν μεθόδων. ὅταν γὰρ διὰ τοῦ τὴν ἐκτὸς τοῦ τριγώνου γωνίαν ἴσην εἶναι δύο ταῖς ἐντὸς καὶ ἀπεναντίας δεικνύηται τὸ τρίγωνον ἴσας ἔχον τὰς ἐντὸς τρεῖς γωνίας δυσὶν ὀρθαῖς, πῶς ἀπ'αἰτίας ἡ ἀπόδειξις αὕτη, πῶς δὲ οὐχὶ τεκμήριόν ἐστι τὸ μέσον; Translation by Morrow in *A Commentary on the First Book of Euclid's Elements*, p. 161: "What is called 'proof' we shall find sometimes has the properties of a demonstration in being able to establish what is sought by means of definitions as middle terms, and this is the perfect form of demonstration; but sometimes it attempts to prove by means of signs. This point should not be overlooked. Although geometrical propositions always derive their necessity from the matter under investigation, they do not always reach their results through demonstrative methods. For example, when from the fact that the exterior angle of a triangle is equal to the two opposite interior angles it is shown that the sum of the interior angles of a triangle is equal to two right angles, how can this be called a demonstration based on the cause? Is not the middle term used here only a sign?" Grynaeus' edition, the one read by Piccolomini, does not show any variation on this passage. See Proclus, *είς τοῦ αὐτοῦ τὸ πρῶτον ἐξηγημάτων Πρόκλου βιβλίον*, p. 52.

110 Dubouclez, *Descartes et la voie de l'analyse*, pp. 11–116. Philoponus, *In Analytica posteriora Commentaria*, p. 49, l. 5–14.

That mathematics receives its necessity not from its demonstrative means, but from its subject-matter, is insisted upon and directly quoted in Greek in the concluding chapter of *De certitudine* (XII):

> Proclus asserts the same in his commentary on the *Elementa*, page 57, where he has these legitimate words: "For geometrical arguments (λόγοι) have everywhere the necessity due to the underlying matter".[111]

Regarding (iii), it seems that Piccolomini interpreted Proclus' rhetorical πῶς ἀπ'αἰτίας ἡ ἀπόδειξις αὕτη not as referring specifically to theorem I.32 just mentioned by Proclus, but to mathematical demonstration generally. Thus, the Sienese arguably read ἀπ'αἰτίας in a "Simplician" manner, seeing in Proclus' words the already known opinion that there is no αἰτία in mathematics.[112] Therefore, as Piccolomini's Proclus fittingly concludes, it follows that, *certissima* as they might be, *all* mathematical demonstrations are "from signs" and none of them is perfect, i.e., *potissima*.

Secondly, Piccolomini points to a locus of Proclus in which it is affirmed that not every mathematical demonstration has the same degree of certainty ([…] *non omnes demonstrationes mathematicas habere eandem certitudinem*). He refers to f. 34, l. 11–13, where Proclus grants different degrees of certainty to mathematical disciplines depending on their subject of study:

> Differences in subject-matter at once produce differences in the sciences and the arts that are concerned with them. Thus, some things are unchangeable, others changing; some simpler, others more composite; some intelligible, others sensible. Even in mathematics we cannot demand the same degree of accuracy (τὴν αὐτὴν ἀκρίβειαν) in all parts.[113]

It is hard to see how this precise point fits in Piccolomini's argument.[114] Piccolomini's philosophy seems to imply a homogeneous degree of certainty in mathematical sciences. Except for this line, no mention of different degrees of mathematical certainty is ever made in *De certitudine*. Some insight into

---

111 Alessandro Piccolomini, *De certitudine*, f. CIX: *Quod idem asseverat Proclus in Elementa pagina 47* [57] *ubi habet haec legiptima* [sic] *verba*: πανταχοῦ [μὲν] γὰρ τὸ ἀναγκαῖον ἔχουσιν οἱ γεωμετρικοὶ λόγοι διὰ τὴν ὑποκειμένην ὕλην.

112 Simplicius, *In De anima*, p. 210, l. 5–9; p. 233, l. 22–30; pp. 276, l. 27–277, l. 35.

113 Morrow's translation. Original Greek in Proclus, *In Euclidem*, p. 34, l. 8–13: Εὐθὺς γὰρ τὰ υποκείμενα ταῖς ἐπιστήμαις ἢ ταῖς τέχναις ποιεῖ διαφοράς, οἷον εἰ τὰ μὲν ἀκίνητα, τὰ δὲ κινούμενα, καὶ τὰ μὲν ἁπλούστερα, τὰ δὲ συνθετώτερα, καὶ τὰ μὲν νοητά, τὰ δὲ αἰσθητά. Οὔτ' οὖν πᾶσαν τὴν μαθηματικὴν τὴν αὐτήν ἀκρίβειαν ἀπαιτήσομεν.

114 That this quotation of Proclus does not contribute to Piccolomini's argument was already pointed out by Barozzi. See Francesco Barozzi, *Opusculum*, p. 30.

this point can be gained by exploring other passages from *In Euclidem* where Proclus deals with the same topic, namely f. 59, l. 15–17. There Proclus writes:

> For one science is more accurate (ἀκριβεστέρα) than another, as Aristotle says; that is, a science that starts from simpler principles than one whose starting point is more complex, or one that states why a fact is so than one which says that it is so, or a science concerned with intelligibles than one that applies to objects in the sense world. According to these criteria of exactness, arithmetic is more precise than geometry, for its principles are simpler (τῆς ἀκριβείας ἀριθμητικὴ μὲν ἀκριβεστέρα γεωμετρίας).[115]

In this section Proclus emphasises, following Aristotle in *Analytica posteriora* I, 27 (87a, l. 31–37), that not every mathematical science shows the same degree of certainty (ἀκρίβεια). Different degrees of certainty, Proclus proceeds, stem from two facts: firstly, (1) the complexity of the principles assumed, which vary from science to science, and which are simpler the more abstract the science is. Secondly, (2) the kind of explanation a given science can provide, whether ὅτι or διότι, may differ: the science able to provide the ὅτι and the διότι would be more certain in the sense that it provides the fact and its explanation. Therefore, Proclus concludes that, because of (1), arithmetic is more certain than geometry, and, because of (2), geometry is more certain than mechanics.[116] Indeed, Proclus is here assuming that in subalternate sciences the *explanandum* (ὅτι) is a sensible phenomenon, whereas the *explanans* (διότι) is mathematical – an assumption unacceptable to Piccolomini.

But, yet again, how do such claims help Piccolomini's views on the homogeneity of mathematical certainty and his tenet that mechanics is a contemplative science dealing with abstract objects and not with sensible ones? If we want to charitably make sense of Piccolomini's use of Proclus' view of the disparity in the certainty amongst mathematical proofs, we can make two claims.

Firstly, (1) it can be argued that Piccolomini understood Proclus' claims on ἀκρίβεια as referring to the possible variety of demonstrations and middle-terms

115 Morrow's translation. Original Greek in Proclus, *In Euclidem*, p. 59, l. 10–16: ἀκριβεστέρα γάρ ἐστιν ἐπιστήμη ἄλλη ἄλλης, ὥς φησιν Ἀριστοτέλης, ἥ τε ποικιλωτέραις ἀρχαῖς χρωμένη τῆς ἐξ ἁπλουστέρων ὑποθέσεων ὡρμημένης καὶ ἡ τὸ διότι λέγουσα τῆς τὸ ὅτι γινωσκούσης καὶ ἡ περὶ νοητῶν πραγματευομένη τῆς τῶν αἰσθητῶν ἐφαπτομένης. Καὶ κατὰ ταύτας τὰς ἀποδόσεις τῆς ἀκριβείας ἀριθμητικὴ μὲν ἀκριβεστέρα γεωμετρίας.

116 Proclus, *In Euclidem*, p. 59, l. 21–23: αὗται γὰρ τὰς αἰτίας ἀποδιδόασι καθόλου τῶν ὑπ'ἐκείνας θεωρημάτων – γεωμετρία δὲ μηχανικῆς ἢ ὀπτικῆς [...]. In Morrow's translation (slightly modified): "They [the subordinating, superior science] furnish the principles of the theorems subordinate to them – geometry is superior to mechanics and optics".

that can be adopted in proving the same theorem. Since some demonstrations for the same proposition are more efficient or flawed, unwieldy or expedient, they consequently differ in ἀκρίβεια, taken here as the accuracy (mathematical elegance) through which they prove the theorem.

Secondly, (2) it might be postulated that Piccolomini understood ἀκρίβεια as referring to the operative moment of mathematical, speculative sciences: since mechanics, in Piccolomini's view, does not speak about sensible reality, when its propositions are brought forth from the abstract *intellectus speculativus* and applied as arts in the use of *intellectus practicus*, mathematics become inexact and only approximate. Nonetheless, it is unclear how this point would contribute to the argumentation in chapter XI.

Because of this, I tend to privilege interpretation (1) over (2). Otherwise, I can only understand the addition of this line from Proclus' *In Euclidem* as part of a rather unthoughtful or incoherent anti-mathematical screed from Piccolomini.

Finally, (3) Piccolomini quotes another Greek line from Proclus' *In Euclidem* whose ambiguous phrasing was destined to be contested by Barozzi. This is Proclus' reference to Amphinomus' and also Aristotle's view, that geometry does not contemplate causes or the "why".[117] In agreement with Priscian's views, Proclus' passage emphasises that Amphinomus and Aristotle, the source of this opinion, held that geometry does not contemplate (θεωρεῖν). That said, Proclus swiftly invokes Geminus' counterargument to shield mathematics against the accusation. He concludes, with Geminus and against Amphinomus and the Stagirite, that, in mathematics, "if the conclusion is universal and applies to all cases, the "why" is thereby made manifest".[118] Although Proclus' view (unlike that of Priscian) is that mathematics contemplates causes, he does not reveal what kind of causes these may be in his comment.

What did Piccolomini make of this passage?

It is hard to tell, but, already in Barozzi's influential answer, Piccolomini was criticized as misattributing Amphinomus' denial of geometrical causation and

117 Proclus, *In Euclidem*, p. 202, l. 9–12: τήν γε μὲν αἰτίαν καὶ τὸ διὰ τί πολλοῖς ἔδοξεν ἡ γεωμετρία μὴ θεωρεῖν. [καὶ] ταύτης γάρ ἐστι καὶ ὁ Ἀμφίνομος τῆς δόξης Ἀριστοτέλους κατάρξαντος. In Morrow's translation: "Many persons have thought that geometry does not investigate the cause, that is, does not ask the question 'why?' Amphinomus is of this opinion, though Aristotle originated it". As Morrow points out, it is hard to pinpoint a place in the Aristotelian corpus where this opinion is stated. See Proclus, *A Commentary on the First Book of Euclid's Elements*, p. 157.

118 Morrow's translation (slightly modified). See Proclus, *In Euclidem*, p. 202, l. 23–25: εἰ δὲ καθ' ὅλον καὶ ἐπὶ πάντων τῶν ὁμοίων, εὐθὺς καὶ τὸ διὰ τί γίγνεται καταφανές.

explanative power *also* to its reporter, Proclus.[119] In accounting for this hermeneutics, it cannot be argued that Piccolomini's text was faulty: Grynaeus' edition contains Proclus' subsequent refutation of Amphinomus in its totality.[120]

At this point, I think, Piccolomini's acquaintance with Priscian's *De anima* plays a decisive role in the entire interpretation of (i), (ii), and (iii). Two most likely explanations can be proposed:

(a) **Inadvertent "Simplician" reading**

Piccolomini read Proclus' *In Euclidem* assuming that Proclus' philosophy of mathematics agreed with that of Pseudo-Simplicius. Finding some doctrinal and linguistical coincidences, he read *In Euclidem* under a Simplician bias, interpreting it as an expansion of the doctrines already known to him from Priscian's *De anima*. Thus, Proclus would have been understood as pointing to the deficiencies of mathematical demonstrations as they would inevitably follow from the core tenet of the insubstantiality and causal idleness of mathematical objects. In reading Proclus in this way, Piccolomini would have found a predecessor to his views regarding *demonstratio potissima*.

(b) **Deliberate "Simplician" reading**

Piccolomini was aware of some doctrinal differences between Proclus and Priscian. Yet he would have found that some isolated comments of Proclus fittingly backed his own views, inspired by Pseudo-Simplicius, that mathematical demonstrations are neither *potissimae*, nor even real demonstrations and, at the same time, that mathematics is the most certain discipline. Thus, he picked those *loci* from Proclus – an impressive authority in the disputed field – so gaining the voice of the only surviving ancient Greek commentator on Euclid's *Elementa* to his side.

It is hard to determine whether Piccolomini engages in (a) or (b) in every case, or a mixture of both, sometimes reading Proclus in light of Pseudo-Simplicius, sometimes forcing Proclus to fit Pseudo-Simplicius.

Regarding the hermeneutics of passage about Amphinomus above, I may suggest that, in that particular case, Piccolomini deliberately remained silent about Geminus' counterargument. The reason for this is that, as mentioned in section 2.1.1.3, the Sienese alludes to Proclus only once in his paraphrase of the *Quaestiones mechanicae*: in the 30th question. This means that Piccolomini was familiar with some contents of Proclus' commentary before 1540. Coincidentally, he invoked *In Euclidem* there to claim that in mathematics the same proposition can be demonstrated "from several assumptions and several

---

119 As said, that this is not Proclus' but Amphinomus' opinion was also pointed by Barozzi in his refutation. See Francesco Barozzi, *Opusculum*, pp. 30–31.

120 Proclus, *εἰς τοῦ αὐτοῦ τὸ πρῶτον ἐξηγημάτων Πρόκλου βιβλίον*, p. 56.

true causes, as Proclus writes" (*pluribusque veris causis in mathematicis haberi posse*).[121] This quotation suggests that Piccolomini was by that time aware of Geminus' counterargument in Proclus' commentary, as that is the only locus in *In Euclidem* where mathematical causation is generally cast in doubt and then defended.

My suggestion is the following: by the time Piccolomini wrote the paraphrase, already endorsing his anti-Mertonian standpoint but not deeply cognizant of Proclus' text, he was still hesitant regarding the issue of mathematical causation. As he paraphrased the *Quaestiones*, he wanted to defend mechanics against the "Simplician" onslaught, and he found resources to do it in Proclus' *In Euclidem*. However, six years later, when he wrote *De certitudine* and the *Praefatio*, he had matured in his philosophy of mathematics. This philosophy allowed him to dignify mechanics while neutralising its menacing power as a discipline potentially emulative of natural philosophy. To do so, he had to embrace and wholly develop the precise position he had targeted in the 30th question some years before. However, Piccolomini did not delete his allusion to Proclus in the paraphrase before publishing it together with his *De certitudine*, whence a token of his philosophical transition has remained to us.

Thus, I am inclined to think that Piccolomini deliberately omitted Geminus' argument when writing *De certitudine* and that he purposefully presented Amphinomus' view vaguely, without neither explicitly attributing it, nor denying that it was also Proclus' persuasion. So, he left the reader with a true but incomplete fact: an opinion reported by Proclus, the opinion of Amphinomus and, its alleged originator, Aristotle.

The existence of this conflicting reference to Proclus in Piccolomini's writings escaped all those who partook in the question subsequently triggered, including Barozzi himself. As far as I know, it has not been detected by modern scholarship. From this, it appears that Piccolomini's philosophical views were received only through *De certitudine*, whereas his other philosophical positions, both in the paraphrase and its *Praefatio*, and his contradictory statement in the 30th question, remained marginalised in the subsequent debate.

Coming back to Piccolomini's doctrine by 1546, the influence of Priscian's *In De anima* has been shown of the utmost importance for grasping the context under which Piccolomini was interpreting *In Euclidem*. The impact of Priscian's text will be further attested in accounting for Piccolomini's take

121 Alessandro Piccolomini, *In mechanicas quaestiones* f. LVIII–LVIIIv: *Non enim inconveniens est eandem complexionem sive (ut ita dicam) conclusionem ex pluribus assumptionibus pluribusque veris causis in mathematicis haberi posse, ut scribit Proclus, et ut quilibet cognoscere potest, qui mathematicas disciplinas quam modice degustaverit.*

on Proclus' philosophy of mathematics, phantasy, and *abstractio* (see 3.5.2 and 3.5.3). It may be added that, in light of this, Barozzi's famous response to Piccolomini in his 1560 *Opusculum* could be rightfully interpreted as an attempt of the Venetian to get the right interpretation of Proclus against an overly "Simplician" reading of Piccolomini.

After this analysis of Piccolomini's denial of mathematical causality, its sources, and possible development, it could be useful to offer a synoptic view of all its influences. The reader will find one in Table 9 in the Appendix.

3.5 Piccolomini's Philosophy of Mathematics

In this section, I will explore in detail Piccolomini's longest account of his philosophy of mathematics and its akribological implications. The study will be divided into three parts. Firstly, I will introduce some of the questions on the philosophy of mathematics as inherited from the ancient world, focusing on Aristotle's legacy. Then I will analyse chapter VII of *De certitudine*, where Piccolomini expounds his ontology of mathematics. Lastly, I will elaborate on the metaphysical and epistemological presuppositions and implications of Piccolomini's theory to account for his akribology of mathematics.

3.5.1 *Aristotle's Riddles of Quantity*

The Latin term *abstrahere*, either in its verbal, participle, or adjectival forms, occurs 12 times in the body of Piccolomini's paraphrase of Pseudo-Aristotle's *Quaestiones mechanicae* and *De certitudine*. It appears once in the passage of the *Quaestiones* discussed in chapter 2.5.3 (*abstrahimus imaginantes*) where Piccolomini established, in open hostility to the Italo-Mertonian Calculators, that mathematical propositions cannot be enacted in sensible reality.[122] The discussion on the abstracted nature of mathematical objects recurs in *De certitudine* where the other 11 mentions are found.

Since the meaning and function of *abstractio* and its relation to mathematical items and knowledge will prove one of the central concepts of Piccolomini's philosophy of mathematics, I shall devote some space to analyse the puzzles that this concept has raised.

The so-called abstractionism of mathematical objects has its most influential proponent in Aristotle. However, the exact process by which ἀφαίρεσις (translated as *abstractio* by Boethius) leads to the understanding of mathematical

122 Alessandro Piccolomini, *In mechanicas quaestiones*, f. XX–XXV.

objects, its connection to other views on the nature of mathematical objects in Aristotle's works, and its precise metaphysical basis have been topics of debate among scholars since the time of the Greek commentators.[123] I will mention here seven *loci* in which Aristotle provides some details on the philosophy of mathematics:

(I) In some passages,[124] Aristotle seems to acknowledge the existence of κυρίως πόσον, which can be translated as "quantity in the strict sense", in opposition to something which is labelled τὸ πόσον κατὰ συμβεβηκός or "accidental quantity". By this distinction, Aristotle seems to suggest that there is a kind of entity which is more properly called "quantity" and another kind consisting of quantities which accidentally belong to other, non-quantitative entities.

(II) In other places,[125] Aristotle introduces the notion of τὸ πόσον κατὰ οὐσίαν or "according to substance" in opposition to quantity as a πάθος or affection of another entity. In this way, he seems to suggest that quantities are substances. In the same passage, he introduces a new concept that is arguably related to the first distinction: that of τοῦ πόσου πάθη καθ'αὑτά or "quantitative affections in themselves". It is unclear whether Aristotle means that passions such as "the large" and "the small" belong in themselves to quantity, or that they are passions that, taken in themselves, constitute quantity itself, or both.[126]

(III) In his most detailed accounts of the philosophy of mathematics in *Physica* II, 2 and in the *Metaphysica*,[127] Aristotle introduces the operator ᾗ or *qua* for referring to the isolated study of quantitative traits found in sensible objects.[128] This logical move allows the epistemic agent to remove the non-quantitative traits of a sensible entity (including matter itself) and consider quantity alone. Thus, this crystal sphere is treated *qua* sphere (and only as such) in doing geometry.[129]

(IV) Aristotle entertains in some passages that there is a logical procedure able to isolate properties which exist together and treat them as if they were

123 A. De Libera, *L'art des généralités: Théories de l'abstraction* (Paris: Aubier, 1999), pp. 25–153; Helmig, *Forms and Concepts*, p. 97.

124 *Categoriae*, ed. I. Bekker, vol. 1, 6, 5a, l. 38 and forth.

125 *Metaphysica*, V, 13, 1020a, l. 7–33.

126 For an elaboration on Aristotle's different views on mathematical objects in *Categoriae*, *Metaphysica*, and *Topica* and their Platonic leanings, see Cleary, *Aristotle and Mathematics*, chap. 3.

127 *Metaphysica*, VI, 1, 1026a, l. 5; VII, 3, 1029a, l. 10 and forth; XI, 3, 1061a, l. 28 and forth; XIII, 2, 1076b, l. 1–1077b, l. 17; 3, 1077b, l. 17–30.

128 Aristotle discusses the operator ᾗ also in *Analytica posteriora*, I, 5.

129 Helmig, *Forms and Concepts*, chap. 3; Cleary, *Aristotle and Mathematics*, chap. 5, IV; J. Lear, "Aristotle's philosophy of mathematics", in *Philosophical Review* 91 (1982), pp. 161–192.

self-subsistent or independent.[130] Hence, entities which are "logically separated" (λόγῳ χωριστόν) but not really independent of sensible items (such as shapes, angles, or lines) would be adopted as objects of study of mathematical sciences.[131] It is to be noted that it is strongly implied (and so was read by the commentators) that the process of separation and that of abstraction are, if not the same, at least closely related operations or parts of the same process: "separating from matter by reason" and "abstracting from matter" can be interpreted as describing the same epistemic operation.

(v) Aristotle sometimes refers to a never clearly defined material principle of geometrical objects which he labels ὕλη νοητή or "intelligible matter" or "noetic matter", perhaps more fittingly translated as "mental matter".[132] Arguably, Aristotle introduced the notion of a matter proper to mathematical items in contrast to the matter of sensible items, i.e., intelligible matter works as the substrate onto which mathematical items are instantiated as multiple (arguably by the epistemic agent doing mathematics).[133]

(vi) Aristotle conceives different degrees of separability in the study of different entities. This is explicitly affirmed in *Physica* II, 2 (193b, l. 36 ff.) where he attributes different degrees of separability to the objects of physics (χωρίζουσιν ἧττον) and those of mathematics. The same topic is also touched upon in other parts of the corpus, such as *De anima* III, 4 (429b, l. 10–23), and amply in *Metaphysica* VII, 5. By "more separable" Aristotle means that the nature of a given thing can be grasped without including its necessary instantiation in a particular kind of matter. A usually repeated example is that of convexity (κοιλότης) and snub-ness (τὸ σιμόν): whereas convexity can be understood without reference to any particular matter (for either the surface of crystal or wood can instantiate convexity), snub-ness, defined as "the convexity of the nose", can only be universally understood as belonging to a determined body part and found in a determined material (namely, flesh). Thus, from the nose of Socrates, both snub-ness and concavity can be separated, for wherever is snub-ness there is also concavity. But the concavity of Socrates' nose (a mathematical property) is more separable than snub-ness: its definition excludes the material substratum in which it has been seen and it must be found.

---

130 *Metaphysica*, VIII, 1, 1042a, l. 29; *De anima*, III, 7, 431a, l. 8, l. 15–19; 432a, l. 20; 10, 433b, l. 24–25.

131 For an account of the use of logical separation and its relation to abstraction in Aristotle's philosophy of mathematics, see Felgner, *Philosophy of Mathematics in Antiquity and in Modern Times*, chap. 3, sec. 4 and 5.

132 *Metaphysica*, VII, 10, 1036a, l. 10–12; 11, 1037a, l. 2–4.

133 On intelligible matter and mathematical objects according to Aristotle, see also Cleary, *Aristotle and Mathematics*, chap. 7, *passim*.

Finally, (VII) Aristotle's characterisation of the position of mathematical items and how they relate to natural entities is dubious. A clear example of this, highly relevant to the purpose of this study, is the first three chapters of *Metaphysica* XIII, 1–3, where Aristotle analyses Plato's and the Pythagorean views. Thus, against Plato, the Stagirite claims that mathematical objects exist unqualifiedly in perceptible things and are only taken as separated by the mathematician.[134] On the other hand, against the Pythagoreans, Aristotle defends that mathematical items cannot be said to exist unqualifiedly in sensible things.[135] In consonance with the first view, Aristotle also proposes in *Physica* II, 2 that the line of the geometer exists in the physical object.[136]

Aristotle's claims about the ontological status of and access to mathematical objects, the ways and degrees of separation, abstraction, intelligible matter, and the link between mathematical and physical objects leave room for many interpretations and variations (even for Platonising ones). Mueller has indexed the three most relevant readings and uses of Aristotle's claims, including Neoplatonic variations and partial adoptions, such as those of Syrianus and Proclus (and we could add also Priscian). The three paradigmatic interpretations are as follows:[137]

(1) **Naturalised full-correspondence view:** Mathematical entities are ontologically the same as physical objects but they are studied in a manner that excludes non-quantitative properties. The natural philosopher and the mathematician explore different aspects of one and exactly the same item; they only differ in the way they approach the object. On this account, to the question "Where are mathematical objects?" the mathematician would point to corporeal reality and say that he speaks only of the quantitative traits out there as if they were unchangeable and immobile. Mathematical propositions are made true and refer to quantitative traits of natural entities, and hence fit external reality.

(2) **Abstractionist full-correspondence view:** Mathematical entities are within the private representations of the mathematicians. Doing mathematics consists of taking the quantitative traits of natural items and, by stripping them from non-quantitative traits and change, duplicating

134 Aristotle, *Metaphysica* XIII, 3, 1077, l. 31–34. For an exposition and discussion of Aristotle's *loci* and the hermeneutic problems raised by them, see Helmig, *Forms and Concepts*, pp. 100–107.

135 Aristotle, *Metaphysica* XIII, 3, 1077b, l. 12–17.

136 *Physica*, II, 2, 194a, l. 8–11: ἡ μὲν γὰρ γεωμετρία περὶ γραμμῆς φυσικῆς σκοπεῖ, ἀλλ᾽ οὐχ ᾗ φυσική, ἡ δ᾽ ὀπτικὴ μαθηματικὴν μὲν γραμμήν, ἀλλ᾽ οὐχ ᾗ μαθηματικὴ ἀλλ᾽ ᾗ φυσική.

137 Mueller, "Aristotle's Doctrine of Abstraction". For clarity, I have rephrased the three interpretative models and assigned them proper names.

them in the mind as refined imitations. Yet, mathematical objects, *qua* copied-pasted quantities staged in the mind, describe also extramental quantities and ontologically depend upon them. Thus, mathematical propositions could be verified by both abstracted items and extra-mental quantities alike – explaining how mathematics can be effectively applied to natural items in sciences such as, say, optics. Asked where mathematical objects are, the mathematician would point first to external objects and then to his own head.

(3) **Discrete non-correspondence view:** Mathematical entities exist only in the mind of the mathematician: things like unchangeable two-dimensional items or perfect equality do not exist in the extra-mental world. Thus, mathematical objects cannot be unqualifiedly instantiated in the extra-mental world. Mathematicians represent in their minds quantitative properties which are suggested by, but always factually absent from, the extra-mental world. Thus, mathematical objects resemble but do not perfectly fit or have their origin in natural quantity, but in a discrete mental power. Mathematical propositions hold true exclusively in discrete mathematical objects. Questioned about the position of mathematical items, the mathematician would just point to his own head.

Different interpreters and commentators have endorsed different variations of these three archetypical views backed by additional theoretical suppositions. In the Latin tradition of commentary on Aristotle the most influential authors tend towards interpretation (2). Ex. gr., Albert the Great points out in his *Super Euclidem* that the treatment of mathematics turns the mind to imagination while it considers the quantities as they are present in sensible items (*Convenit enim geometria cum physica*).[138] Nevertheless, when he asks himself about the origin of the certainty of mathematical objects (which cannot be found in natural bodies), he leans towards (3), opening a gap between natural and

138 Albert the Great, *Super Euclidem*, ed. P. M. J. E. Tummers (Münster: Aschendorff, 2014), p. 1: [...] *et considerans formam quae est in materia mobili nec tamen est concepta cum ipsa secundum rationem diffinitivam* (*quae dicitur mathematica*) [...]; [...] *secunda autem mentem ad imaginationem conversam ornat* [...]. A clear example of Albert's theory of the mathematical object of study as immersed in sensible matter and, thus, a case of overlapping in the object of study of physics and mathematics is found in his commentary on *Physica* II, 1. See Albert the Great, *In libros Physicorum Aristotelis*, II, trac. 1, cap. 8, p. 107: *Nos enim videmus quod superficies planas et soliditates formas et longitudines et puncta habent physica corpora, et de his intendit mathematicus in geometria* [...]. ***Convenit enim geometria cum physica in subiecto suarum quaestionum,*** *licet differat in praedicato. Utraque enim inquirit de corpore et superficie et longitudine, sed geometria dicit de recta vel obliqua, physicus de calida vel alba, vel simplici vel commixta.*

mathematical quantity.[139] Aquinas also seems to have endorsed a version of (3) in his Aristotelian commentaries, claiming that the term *abstrahere* means in this context the separation of quantity from natural movement and its consideration in intelligible matter, where the mathematical objects possess *sui generis* traits.[140]

As explained in section 2.1, Syrianus, Proclus, and Priscian upheld a variation of (3) within a Neoplatonic framework. Their version of (3) includes a sophisticated theory of phantasy, projection of ensouled reasons, and, hence, a theory of Platonic recollection dependent on a psychological theory. One of the most emphatic claims of Proclus (and Syrianus) against Aristotle's theory of abstraction is that mathematical items are not ὑστερογενεῖς regarding

---

139 For an analysis on the variations of Albert's philosophy of mathematics, see Molland, "Mathematics in the Thought of Albertus Magnus", in James A. Weisheipl (ed.), *Albertus Magnus and the Sciences*, pp. 463–478.

140 Thomas Aquinas, *Phyiscorum Aristotelis expositio*, ed. Leonina, vol. 2 (Paris: Commissio Leonina, 1884), II, lec. 3: *Quia enim mathematicus considerat lineas et puncta et superficies et huiusmodi et accidentia eorum non inquantum sunt termini corporis naturalis, ideo dicitur abstrahere a materia sensibili et naturali. Et causa quare potest abstrahere, est ista: quia secundum intellectum sunt abstracta a motu.* See also Thomas Aquinas, *Sententia libri Metaphysicae*, VII, lec. 11: *Videtur enim eis, quod sicut materia sensibilis non est pars speciei in naturalibus, ita etiam quod materia intelligibilis non sit pars speciei in mathematicis. Materia autem figurarum mathematicarum intelligibilis, est continuum, ut linea vel superficies.* See also Thomas Aquinas, *Sententia libri De anima*, ed. Leonina, vol. 45.1, (Paris: Commissio Leonina, 1984), III, lec. 8: *Quaedam vero sunt formae, quae non exigunt materiam sub determinata dispositione sensibilium qualitatum, tamen requirunt materiam sub quantitate existentem: sicut triangulus, et quadratum, et huiusmodi: et haec dicuntur mathematica; et abstrahunt a materia sensibili, sed non a materia intelligibili, inquantum in intellectu remanet continua quantitas, abstracta a sensibili qualitate.* See also Thomas Aquinas, *In Boethii de trinitate*, q. 6, 2: *Quaedam vero sunt, quorum iudicium non dependet ex his quae sensu percipiuntur, quia quamvis secundum esse sint in materia sensibili, tamen secundum rationem diffinitivam sunt a materia sensibili abstracta. Iudicium autem de unaquaque re potissime fit secundum eius diffinitivam rationem. Sed quia secundum rationem diffinitivam non abstrahunt a qualibet materia, sed solum a sensibili et remotis sensibilibus condicionibus remanet aliquid imaginabile, ideo in talibus oportet quod iudicium sumatur secundum id quod imaginatio demonstrat. Huiusmodi autem sunt mathematica. Et ideo in mathematicis oportet cognitionem secundum iudicium terminari ad imaginationem, non ad sensum, quia iudicium mathematicum superat apprehensionem sensus. Unde non est idem iudicium quandoque de linea mathematica quod est de linea sensibili, sicut in hoc quod recta linea tangit sphaeram solum secundum punctum, quod convenit rectae lineae separatae, non autem rectae lineae in materia, ut dicitur in I De anima.* Amongst modern interpreters, Cleary's interpretation leans to (2), Mueller has endorsed (2) as the most coherent view, and Lear has attributed (1) to Aristotle. See Cleary, *Aristotle and Mathematics*, chap. 7; Mueller, "Aristotle on geometric objects", in J. Barnes, M. Schofield, Richard Sorabji (eds), *Articles on Aristotle*, vol. 3 (London: Duckworth, 1979), pp. 96–107; Lear, "Aristotle's philosophy of mathematics".

physical entities.[141] In other words, mathematical entities are not produced by abstraction from, nor do they ontologically depend on, physical quantities. Rather, the situation is reversed: the physical world and its mathematical-like properties depend ontologically on the Soul and, consequently, on ensouled mathematics.[142]

Lastly, we have the concept of ἀφαίρεσις as received by the philosophical tradition following Alexander of Aphrodisias, for the Peripatetic commentator introduced an epistemic aspect to this notion which persisted ever since. In his *De intellectu*, Alexander equated ἀφαιρεῖν (understood as synonymous with χωρίζειν) with the process of grasping a universal form from numerous particulars of the same kind.[143] Thus, he assimilated abstraction/separation to the operation of ἐπαγωγή (translated as *inductio*) as it appears in the Aristotelian corpus. From this, a double understanding of *abstractio* was present in the Latin semantics of the term: one (i) related to the obtention of mathematical items, another (ii) related to the obtention of separated universal forms via induction.[144]

141 Proclus, *In Euclidem*, pp. 12, l. 2–18, l. 4; Helmig, *Forms and Concepts*, chap. 5; Helmig, "Proclus' Criticism of Aristotle's Theory of Concept Formation in *Analytica Posteriora* II.19".

142 Mathematical properties as manifested in the sensible world through the Soul are called by Proclus ἀριθμοὶ φυσικοί or "physical numbers". See *In Timaeum*, I, p. 255, l. 7–26; III, p. 85, l. 23–31; and p. 89, l. 7–13. For an account of physical number see O'Meara, *Pythagoras Revived*, pp. 134–135.

143 Alexander of Aphrodisias, *De intellectu*, in I. Bruns (ed.), *De anima cum Mantissa* (Berlin: Raimer, 1887), Vol. 2, p. 110, l. 17–20: ὁ γὰρ νοῦς ὁ ἡμέτερος νοεῖ τὰ αἰσθητὰ δυνάμει ὄντα νοητά, καὶ ὑπὸ τοῦ νοῦ νοητὰ ταῦτα γίνεται. αὕτη γὰρ ἐνέργεια τοῦ νοῦ, τὰ ἐνεργείᾳ αἰσθητὰ τῇ αὐτοῦ δυνάμει **χωρίσαι καὶ ἀφελεῖν** τούτων, σὺν οἷς ὄντα ἐστὶν αἰσθητά, καὶ ὁρίσασθαι καθ' αὐτά;' and p. 111.15–19: 'πρότερον γὰρ ποιεῖ **τῇ ἀφαιρέσει** νοητόν, εἶθ' οὕτως λαμβάνει τούτων τι ὃ νοεῖ τε **καὶ** ὁρίζεται, ὅτι τόδε τί ἐστι. καὶ γὰρ εἰ ἅμα χωρίζεται καὶ λαμβάνει, ἀλλὰ **τὸ χωρίζειν** προεπινοεῖται· τοῦτο γάρ ἐστιν αὐτῷ τὸ ληπτικῷ εἶναι τοῦ εἴδους. See Helmig, *Forms and Concepts*, pp. 155–157; De Libera, *L'art des généralités*, pp. 25–153.

144 A clear example of the persistence of this double semantics of *abstractio* in Latin, as both the separation from matter and the obtention of the universal, is found in Albert the Great, *In De anima*, II, 3, cap. 4, p. 238a: *Similiter autem est quando dicitur, quod abstractio fit a materia, intelligitur quod est quoddam particulare quod particularitatem habet a materia. Hoc enim est primum subiectum formae, et per ipsum substat particulare et naturae communi quae abstrahitur universaliter ab intellectu* [...]. Helmig also proposes Aquinas to illustrate this point. See Thomas Aquinas, *In De anima*, IV, 48: *Habemus autem duplicem modum abstractionis per intellectum: unum qui est a particularibus ad universalia; alium per quem abstrahimus mathematica a sensibilibus*. See Helmig, *Forms and Concepts*, p. 97, note 55.

3.5.2 Phantasia efficiens: *Piccolomini's Reading of Proclus' Ontology of Mathematics*

Piccolomini expounds on his ontology of mathematics in chapter VII of *De certitudine: Quae praelibanda sunt ad mathematicas disciplinas pertinentia. Et primum de materia ipsarum.*[145] In this section, the non-correspondence intuitions we encountered in the 1540 paraphrase will be developed by using Proclus' *In Euclidem*. The section runs as follows:

> [1] In discussing the object of mathematical entities in book II of the commentary on the *Elementa*, Proclus asserts that mathematical entities, being nothing more than abstracted quantities themselves, cannot be entirely classified among sensible things.
> [2] Firstly, he questions how entities like indivisible points, non-deep surfaces, non-broad lines, equalities of lines and surfaces, and the like, could be placed among sensible things when there is no natural quantity that does not possess three dimensions, neither [do natural quantities possess] true equality, or any precise and absolute figure, whether circular, triangular, or of any other kind. Furthermore, how will universal and indubitable reasons be possible in mathematics if they will deal only with continuously flowing quantities and in change? Conversely, if such mathematical entities, according to Proclus, are entirely separated from sensible matter, how can such forms, already detached from sensible matter, undertake divisions, contractions, extensions, intervals, and many others of this kind?
> [3] Proclus concludes, following Plato, that these mathematical entities, on which demonstrations are based, are neither entirely sensible in the subject, nor completely freed from it. Instead, these mathematical figures are found in phantasy with the occasion provided by quantities found in sensible matter. The intellect then gathers universal reasons from the quantities present in the phantasy.
> [4] Therefore, the subject-matter of these sciences is quantity itself in this way, if I may say, phantasised (*quantum ipsum* [...] *phantasiatum*). This is why, albeit not entirely properly, it [*quantum phantasiatum*] is referred to by most people as "intelligible matter". Hence, it is clear why mathematical disciplines are termed middle sciences (*mediae scientiae*) by Proclus, Simplicius in *In De anima*, I, and Ptolemy in the preface to the

145 Alessandro Piccolomini, *De certitudine*, f. LXXXXVII.

> *Almagestum*. Their matter exists in a middle state, not entirely without sensible matter and not completely immersed in it.
> [5] Its producer [i.e., of the matter of mathematics], namely phantasy, is a certain power of the soul that holds a middle ground between sense and intellect. For this reason, and to encompass both extremes, Aristotle calls it "passive intellect". Thus, it becomes evident what is the matter of mathematics and which is the producer (*efficiens*) of mathematical entities.[146]

The core argument in this chapter consists of a Latin paraphrase conflating some passages from Proclus' *In Euclidem*, adapted and enriched by Piccolomini.

The section which Piccolomini quotes (and partially paraphrases in Latin) is found in the second prologue of Proclus' *In Euclidem*.[147] In this passage,

---

146 Alessandro Piccolomini, *De certitudine*, f. LXXXXVII–LXXXXVIIv: [1] *De materia mathematicarum facultatum disputans Proclus In I Elementorum, libro II, dicit primum entia ipsa mathematica, cum nihil aliud sint quam quantitates ipsae abstractae, non posse inter ipsa penitus sensibilia connumerari.* [2] *Primo enim quomodo puncta omnino impartibilia, superficies non profundas, lineas non latas, linearum et superficierum aequalitates, et alia id genus, inter sensibilia existentia poni posse dicendum est, cum non detur inter quanta naturalia quod trinam dimensionem non recipiat, nec vera aequalitas, nec tandem quaevis figura, circularis, triangularis, aut quaevis alia, exacta penitus et absoluta. Dehinc, quomodo in mathematicis fieri poterunt rationes ipsae universales et inconvincibiles si circa singularia ipsa quanta continuo fluentia et mutationem recipientia tantummodo versabuntur? Si vero e contra huiuscemodi entia mathematica, dicit Proclus, omnino a materia sensibili semoventur, quanam ratione species illae a materia penitus iam semotae, partitiones, contractus, protensiones, intervalla, et huiuscemodi generis multa suscipere poterint?* [3] *Concludit ergo Proclus ex Platone quod res ipsae mathematicae, de quibus fiunt demonstrationes, nec omnino in subiecto sensibiles sunt, nec penitus ab ipso liberatae, sed in ipsa phantasia reperiuntur figurae illae mathematicae, habita tamen occasione a quantitatibus in materia sensibili repertis. Intellectus autem ex iis, quae in phantasia sunt quantitatibus, rationes illas universales colligit.* [4] *Materia ergo harum scientiarum erit quantum ipsum, hoc modo, ut ita dicam, phantasiatum. Et id a plerisque, quamvis non satis proprie, materia intelligibilis nuncupatur. Ex hoc patet quare mathematicae disciplinae a Proclo, a Simplicio in I De anima, et a Ptolemaeo in praefatione Almagesti mediae scientiae nuncupantur. Sua enim materia medio modo se habet, non omnino sine materia sensibili, nec etiam in illa penitus immersa.* [5] *Eius vero efficiens, phantasia scilicet, est animae quaedam vis quae inter sensum et intellectum medium locum tenet, et idcirco ut duo extrema comprehendat eius nomen, ab Aristotele intellectus passivus appellatur. Et sic patet quaenam sit materia in mathematicis et quid mathematicarum rerum efficiens.* The division into paragraphs and numeration is mine. I will refer to these paragraphs by the assigned numbers.

147 Proclus, *In Euclidem*, pp. 49, l. 7–50, l. 9: εἴτε γὰρ ἐν τοῖς αἰσθητοῖς ἐστι τὰ σχήματα, περὶ ὧν ὁ γεωμέτρης διαλέγεται, καὶ ἀχώριστα τῆς ὕλης, πῶς ἔτι τὴν γεωμετρίαν τῶν αἰσθητῶν ἡμᾶς ἀπολύειν φήσομεν καὶ περιάγειν εἰς τὴν ἀσώματον ὑπόστασιν καὶ συνεθισμὸν εἶναι πρὸς τὴν θέαν τῶν νοητῶν καὶ προευτρεπίζειν εἰς τὴν κατὰ νοῦν ἐνέργειαν, ποῦ δὲ καὶ τεθεάμεθα ἐν τοῖς αἰσθητοῖς

τὸ ἀμερὲς σημεῖον ἢ τὴν ἀπλατῆ γραμμὴν ἢ τὴν ἀβαθῆ ἐπιφάνειαν ἢ τὴν ἰσότητα τῶν ἐκ τοῦ κέντρου γραμμῶν ἢ ὅλως τὰ πολύγωνα καὶ πολύεδρα σχήματα πάντα, περὶ ὧν ἡ γεωμετρία διδάσκει; πῶς δὲ καὶ οἱ λόγοι τῆς ἐπιστήμης ταύτης ἀνέλεγκτοι μένουσι τῶν αἰσθητῶν σχημάτων καὶ εἰδῶν τὸ μᾶλλον καὶ ἧττον ἐπιδεχομένων καὶ κινουμένων πάντη καὶ μεταβαλλομένων ἁπάσης τε ἀοριστίας ὑλικῆς ἀναπεπλησμένων καὶ τῆς μὲν ἰσότητος μετὰ τῆς ἐναντίας ἀνισότητος ὑφεστηκυίας, τῶν δὲ ἀμερίστων κατὰ μερισμὸν καὶ διάστασιν προεληλυθότων; εἴτε ἔξω τῆς ὕλης ἔστι τὰ ὑποκείμενα τῇ γεωμετρίᾳ καὶ λόγοι καθαροὶ καὶ χωριστοὶ τῶν αἰσθητῶν, ἀμέριστοι πάντες ἔσονται καὶ ἀσώματοι καὶ ἀμεγέθεις. ἔκτασις γὰρ καὶ ὄγκος καὶ ὅλως διάστασις τοῖς λόγοις διὰ τὴν ὑλικὴν ὑποδοχὴν παραγίνεται, τὰ μὲν ἀμέριστα μεριστῶς, τὰ δὲ ἀδιάστατα διαστατῶς, τὰ δὲ ἀκίνητα κινουμένως δεχομένην. πῶς οὖν ἔτι τὴν εὐθεῖαν τέμνομεν καὶ τὸ τρίγωνον καὶ τὸν κύκλον; πῶς δὲ γωνιῶν διαφορὰς λέγομεν καὶ αὐξήσεις αὐτῶν καὶ μειώσεις σχημάτων, οἷον τριγωνικῶν ἢ τετραγωνικῶν; πῶς δὲ τὰς ἀφὰς τῶν κύκλων ἢ τῶν εὐθειῶν; πάντα γὰρ ταῦτα μεριστῶς εἶναι τὴν γεωμετρικὴν ὕλην ἐπιδεικνύουσι καὶ οὐκ ἐν ἀμερέσι λόγοις ὑφεστηκυῖαν. In Morrow's translation: "If we regard the figures that the geometer talks about as belonging to the sense world and inseparable from matter, how can we any longer say that geometry emancipates us from sensible things, converts us to the realm of bodiless existence, habituates us to the sight of intelligibles, and prepares us for activity in accordance with Intellect? And where amongst sensible things have we seen the point without parts, the line without breadth, the surface without thickness, the equality of the lines from the center, or in general any of the polygonal and polyhedral figures about which geometry teaches us? And how can the propositions of this science remain irrefutable when the figures and forms of sensible things are only more or less what they are, moving and changing in every way and full of all the indeterminateness of matter, when equality is composed of its opposite inequality, and indivisibles parade as divisible and separated? But if the objects of geometry are outside matter, its ideas pure and separate form sense objects, then none of them will have any parts or body or magnitude. For ideas can have magnitude, bulk, and extension in general only through the matter which is their receptacle, a receptacle that accommodates indivisibles as divisible, unextended things as extended, and motionless things as moving. How, then, can we still bisect the straight line or the triangle or the circle? Or speak of the difference between angles, or of increase and decrease of figures such as triangles or squares? How can we talk of contact between circles and straight lines? All these things indicate that the subject-matter of geometry is divisible and not composed of partless ideas". A second passage of Proclus' commentary which very closely resembles the one quoted by Piccolomini is Proclus, *In Euclidem*, pp. 12, l. 9–13, l. 8: πρῶτον μὲν οὖν εἰ ἀπὸ τῶν αἰσθητῶν τὰ μαθηματικὰ εἴδη λέγομεν ὑποστῆναι, τῆς ψυχῆς ἀπὸ τῶν ἐν ὕλῃ τριγώνων ἢ κύκλων τὸ εἶδος τὸ κυκλικὸν ἢ τὸ τριγωνικὸν ὑστερογενῶς ἐν ἑαυτῇ μορφούσης, πόθεν ἡ ἀκρίβεια καὶ τὸ ἀνέλεγκτον ὑπάρχει τοῖς λόγοις; [...] ποῦ γὰρ ἐν τοῖς αἰσθητοῖς τὸ ἀμερὲς ἢ τὸ ἀπλατὲς ἢ τὸ ἀβαθές, ποῦ δὲ ἡ τῶν ἐκ τοῦ κέντρου γραμμῶν ἰσότης, ποῦ δὲ οἱ ἑστῶτες ἀεὶ λόγοι τῶν πλευρῶν, ποῦ δὲ αἱ τῶν γωνιῶν ὀρθότητες; οὐχ ὁρῶμεν, ὡς ἐν ἀλλήλοις πάντα τὰ αἰσθητὰ συμμέμικται καὶ ὡς οὐδὲν ἐν τούτοις εἰλικρινὲς οὐδὲ τοῦ ἐναντίου καθαρεῦον, ἀλλὰ μεριστὰ πάντα καὶ διαστατὰ καὶ κινούμενα; πῶς οὖν τοῖς ἀκινήτοις λόγοις ἐκ τῶν κινουμένων καὶ ἄλλοτε ἄλλως ἐχόντων αὐτὴν τὴν μόνιμον οὐσίαν δώσομεν; [...] πῶς δὲ τοῖς ἀκριβέσι καὶ ἀνελέγκτοις εἴδεσιν ἀπὸ τῶν μὴ ἀκριβῶς τὴν ἀκρίβειαν προσθήσομεν; πᾶν γὰρ τὸ τῆς ἀκινήτου γνώσεως αἴτιον μειζόνως ἐστὶν αὐτὸ τοιοῦτον. ψυχὴν ἄρα τὴν γεννητικὴν ὑποθετέον τῶν μαθηματικῶν εἰδῶν τε καὶ λόγων. In Morrow's translation (slightly modified): "In the first place, if we say that mathematical forms are derived from sense objects – that the soul, from seeing material circles and triangles, afterwards shapes in herself the form of circle and the form of triangle – whence come the exactness and irrefutability that belong

Proclus examines the ontological status of geometrical objects and delves into their unique characteristics to provide a coherent account of their precise metaphysical nature. He presents a series of rhetorical questions ultimately aimed at challenging the Aristotelian views that (i) geometrical objects exist within sensible matter, and (ii) they are derived through abstraction and, consequently, are secondary to sensible quantity.

Proclus begins by noting that certain characteristics of geometrical objects cannot be found amongst sensible objects. He identifies two types of these traits: (a) ontological traits related to spatial or visual properties, such as the indivisibility of a point, the absence of three-dimensionality, and the relations of identity that hold amongst geometrical objects; and (b) epistemological traits concerning the discourse about these objects, such as the irrefutability and timeless truth of geometry. Conversely, he also acknowledges that geometrical objects can exhibit body-like traits (c), such as divisibility, enlargement, reduction, and contact, suggesting that it would be paradoxical to assume they are entirely devoid of sensible-like characteristics. Through this reasoning, Proclus is preparing the reader for his concept of φαντασία as the intermediary between the intellectual and sensible levels of reality.[148]

---

to our ideas? […] Where amongst sensible things do we find anything that is without parts, or without breadth, or without depth? Where do we see the equality of the lines from centre to circumference? Where the fixed ratios of the sides? Where the rightness of angles? Do we not see that all sensible things are confused with one another and that no quality in them is pure and free of its opposite, but that all are divisible and extended and changing? How, then, can we explain that very stability which unchangeable ideas have, if they are derived from things that are ever changing from one state to another? […] And how can we get the exactness of our precise and irrefutable concepts from things that are not precise? For whatever yields knowledge that is steadfast has that quality itself in greater degree. We must therefore posit the soul as the generatrix of mathematical forms and ideas".

148 Proclus had already pointed out the hybrid nature (intellectual-sensible) of mathematical entities in the incipit of the first prologue to indicate their in-betweenness (μέση χώρα). Proclus, *In Euclidem*, p. 3, l. 1–14: τὴν μαθηματικὴν οὐσίαν οὔτε τῶν πρωτίστων ἐν τοῖς οὖσι γενῶν οὔτε τῶν ἐσχάτων εἶναι καὶ παρὰ τὴν ἁπλῆν διῃρημένων, ἀναγκαῖον, ἀλλὰ τὴν μέσην χώραν ἀπειληφέναι τῶν τε ἀμερίστων καὶ ἁπλῶν καὶ ἀσυνθέτων καὶ ἀδιαιρέτων ὑποστάσεων καὶ τῶν μεριστῶν καὶ ἐν συνθέσεσιν παντοίαις καὶ ποικίλαις διαιρέσεσιν ἀφωρισμένων. τὸ μὲν γὰρ ἀεὶ κατὰ τ'αὐτὰ ἔχον καὶ μόνιμον καὶ ἀνέλεγκτον τῶν περὶ αὐτὴν λόγων ὑπερανέχουσαν αὐτὴν ἀποφαίνει τῶν ἐν ὕλῃ φερομένων εἰδῶν, τὸ δὲ διεξοδικὸν τῶν ἐπιβολῶν καὶ τὸ ταῖς διαστάσεσι τῶν ὑποκειμένων προσχρώμενον καὶ τὸ ἀπ'ἄλλων ἀρχῶν ἄλλα προκατασκευάζον καταδεεστέραν αὐτῇ δίδωσι τάξιν τῆς ἀμερίστου καὶ ἐν αὑτῇ τελείως ἱδρυμένης φύσεως. In Morrow's translation: "Mathematical being necessarily belongs neither amongst the first nor amongst the last and least simple of the kinds of being but occupies the middle ground between the partless realities – simple, incomposite, and indivisible – and divisible things

From this section of Proclus' commentary, Piccolomini produces paragraphs [1] and [2] of his exposition quoted above.

In [1] Piccolomini invokes Proclus to sustain that mathematical entities are what he calls *quantitates abstractae*. It is noteworthy that, whereas Proclus speaks only of geometrical objects, Piccolomini generalises his first claim and applies it to all mathematical objects, not just to those of geometry. Also important is that the term *abstractum* should be read here as "separated". That *abstractae* means "separated" in this context is further confirmed by the fact that, when referring to intellect, Piccolomini also labels it as *abstractus* to signal its separateness.[149] Therefore, the reader must be warned that the participle *abstractum* in Piccolomini's text can mean either just "separated" (as the intellect) or "abstracted and, thus, separated". This shows how the original Greek terms ἀφαιρεῖν and χωρίζειν and their Latin renderings *abstrahere* and *separare* came to be used interchangeably.

In paragraph [2], Piccolomini provides reasons for sustaining that mathematical and sensible quantities cannot be the same kind of beings and, thus, that they are entities separated and distinct from natural matter. Much like Proclus, he offers a chain of rhetorical questions stating some traits of mathematical objects that cannot be found amongst sensible quantities (*quantitates naturales*): the ontological objectual precision or sense (b) of *certitudo*, and irrefutability or sense (a) of *certitudo*.[150] The Sienese argues that, nevertheless, separated quantities share some features with natural quantities. Therefore, they can be neither the same nor completely alien to sensible-like properties. Piccolomini paraphrases again Proclus' argument: if separated quantities were not some sort of sensible-like subject, then geometrical operations such as elongation or division would be impossible. In doing this, he openly refers to Proclus (*dicit Proclus*).

Therefore, as seen, Piccolomini endorses in his argumentation all three points (a), (b), and (c) expounded on by Proclus.

---

characterized by every variety of composition and differentiation. The unchangeable, stable, and incontrovertible character of the propositions about it shows that it is superior to the kinds of things that move about in matter. But the discursiveness of [mathematical] procedure, its dealing with its subjects as extended, and its setting up of different prior principles for different objects – these give to mathematical being a rank below that indivisible nature that is completely grounded in itself".

149 Alessandro Piccolomini, *De certitudine*, f. CIXV: *Intellectus vero, qui sua natura* ***abstractus*** *est, sine sensibili materia non possit declarari nec diffiniri.* I want to thank Professor Dragos Calma for his thoughtful suggestions and the inspiring conversation regarding this point.

150 See section 2.4.1 above.

Piccolomini concludes in paragraph [3] that mathematical entities are found in some in-between region called "*phantasia*". This allows mathematical beings to exist neither immersed in a sensible matter nor completely freed (or diverse) from it (*nec penitus ab ipso liberatae*). What does this puzzling affirmation mean? The answer is that, much like Proclus, Piccolomini holds that phantastic extension and multiplicity are the very "matter" of mathematical items and, in that sense, mathematical items are not freed from "matter" as universal concepts, the objects of intellect, are. Piccolomini's statements in this section follow Proclus' account of the double-matter in the second prologue:

> If we assume two classes of things that participate in the universal, namely, sense objects and objects that have existence in the phantasy (for matter likewise is twofold, as Aristotle somewhere says: the matter of things tied to sensation and the matter of phantastic objects), we shall admit that the corresponding universals are of two kinds: one perceptible, since it is participated in by sense objects, and the other phantastic, as existing in the plurality of pictures onto phantasy.[151]

Here Proclus sustains that, to explain the nature of geometrical objects, we need to posit the existence of two matters. Proclus claims, by invoking Aristotle's view, that universals (τὸ καθόλου) exist as enmattered individuals in two ways.[152] This means that there are two forms of being-a-particular. From this, it necessarily follows that there are also two kinds of substrate or "matter" by virtue of which something is a particular thing in one or the other way – for these two kinds of matter must differ in some aspect for the object instantiated in them to be regarded as two kinds instead of one. Proclus then declares that particulars are either sensible, i.e., perceived by external sense-organs, or phantastic, i.e., perceived in phantasy.

Piccolomini establishes in [3] that mathematical items exist neither as sensible in sensible subjects, nor totally freed from it [i.e., from a sensible-like subject]. Although the second part of the affirmation in Latin is obscure, the following conclusion, sc., that mathematical items are found (exclusively)

151 In Morrow's translation (modified). Original Greek in Proclus, *In Euclidem*, p. 51, l. 13–20: καὶ τὰ μετέχοντα αὐτὰ διττὰ θέμενοι, τὰ μὲν αἰσθητὰ τὰ δὲ ἐν φαντασίᾳ τὴν ὑπόστασιν ἔχοντα – καὶ γὰρ ἡ ὕλη διττή, καὶ ἡ μὲν τῶν αἰσθήσει συζυγούντων ἡ δὲ τῶν φανταστῶν, ὥς που καὶ Ἀριστοτέλης φησὶ διττὸν εἶναι τὸ καθόλου – τὸ κατατεταγμένον συγχωρήσομεν, τὸ μὲν αἰσθητὸν ὡς μετεχόμενον ὑπὸ τῶν αἰσθητῶν, τὸ δὲ φανταστὸν ὡς ἐν τοῖς τῆς φαντασίας πλήθεσιν ὑφεστηκός.

152 *Metaphysica*, VII, 10, 1036a, l. 10–12; 11, 1037a, l. 2–4.

in *phantasia*, allows for this most likely reading: by saying this, Piccolomini means that mathematical objects exist as non-sensible items although they exist in a subject which is not deprived of sensible-like traits.[153] This opinion is attributed to Proclus as an interpreter of Plato (*ex Platone*).[154] It must be underscored that, in this passage, Piccolomini uses the passive form of the verb *reperire* in the third-person plural (*reperiuntur*) for locating mathematical entities in phantasy: "they are found" or "they are discovered" in phantasy. This seems to suggest that mathematical items are endowed with some "being" of their own – even before they are found in phantasy. I will address this question in the next section.

Also, in paragraph [3] Piccolomini adds a detail that is absent from Proclus' source passage and that reveals further details regarding his sources. He affirms that mathematical entities are found "*habita tamen occasione a quantitatibus in materia sensibili repertis*". It is possible that this line was suggested to Piccolomini by Proclus in 54, l. 16–28, where the Diadochus describes how the mathematician, in attempting to separate the circle from sensible matter (χωρίζειν γὰρ ἀπὸ τούτων ἐπιχειρεῖ), triggers the projection of the phantastic circle upon which he will work out. Nonetheless, I may suggest that a more likely source for this line is the elaborations of the *Simpliciani* on the doctrine of the impassibility of the soul activities. In Genua De' Passeri commentary on *De anima* we find some revealing *loci*, inspired by his reading of Pseudo-Simplicius:

---

153 In his comment on *De anima*, III, 427a, l. 9 Priscian elaborates on the judging operation (ἐνέργεια κριτική) and how it is exercised on received perceptible input. There he writes: "it is on its own in the judging activity, **which is not altogether separate from bodies**, since in the end it comes to the soul through the affection of the organ" (οὐκ οὔσῃ πάντῃ σωμάτων χωριστῇ, ἐπειδὴ καὶ εἰς ταύτην διὰ τῆς τοῦ ὀργάνου πείσεως τελευτᾷ, οὐ πάθην οὖσαν οὐδὲ κίνησιν οὐδὲ τῶν ἔξωθεν εἰσδοχήν). That the first period 'οὐκ οὔσῃ πάντῃ σωμάτων χωριστῇ' has a particular resemblance to Piccolomini's Latin *nec penitus ab ipso liberatae* and that this precise phrase has no parallel in Proclus' text may be a hint that Piccolomini was using Priscian's section to read Proclus' φαντασία. The fact that in the following paragraph, Piccolomini introduces another passage without a straight parallel in Proclus' text but that it is found and repeated in Priscian's text, might also reinforce this conjecture. Additionally, as put in section 2.1.5, Priscian also discusses ἐνέργεια κριτική to show that the results of mathematical projection are rather an inner activity of the soul. See 'Simplicius', *On Aristotle On the Soul 3.6–13*, p. 78. See Simplicius, *In De anima*, comm. in III, 431b, l. 12, p. 277, l. 24–33.

154 Piccolomini was likely familiar with the intermediary position of mathematics explicitly attributed to Plato by Aristotle in *Metaphysica*, II, 987b, l. 14–17; III, 997b, l. 12–24; and XIII, 1076b, l. 39–1077a, l. 9. Thus, he readily aligned Proclus' position with Plato's view as conveyed by Aristotle.

TABLE 6 Genua's occasionalism

Genua De' Passeri, *In De anima*	Original Latin	English translation
p. 148v	**Occasionem** tamen sumit a naturalibus mathematicus problemata illa considerandi quamvis **intus sint**, ut **phantasticae** tamen **formulae** sunt; non autem ut optima cognoscibilia aut intelligibilia quoniam et quae in phantasia formula est, non veluti forma est, sed veluti formatum et compositum est in divisibilitate, magnitudines sunt.[a]	The mathematician, however, takes the occasion to consider those problems from natural things, although they [mathematical items] are internal, as they are phantastic little forms; but [they are internal] not as the best knowable or intelligible things, because even the [mathematical] little-form in phantasy is not like a Form, but, as something formed and composed in divisibility, they are magnitudes.
p. 181v	Quia **sensatio non fit a sensibilibus**, nec scientia a scibilibus, quamvis nec sine scibilibus scientia, neque sine sensibilibus sensatio fiant; quia utrumque fit ab anima tametsi **occasione obiectorum**, unde concurrunt obiecta **veluti causa sine qua non.**	Because sensation does not happen from sensible things, nor does knowledge come from knowable things, even though knowledge is not without knowable things, and sensation does not occur without sensible things; for both are produced by the soul, although with the occasion of the objects, whereby the objects concur as a cause without which it does not happen.
p. 80	Addit Simplicius quod tametsi contingat in hac operatione sensitorium pati, in ipsa tamen vita sensitiva secundum proprium actum fit sensibilium apprehensio, **non extra capiens, sed occasione specierum sensibilium constituta**, non patiens ab illis, sed secundum eas operatur [...].	Simplicius adds that even though it happens in this operation that the sensitive organ is acted upon, in the very sensitive life, though, the apprehension of sensible things occurs according to its proper act, not receiving anything from outside, but with the occasion of sensible species, it is not acted upon by them, but operates according to them [...].

a The phrase "[...] *quae in phantasia formula est, non veluti forma est, sed veluti formatum et compositum est* [...]" arguably paraphrases Priscian's com in III, 429b, l. 18, p. 233, l. 24–25: ἐπεὶ καὶ ὁ ἐν φαντασίᾳ τύπος οὐχ ὡς τὸ εἶδος ἀλλ᾽ὡς τὸ εἰδοπεποιημένον.

Those three *loci* provide highly relevant linguistic and doctrinal insight. First of all, these three passages and many others found in Genua's commentary convey different applications of the same doctrine. As said, this is the Neoplatonic idea that, in the cognition-related powers of the soul, from perception to intellection, the stimulus initiating the process does not strictly alter or affect the activity itself, but it affords the power activation.[155] For instance, in enacting perception, the external sensible species work as affordances initiating the internal process of cognitive apprehension of sensible species. This means that, in a sense, every cognitive activity is a kind of recollection or self-recognition of the intellect, for different kinds of stimuli allow the intellect (and its inner reasons) to be manifested and hence understand itself at different levels.

Secondly, in the Latin phrasing of Genua this particular sort of intervention of a stimulus which triggers a parallel, inner cognitive activity was described with the term *occasio*; the adverbs *occasionaliter* and *excitative* are also introduced for describing this parallel process.[156] These terms, meaning "by occurrence" and "by arousing" respectively, clearly capture what Genua wanted to express as he interpreted Priscian, namely that the cognitive powers (*vires*) of the soul are not directly affected by external sensed stimuli. In other words, in the same way that intellection happens with a *phantasma*, but the *phantasma* does not act upon the intellect (which is superior, active, and nobler), but rather the *phantasma* affords the manifestation of the intellect as a power of the soul, so, generally, external stimuli are conditions, concurrent events, or affordances providing the chance for inner cognitive activities to be triggered. Yet, such occasional causes do not act directly upon the soul's powers and so, they are described as *causa sine qua non*.

Lastly, Genua also holds that the power which grants access to mathematical items is phantasy. How he phrases this particular point is extremely close to Piccolomini's additional clause to his paraphrase of Proclus: we are told by

155 On Genua De' Passeri's and Priscian's noetics, see respectively Spruit, *Species Intelligibilis*, pp. 163–164 and Steel, *The Changing Self*, pp. 132–142. For some details on the impassibility of the soul's activities and epistemic "occasionalism" in the history of ancient and renaissance Neoplatonism see S. Fellina, *Modelli di episteme neoplatonica nella Firenze del '400. Le gnoseologie di Giovanni Pico della Mirandola e di Marsilio Ficino* (Florence: Olschki, 2014), pp. 9–26. The occasionalist view stems from the later Platonists' interpretation of Plato's *Timaeus*, 72a, l. 2–7; *Theaetetus*, 186b ff. and Aristotle's position in *De Anima*, II, 8, 416b, l. 32–417b, l. 25. According to this view, the soul's perceptive activities, in and of themselves, are not altered or affected by physical sensory stimuli; rather, only the bodily organs are, strictly speaking, passive. See, for instance, Plotinus, *Enneades*, III, 6, 2–3.

156 Marco Antonio (Genua) De' Passeri, *In De anima*, Index: *Ad intellectionem explendam non intellectus solus agens cum possibili concurrit, sed phantasma quoque* ***occasionaliter et excitative*** *requiritur.*

both authors and in almost the same words that mathematical items (*quantum phantasiatum* for Piccolomini, *formulae phantasticae* or *magnitudines* for Genua) are obtained in *phantasia* with the *occasio* of quantity found in sensible items.

Thus, it seems a likely conjecture that, in interpreting Proclus, Piccolomini added a clause coming from his Simplician background, namely that the process of obtaining *quantum phantasiatum* (which he calls *abstractio*) must not be understood as a copy-paste affect of phantasy when it is acted upon by a perceived natural quantity. In other words, the perceived quantity is not "imprinted" on a passive phantasy, whereby it becomes a mathematical object. Rather, when quantity is perceived and the mathematical object which it resembles is sought by the perceiver, mathematical phantasy triggers generating by itself a "mathematical image" that possesses its own distinct characteristics, only loosely connected to the sensory quantity that afforded the activity.

The last line of [3] closes with a mention of intellect. We are told that *intellectus*, which Piccolomini considers as separated (*abstractus*), gathers its *rationes universales* from the quantities of phantasy (*Intellectus autem ex iis, quae in phantasia sunt quantitatibus, rationes illas universales colligit*). It is noteworthy that Piccolomini uses here *rationes* to designate the universal form, seemingly as a translation of λόγοι as he found them in the texts of Proclus and Priscian. Nevertheless, Piccolomini remains silent on whether, in the manner of Genua,[157] he accepts the existence of the ensouled *rationes* of which Priscian and Proclus speak. In section 3.5.3.1 below, I will present an alternative theory that would allow Piccolomini to explain the production of mathematical objects without accepting the Neoplatonic ensouled reasons.

In paragraph [4], Piccolomini introduces a syntagm of his own invention to refer to abstracted quantities: *quantum phantasiatum*.[158] It is also noteworthy that, in *De certitudine*, Piccolomini always adopts the transliterated from *phantasia* and related terms instead of *imaginatio* – which he used in the paraphrase. This terminological choice might suggest that Piccolomini gradually and increasingly endorsed Proclus' philosophy of mathematics from 1540 to 1546.

157 See, for instance Marco Antonio (Genua) De' Passeri, *In De anima*, p. 148v: [...] *ut Pythagorici opinabantur, qui formulas [mathematicas] illas substantiales opinabantur, scilicet quia ipsi respiciebant in* ***vitales mathematicarum rationes***. See also Marco Antonio (Genua) De' Passeri, *In tres libros Aristotelis De anima*, p. 168v: [...] *quia, scilicet, de potentia ad actum educitur imperfecta a se ipsa perfecta, cum in se ipsam recurrat; ac se ipsam intueatur* ***veris plenam rationibus***.

158 The passive participle *phantasiatum* was already in use, see section 2.2.4.

Piccolomini then identifies *quantum phantasiatum* with intelligible matter (*materia intelligibilis*), sustaining that both terms are synonyms. In doing this, he is seemingly following two other passages from Proclus' second prologue, very close in their position to the passages he has already used, namely:

> For this reason, everything that it [discursive reason] thinks is a picture or a shape of its thought. It thinks the circle as extended, and although this circle is free of external matter, it possesses intelligible matter provided by phantasy itself. This is why there is more than one circle in phantasy, as there is more than one circle in the sense world.[159]

And

> Let us, then, think of the universal in its instances as of two sorts, the universal in sense objects and the universal in objects of phantasy, and likewise the Form of the circle, or of the triangle or of the figure itself, as of two kinds, one presiding over intelligible matter, the other over perceptible.[160]

In these sections, Proclus strongly relates the two Aristotelian terms φαντασία and ὕλη νοητή. He addresses the question of intelligible matter by treating it as the counterpart of sensible matter in phantastic space, being the "matter" of phantastic entities. In this manner, Proclus develops Aristotle's passages in *Metaphysica* (1036a, l. 9–10; 1037a, l. 4; 1045a, l. 33–35) regarding a second type of matter which allows the universal Forms of geometrical entities to be instantiated and multiplied as something other than sensible items.

Piccolomini translates these views into his Latin paraphrase, rendering *quantum phantasiatum* and *materia intellegibilis* total equivalents: "intelligible matter" is another name for the sensible-like properties displayed by phantastic quantities. As already said above, Piccolomini adds the caveat that the

---

159 Morrow's translation. Original Greek in Proclus, *In Euclidem*, pp. 52, l. 25–53, l. 3: [...] καὶ διὰ τοῦτο πᾶν, ὅπερ ἂν νοῇ, τύπος ἐστὶ καὶ μορφὴ νοήματος, καὶ τόν τε κύκλον διαστατῶς νοεῖ τῆς μὲν ἐκτὸς ὕλης καθαρεύοντα νοητὴν δὲ ὕλην ἔχοντα τὴν ἐν αὐτῇ, καὶ διὰ τοῦτο οὐχ εἷς ἐν αὐτῇ κύκλος, ὥσπερ οὐδὲ ἐν τοῖς αἰσθητοῖς.

160 Morrow's translation. Original Greek in Proclus, *In Euclidem*, p. 53, l. 18–22: διττὸν οὖν σοι νοείσθω τὸ καθόλου τὸ ἐν τοῖς πολλοῖς, τὸ μὲν ἐν τοῖς αἰσθητοῖς, τὸ δὲ ἐν τοῖς φανταστοῖς. καὶ ὁ κυκλικὸς λόγος διττὸς καὶ ὁ τριγωνικὸς καὶ αὐτὸς ὁ τοῦ σχήματος, ὁ μὲν ἐπὶ τῆς νοητῆς ὕλης, ὁ δὲ ἐπὶ τῆς αἰσθητῆς. Intelligible matter is introduced a third time (without direct reference to phantasy) in explaining the space within which theorems are realised. Proclus, *In Euclidem*, p. 78, l. 19–20.

label *materia* is improperly said of *materia intelligibilis* (*non satis proprie*) – a view argued in his denial of mathematical causation and that we encountered in Boccadiferro's lectures (2.4.3). As said, since *materia intelligibilis* is nothing but actual "quantity itself located in phantasy", as also phrased in *De certitudine*,[161] the rejection of the unequivocal materiality of intelligible matter fits Piccolomini's denial of mathematical causality.

Still in paragraph [4], Piccolomini relates the name of the *scientiae mediae* to their ontological substratum in *phantasia*. This is another move of Piccolomini's own with no equivalent in Proclus' text. He tells us that the appellative *mediae* was endorsed by ancient commentators such as Ptolemy,[162] Simplicius, and Proclus for referring to subalternate sciences. The reason for this is, he claims, that the objects of mathematical disciplines exist in a "middle way" (*medius modus*) between intellectual and sensible realities, neither completely immersed in matter, nor without (intelligible) matter. Therefore, the label *scientiae mediae* stands for "the sciences whose objects of study are in-between".

This statement completes Piccolomini's views on the precise status of *scientiae mediae* expressed in the *Praefatio* to the paraphrase of the *Quaestiones mechanicae*. For one thing, every *scientia media* is a contemplative science from the very fact that it treats mathematical objects. Hence, in Piccolomini's view the *scientiae mediae* and the superior, subalternating mathematical sciences, geometry and arithmetic, do not differ regarding the ontological status of their subject-matter: both the object of study of geometry and mechanics are both *quantum phantasiatum*.

Thus, according to Piccolomini, the two key differences between mechanics and geometry are: (1) the level of generality in their propositions and principles, with geometry dealing with more general principles than mechanics and, thus, subordinating it. Secondly (2), the extent to which the *quantum phantasiatum* which is the object of study in mechanics relates to sensible items. This suggests that the phantastic balance of the mathematician, although a distinct mathematical entity with its own principles, is more tightly grounded on the empirical balance – whose movements it resembles – than a line or a point does. This reasoning is also evident in the *Praefatio* and the paraphrase, where

161 Alessandro Piccolomini, *De certitudine*, f. CIII: *Materia autem intelligibilis quantitas ipsa est in phantasia collocata, ut declaravimus superius de materia mathematica pertractantes.*

162 See Claudius Ptolemy, *Claudii Ptolemaei opera quae exstant omnia*, ed. J. L. Heiberg (Leipzig: Teubner, 1898), Vol. 1, *Syntaxis astronomica*, p. 6, l. 3–6: [...] ἔτι τε τόπου καὶ χρόνου καὶ τῶν ὁμοίων ζητητικὸν ὑπάρχον ὡς μαθηματικὸν ἂν ἀφορίσειε τῆς τοιαύτης οὐσίας μεταξὺ ὥσπερ ἐκείνων τῶν δύο πιπτούσης οὐ μόνον τῷ καὶ δι᾽αἰσθήσεως καὶ χωρὶς αἰσθήσεως δύνασθαι νοεῖσθαι [...].

Piccolomini has emphasized that, although mechanics is a mathematical science, its object is closer to natural phenomena.[163]

Nonetheless, the ontological status of the objects of study of geometry and mechanics is the same: both disciplines study phantastic, intermediary items. Moreover, from this position follows that arithmetic and geometry are also *scientiae mediae*. Piccolomini seemingly reads the traditional adjective *mediae* as referring to subalternate mathematical sciences through the Neoplatonic in-betweenness to locate all of them in the same ontological level, between intellect and sensible items.

Finally, Piccolomini offers some more details about *phantasia*. The source of paragraph [5] is another passage from Proclus' second prologue:

> For phantasy, both thanks to its formative movement and because it has existence with and in the body (τὸ μετὰ σώματος καὶ ἐν σώματι), always produces individual pictures that have divisible extension and shape, and everything that it knows has this kind of existence. For this reason, a certain person has ventured to call it passive intellect (νοῦν παθητικόν).[164]

Proclus makes three points regarding phantasy: (i) φαντασία exists "with the body and in the body;" (ii) phantasy knows things by producing entities which are divisible, extended, and multiple, and (iii) phantasy has been called "passive intellect" by someone, namely Aristotle.[165]

As we saw in section 2.2.3, by (i) Proclus means two things: (1) that the activity of phantasy is dependent on body perception as it is dependent on perception and, (2) that phantasy, as located in the inferior vehicle of the soul, is bound to act together with the body.

By (ii) Proclus means that phantasy gives shape to all the objects it treats or (the other way around) that it only deals with objects endowed with shape. As we explored in section 2.2.3, this can happen in a variety of ways, including the poetic and mathematical activities of phantasy.

With (iii) Proclus makes the move of identifying φαντασία with νοῦς παθητικός. The identification of these two powers recurs in the body of *In Euclidem*,

---

163 See Alessandro Piccolomini, *In mechanicas quaestiones*, f. IIv and VI.

164 Morrow's translation (modified). Original Greek in Proclus, *In Euclidem*, p. 51, l. 20–52, l. 4: καὶ γὰρ ἡ φαντασία διά τε τὴν μορφωτικὴν κίνησιν καὶ τὸ μετὰ σώματος καὶ ἐν σώματι τὴν ὑπόστασιν ἔχειν μεριστῶν ἀεὶ καὶ διῃρημένων ἐστὶν καὶ ἐσχηματισμένων τύπων οἰστική, καὶ πᾶν ὃ γιγνώσκει τοιαύτην ἔλαχεν ὕπαρξιν. ὅθεν δὴ καὶ νοῦν παθητικόν τις αὐτὴν προσειπεῖν οὐκ ὤκνησεν.

165 Aristotle, *De anima*, III, 5, 430a, l. 24.

namely in the second prologue,[166] and in the commentary on postulates I–III.[167] The Diadochus justifies this view by claiming that νοῦς, if properly understood, cannot be passive in any sense, for it is an active principle. The oxymoron νοῦς παθητικός or "intellect able to be acted upon" is meant to signify nothing other than the ontological and epistemic in-betweenness of phantasy as an intellect-related power which is, nonetheless, able to be passive. This is to say, phantasy is characterised by both the passive and receptive powers of body perception and by the active and science-related powers of the intellect, being able, at its highest, to unfold the ensouled images of the noetic Forms.[168]

Piccolomini expresses all (i), (ii), and (ii) in paragraph [5].

He firstly indicates (ii), namely that phantasy is the producer of the objects of study of the *scientiae mediae*, sc., mathematical items. This is the first explicit mention in Piccolomini's account of an activity or process that takes place in phantasy.[169] So far, phantasy has been described as matter-like stuff where separated quantities are placed but no activity or movement has been explicitly attributed to it. Now we are told that *phantasia* is the *efficiens* of the matter of mathematics, i.e., of phantastic quantity. Two observations are in order in relation to this objectivation.

Firstly, (1) this closes the circle concerning the "Simplician" note that natural quantities are the "occasion" for mathematical items to be obtained. Now we are explicitly told that the power parallelly triggered in producing *quantum phantasiatum* is *phantasia* itself.

Secondly, (2) it must be highlighted that in Piccolomini's exposition, there is no term equivalent to Proclus' central concept of προβάλλω/προβολή. Nevertheless, although the term is also found (and abundantly so) in Priscian's *In De anima*,[170] it has not a definite and systematically adopted equivalent in Latin – not even in Genua's philosophy. The projectionist language of Proclus and Priscian was perceived by Piccolomini as a description of the discrete

166 Proclus, *In Euclidem*, p. 56, l. 17–18.

167 Proclus, *In Euclidem*, p. 186, l. 7. Outside *In Euclidem*, Proclus adopts this expression for referring to phantasy in other works. See *In Timaeum*, I, p. 244, l. 20–21; III, p. 158, l. 8–9. Proclus coins the syntagm νοῦς φανταστικός in *In Rempublicam*, II, f. 107, l. 17. For an elaboration on the topic see Chlup, *Proclus, An Introduction*, p. 154.

168 For Proclus' account see Proclus, *In Euclidem*, p. 52, l. 4–20.

169 For an account of the active capacities of Proclus' φαντασία, see MacIsaac, "Phantasia between soul and body in Proclus".

170 Simplicius, *In De anima*, p. 6, l. 4; p. 7, l. 9, l. 16, and l. 23; p. 58, l. 32; p. 60, l. 35; p. 67, l. 31; p. 77, l. 12; p. 98, l. 18; p. 119, l. 8 and l. 17; p. 121, l. 33; p. 123, l. 19; p. 124, l. 35; p. 152, l. 38; p. 166, l. 6 and l. 16; p. 183, l. 9; p. 190, l. 25; p. 202, l. 18; p. 206, l. 14 and l. 18; p. 208, l. 21; p. 216, l. 30; p. 225, l. 4; p. 264, l. 33; p. 290, l. 31; p. 311, l. 18; p. 250, l. 7; p. 263, l. 32.

cognitive activity generating mathematical phantasy. Therefore, it was rendered in a more general language which points towards the discrete efficiency of phantasy – plausibly reflecting Proclus' summary claim that phantasy is the "genetrix of mathematical forms and ideas".[171]

That *phantasia* possesses efficient powers is a claim that could cast grave doubts on the coherence of *De certitudine*. At the heart of Piccolomini's argumentation lies the view that there is no causation at all in mathematics. Moreover, as we saw, Piccolomini explicitly denies that *causa efficiens* takes place in mathematics and that all mathematical movement is metaphorical.[172] Nevertheless, a solution to this apparent incoherence can be found.

In (partially) accepting Proclus' view (i), that φαντασία exists "with the body and in the body", Piccolomini restresses the active aspect of phantasy calling it a "power" (*vis*) of the soul which is intermediary between the powers of sense perception and intellect (*inter sensum et intellectum medium locum tenet*). Piccolomini's hinted rationale seems to be the following: phantasy is the producer of abstracted quantity, but abstracted quantity itself is causally idle. In other words, whereas phantasy is an active power, the products of phantasy are completely inert. From a disciplinary point of view, the same delimiting difference could be expressed as follows. Take the following two statements:

(α) Phantasy causes the triangle.
(β) The triangle causes Pythagoras' theorem.

Note that statements (α) and (β) do not belong to the same discipline: (α) is a claim within the field of psychology, explaining one of the soul's powers; (β), instead, is a mathematical claim. Piccolomini's *De certitudine* only denies causality as in the case of (β).

Piccolomini also assumes Proclus' tenet (iii), i.e., the identification of φαντασία with νοῦς παθητικός as a label introduced by Aristotle for describing the intermediary soul-power of phantasy. He does so with great ease since two of his major philosophical authorities besides Proclus agree.

Firstly, as shown in section 2.2.5, Averroes identified some of the uses of *phantasia* (التوهم, *at-tawahhum*) with *intellectus passibilis* (العقل المنفعل, *al-ʿaql*

171 This claim is found in the first prologue, Proclus, *In Euclidem*, p. 13, l. 8: ψυχὴν ἄρα τὴν γεννητικὴν ὑποθετέον τῶν μαθηματικῶν εἰδῶν τε καὶ λόγων. In Morrow's translation: "We must therefore posit the soul as the generatrix of mathematical forms and ideas". See G. Watson, *Phantasia in classical thought*, chap. 5; MacIsaac, "Phantasia between soul and body".

172 Alessandro Piccolomini, *De certitudine*, f. CIIv.

al-munfaʿil), understanding that the later syntagm in *De anima* III, 5 refers to the individual cognitive power providing the *phantastmata* necessary to enact intellection. So, in Averroes' reading, an aspect of *ymaginatio* is equivocally labelled as *intellectus* as it is an intellection-related power whose activity is required for conjunction with the material intellect – العقل الهيولاني, *al-ʿaql al-hiyūlānī*. But unlike the eternal, separated substances which the agent and the material intellects are, *intellectus passibilis* is perishable and able to undergo alteration and corruption, as *ymaginatio* is tied to the individual body as one of its powers and, thus, it is *passivus*. Averroes' classification of intellects was inherited and discussed by the long-standing Latin Averroistic tradition; the identification of intellection-related activity *phantasia* with *intellectus passivus* was perhaps one of its less problematic views and generally endorsed by 16th-century philosophers in interpreting *De anima*.[173]

Secondly, Priscian also stresses a similar point. For instance, in commenting on section 403a, l. 8, he states that "Neither is phantasy unqualifiedly intellect, but the entirety of it is a passive and formative intellect".[174] In doing this, Priscian identifies the extension-giving power of phantasy with νοῦς παθητικός, for this power is the one able to unfold the inner λόγοι.

Lastly, another of Piccolomini's sources, Themistius, supports the (at least partial) identification of νοῦς παθητικός with φαντασία as part of his effort to reconcile Plato and Aristotle.[175]

That being so, it is unsurprising that Piccolomini easily accepted Proclus' identification of phantasy, intelligible matter, and passive intellect: he just found a token of a widely accepted philosophical theory confirmed in Proclus' text. In doing this, he was reading Proclus' theory of phantasy through his own intellectual background, totally alien to Proclus' theory of the two ὀχήματα of the soul and the πνεῦμα, a doctrine absent from *In Euclidem* (see 2.2.3). Therefore, Piccolomini is arguably interpreting that the power of *phantasia* is

173 Spruit, "Introduction" in Agostino Nifo, *De intellectu*, ed. L. Spruit (Leiden-Boston: Brill, 2011), pp. 1–32; Spruit, "The Pomponazzi Affair. The Controversy over the Immortality of the Soul", in H. Lagerlund, B. Hill (eds), *The Routledge Companion to Sixteenth Century Philosophy*, pp. 225–246; C. Leijenhorst, "The Nature of the Understanding: Intellect, Conception, and Concepts", in H. Lagerlund, B. Hill (eds), *The Routledge Companion to Sixteenth Century Philosophy*, pp. 475–492; S. Boer, "The (human) soul", in H. Lagerlund, B. Hill (eds), *The Routledge Companion to Sixteenth Century Philosophy*, pp. 411–435; G. Di Napoli, *L'immortalità dell'anima nel Rinascimento* (Turin: Società Editrice Internazionale, 1963).

174 Simplicius, *In De anima*, p. 17, l. 4–5: οὔτε ἁπλῶς νοῦς ἡ φαντασία, ἀλλὰ τὸ ὅλον τοῦτο παθητικὸς καὶ μορφωτικὸς νοῦς.

175 Themistius, *In libros Aristotelis De anima paraphrasis*, pp. 88, l. 26–91, l. 29. See also Themistius, *On Aristotle On the Soul*, ed. R. B. Todd (London: Bloomsbury Academic, 1996); M. Gabbe, "Themistius as a Commentator on Aristotle: Understanding and Appreciating his Conception of Nous Pathetikos and Phantasia", in *Dionysius* 26 (2008), pp. 73–92.

(ontologically) an activity of the individual body, inseparable from it, and able to engage in intellection-related activities.

To conclude, Piccolomini's tenets concerning the philosophy of mathematics as expressed in chapter VII of *De certitudine* can be summarised in nine points:

(i) Mathematical items exist as separated quantitative items in phantasy (*quantum phantasiatum*).
(ii) Mathematical items are not totally freed from sensible-like traits; they manifest ontological and epistemological traits impossible in sensible matter.
(iii) Mathematical items, therefore, exist in a middle way (*modus medius*) in-between sensible matter and intellect, i.e., in phantasy.
(iv) Mathematical items are produced by phantasy alone (*phantasia efficiens*).
(v) The affordance (*habita occasione*) of sensible quantity triggers phantasy's production of mathematical objects.
(vi) The intellect gathers universal mathematical concepts from phantastic quantities.
(vii) The object of study of middle or subalternate sciences (*scientiae mediae*) is phantastic quantity – and nothing else.
(viii) Phantasy is a power of the soul between perception and intellect (*medium locum tenet*) – but tied to the body.
(ix) Therefore, phantasy is improperly called "intelligible matter" and also "passive intellect".

3.5.2.1 *Coda*: Is Piccolomini an "Aristotelian"?

As shown by his account in chapter VII of *De certitudine*, following Proclus' text in many respects, Piccolomini's philosophy of mathematics emerges as a species of the third (3) interpretation of abstraction:

> **Discrete resemblance-correspondence view:** Mathematical entities exist only in the mind (*phantasia*) of the mathematician (for things like two-dimensional items or perfect equality do not exist in the extra-mental world). Mathematical objects cannot be unqualifiedly instantiated or found in the extra-mental world (i.e., in sensible matter). The mind autonomously generates quantitative items which are suggested by, but absent from, the extra-mental world. Thus, mathematical objects resemble but do not perfectly correspond or straightforwardly derive from natural realities as copy-paste replicas. Questioned about the ontological position of mathematical entities, the mathematician would point to his own head.

At this point, it must be emphasised that the theory of mathematical objects expounded in *De certitudine* is incompatible with the views expressed by Piccolomini in later works, such as *Prima* and *Seconda parte della filosofia naturale* of 1551 and 1554 respectively (= *Filosofia naturale*). Indeed, in his later vernacular production, when elaborating on the distinction between the mathematician and the natural philosopher in addressing the crucial passage in *Physica* II, 2, Piccolomini affirms that mathematicians and natural philosophers speak about the very same items, but consider them differently:

> I say that the mathematician and the natural philosopher, although they can consider **one and the same** line or surface, will treat them differently: the mathematician as a line or surface. ([…] *dico che'l matematico e'l filosofo naturale, se ben possano una* ***medesima*** *linea ò superficie considerare, tuttavia il matematico in quanto linea e superficie le tratterà* […]).[176]

Thus, in this view, the mathematician considers one and the very same object but only *qua* quantity. Piccolomini explains here that mathematical abstraction (*astrattion*) is a logical operation or a way of considering or paying attention to a natural quantity or, in other words, a manner of understanding things which exist together in natural matter and are not separated in any sense as if they were separated. Therefore, Piccolomini endorses in his *Filosofia naturale* (or

176 Alessandro Piccolomini, *Prima parte della filosofia naturale* (Venice: Lorenzini, 1560), II, f. 60v: *Per la qual cosa, non senza ragione, matematici possan senza falsità e senza inganno alcuno separar con l'intelletto dalle sostanze materiali la quantità como soggetto loro, et d'attorno a quelle far le demostration loro, senza mention alcuna di sostanza che al mondo sia. Tornando dunque à proposito dico* ***che'l matematico e'l filosofo naturale, se ben possano una medesima linea ò superficie considerare, tuttavia il matematico in quanto linea e superficie le tratterà, senza considerare se in oro, od in argento, od in terra, od in acqua, od in Cielo, od in altra sostanza si trovino.*** See also Alessandro Piccolomini, *Prima parte della filosofia naturale* ff. 59v–60: *Per essempio, farem noi, quando le superficie e le linee comprenderemo separatamente dall'oro, ò dall'argento, ò da qualsi voglia altra sostanza sensata, non in modo però che giudichiamo poter essere quelle, per loro stesse disgiunte da cosi fatte sostanze, ma solamente considerando quelle, non facendo consideratione di queste.* See also Alessandro Piccolomini, *Seconda parte della filosofia naturale*, I, f. 33: *Hor se ben queste due linee, cioè la retta e la circolare senza altra materia che la pura quantità loro con astrattion matematica considerate, à nissuna natural mutatione, ò sito determinato si sottopongano.* De Pace has used part of the section in f. 24v from *Prima parte della filosofia naturale* in interpreting Piccolomini's views on chapter VII of *De certitudine*. See De Pace, *Le matematiche e il mondo*, p. 44 and forth. Higashi has also interpreted Piccolomini's philosophy of mathematics as "anti-realist" (Higashi's label) meaning two things, namely (a) that Piccolomini denies the substantiality of phantastic mathematical items (which is true of both Piccolomini and Proclus), and (b) that Piccolomini asserts that phantastic quantity is ontologically derived from natural quantity. My interpretation differs from that of Higashi (and De Pace's) in point (b). See Higashi, *Penser les mathématiques*, p. 136.

attributes to Aristotle as he paraphrases him in Italian) a full-correspondence reading of mathematics, oscillating between the view that mathematical objects are exactly the same thing as natural objects and the view that mathematical objects are intellectual simplifications derived from natural quantity as stripped replicas. Moreover, in *Filosofia naturale*, the process that grants access to mathematical concepts is attributed to the intellect (*intelletto*) – a view alien to *De certitudine*'s philosophy. Additionally, *fantasia* or *imaginatione* is not described as an active power but rather plays a subordinate role to the intellect, which is the faculty responsible for performing abstraction by supposing that accidental natural quantities exist independently from their subjects.[177] Proclus is absent from these accounts, which are focused on explaining Aristotle's *Physica* in its own terms (and with space for some innovation and interpretation).

The conclusion is that Piccolomini's theory in the later *Filosofia naturale* and the views held in *De certitudine* are hard (if not impossible) to reconcile. Because of this, the interpretation of Piccolomini's *De certitudine* which I have offered takes its claims in the opusculum at face value and avoids interpreting his philosophy by using his later vernacular production. The approach of conflating Piccolomini's views in chapter VII of *De certitudine* with his later Italian paraphrases (where he was, by his own confession, faithfully explaining the Peripatetic philosophy)[178] has produced an overly "Aristotelian" reading of *De certitudine*'s ontology of mathematics. In this reading, Piccolomini would hold the view that mathematical items are posterior derivations of natural quantity. Additionally, he would use Proclus to illustrate such a view – without accepting much of his philosophy.

In the previous sections, I have aimed to present a different take on Piccolomini's philosophy of mathematics in *De certitudine*. By considering his educational background as a "Simplician" Averroist and the philosophical ideas on mathematics found in the texts published alongside the opusculum,

---

177 See, ex. gr., Piccolomini's account of time and movement as *continua* in Alessandro Piccolomini, *Prima parte della filosofia naturale* (1560), ff. 113v–114. Here the term *fantasia* is invoked, but Piccolomini never attributes the power of generating imagined movement solely to it. Instead, this power belongs to the pair of *fantasia* and *intelletto*. Moreover, when it refers to the process of mathematical abstraction, *fantasia* is also put together with *intelletto*, something wholly alien to Piccolomini's ontology of mathematics in chap. VII of *De certitudine*. Overall, it appears that, in these writings, *fantasia* plays a secondary role, yielding to intellect in granting access to mathematical objects. In some cases, faithfully to Aristotle's text, *fantasia* is even omitted from the operation of mathematical abstraction (*separar con l'intelletto*, see quotation from f. 60v above).

178 See Piccolomini's testimony in Sgarbi, *The Italian Mind*, p. 181: [...] *secondo la via dei peripatetici; non traducendo, ma ampliando dove bisogna, di maniera però che io non mi parti dal parer primariamente di Aristotile e di poi dei primi suoi greci espositori.*

the *Praefatio* to the paraphrase, I have tried to offer a more nuanced understanding of his views. As seen, in chapter VII, Piccolomini integrates Proclus' ontology of mathematics into a unique intellectual framework shaped by the 16th-century interpretations of Averroes and Pseudo-Simplicius. Regarding his denial of causality in mathematics, evidence has been provided that its origin is not fully Aristotelian: rather it is a full-fledged unfoldment of a theory he received from the Neoplatonist Priscian and which the Siense found confirmed in Proclus' *In Euclidem*. Moreover, *De certitudine* also blends Piccolomini's idiosyncratic theory of *regressus* with a philosophy of mathematics influenced by the contemporary debates on mechanics and Mertonian quantification.

We can conclude that, while Piccolomini was undoubtedly a reader and interpreter of Aristotle (as anyone with a philosophical education at the time), his philosophy of mathematics in *De certitudine* resists being labelled simply as "Aristotelianism". In sum, Piccolomini's opusculum is such a variegated and unique theoretical work that it perhaps defies any overly broad historiographical categorization.

3.5.3 *Abstraction and Akribology in* De certitudine

Piccolomini has accepted (1) that natural quantity plays the role of a triggering condition in the production of separated quantity, and (2) that *phantasia* alone is the producer of mathematical items. Nonetheless, the details of Piccolomini's process of *abstractio* and how exactly mathematical objects are found (*reperiuntur*) in phantasy are aspects of his theory still unclear.

One relevant issue is that, unlike Proclus, the Sienese has not explicitly introduced in chapter VII any sort of recollection-like theory which would allow phantasy to find quantities from some substantial reasons as it is spurred by the stimuli of sense perception. Thus, a paramount question remains still open: without some kind of inborn knowledge, and denying that perceived natural quantity is the source of phantastic quantity, whence does *phantasia* get its *quantum phantasiatum*?

3.5.3.1 Co-eternal Quantity and Quantitative Isomorphism

Piccolomini only twice mentions something akin to an explanation of how separated quantities are produced by phantasy. There, he invokes the notion of co-eternal quantity (*quantitas coaeterna*) or the co-eternal inherence of mathematical forms to naked matter. As Higashi has documented in detail, this notion was originally put forward by Averroes and taken up in the Latin-Averroistic tradition up to Zimara.[179] These two authors are the most likely sources of Piccolomini's development of this doctrine.

179 Higashi, *Penser les mathématiques*, pp. 58–80.

Averroes' view was that prime matter must be considered as endowed with indeterminate dimensions. These dimensions were only determined as a body of this or that particular quantity (size) by a substantial form. This means that, according to Averroes, quantity cannot be stripped out from prime matter, nor can prime matter be taken off from quantity (or, at least, continuous quantity). For instance, in *De substantia orbis*, Averroes claims:

> And because [Aristotle] found that all forms share in indeterminate dimensions, he knew that prime matter is never stripped of indeterminate dimensions. Because, if it were stripped, then the body would be from non-body, and dimension from non-dimension.[180]

Thus, this sort of quantity inherent to matter is undetermined in such a way as to lack a particular shape or figure or any determined dimensions (being, thus, impossible even to imagine). According to Averroes, such shapeless, pure dimensionality enacts the actual extension of a determined figure by the impression of a substantial form which dictates some determined measure and shape.[181] Therefore, co-eternal quantity stands as the pure quantitative potential for quantitative traits within prime matter.

In explaining the process of *abstractio* and the origin of mathematical certainty, Piccolomini appeals to this doctrine and elaborates on it in a (perhaps surprising) direction, intended to be explicative of the being of mathematical objects. One of the two clarifications on abstraction in which a reference to co-eternal quantity is invoked reads as follows:

> Quantity, therefore, due to its imperfection, being co-eternal with prime matter and not bound to any limited subject, consequently, is easily abstracted from all matter. Thus, mathematical sciences are granted their certainty from a familiar and everyday subject to the senses.[182]

---

180 Averroes, *De substantia orbis*, in *Aristotelis opera cum Averrois commentariis*, vol. 9, I, f. 4: *Et quia [Aristoteles] invenit omnes formas communicari in dimensionibus non terminatis, scivit quod prima materia nunquam denudatur a dimensionibus non terminatis; quia, si denudaretur, tunc corpus esset ex non corpore, et dimensio ex non dimensione* [...].

181 For an account on the 13th- to 17th-century Latin views on extension as inherent to prime matter (including Averroes, Ockham, and their heirs), see R. Pasnau, *Metaphysical Themes 1274–1671* (Oxford: Oxford University Press, 2013), chap. 1, sec. 4 and chap. 4.

182 Alessandro Piccolomini, *De certitudine*, f. CIX: *Quantitas igitur, ob imperfectionem suam, cum coaeterna sit primae materiae, et nulli limitato subiecto addicta sit, ex consequenti, ab omni materia facillime abstrahitur, et ita certitudinem ex subiecto sensibus familiari et domestico mathematicae scientiae sortitae sunt.* Piccolomini repeats that quantity as such is a passion of prime matter in Alessandro Piccolomini, *De certitudine*, f. CIXv: [...] *quantitas, quae passio est primae materiae* [...].

Piccolomini, following the Averroistic tradition, establishes here that (i) quantity is co-eternal to matter and, therefore, (ii) that quantity does not require a defined subject to be instantiated. A second passage regarding co-eternal quantity as a trait of prime matter shall help to clarify what Piccolomini had in mind:

> The circle does not ascribe to itself gold, wood, or any other determined stuff. Therefore, since the [greater or lesser] easiness of abstraction arises from a greater or lesser determination and limitation to matter, it follows that things which are not determined to any matter in actuality but are co-eternal with bare matter, will be most abstractable. Consequently, they will be easily knowable, certain, and manifest.[183]

Piccolomini's theory is partially fleshed out here. He proposes the circle as a token of co-eternal quantity and tells us that this figure does not require any particular material to be enacted, for it could be equally instantiated by gold or wood. Thus, quantitative entities only need a quantitative substrate to be (to some degree) instantiated, regardless of the material the particular substrate happens to be. This is the reason why quantities are more abstractable and, thus, more knowable: grasping a quantity does not require understanding any precise material nature, but phantasising it as separated from every natural matter; on the other hand, understanding a substance would require knowing the cause why and how such-and-such a form is instantiated in such-and-such matter (see below). The entire passage arguably follows Aristotle's discussion on the same topic in *Metaphysica* VII, 7, 1033a, and 11, 1036a–1036b where Aristotle explains that, whereas geometrical figures can be instantiated by many materials, substances, such as human beings, must be instantiated in a determined type of matter (i.e., flesh and bones). Thus, whereas a human being made of gold is not a human being, a gold circle is as much of a circle as a wooden circle is.

This doctrine also extends to the dimensions of the geometrical items themselves. Piccolomini tries to make this point by invoking Proclus' *In Euclidem*. The Sienese's argument can be succinctly explained as follows: the measure of a given geometrical entity, say a line or a sphere, could be (by the works of

183 Alessandro Piccolomini, *De certitudine*, f. CVIIIv: *Circulus enim neque aurum, neque lignum, neque aliquid aliud determinatum sibi ascribit. Cum igitur abstractionis facilitas ex maiori vel minori ad materiam determinatione et limitatione oriatur, sequitur quod res illae, quae ad nullam materiam in actu determinantur, sed cum denudata materia coaeternae sunt, abstrahibiles erunt maxime, et idcirco faciles cognitu, certae, ac manifestae.* I will deal with the puzzles of abstraction in the next section.

phantasy) infinitely increased without doing violence to the nature of the mathematical entity. On the contrary, Piccolomini claims that real, natural substances have their quantity (size) restricted within certain limits dictated by their own substantial form: the size of a human being could not but unnaturally (i.e., by something alien to its substantial form) be multiplied by ten. Thus, another consequence of the absence of substantial forms in quantity (*nulli addicta est formae substantiali*) is that mathematical objects possess no fixed size and can be expanded or reduced at the discretion of the mathematician.[184]

Piccolomini then asserts that the things (*res*) that are not determined to be instantiated in any type of matter (i.e., insubstantial mathematical entities) are also co-eternal to bare matter (*cum denudata materia coaeternae*). In this way, Piccolomini locates mathematical items with (some of) the quantities pre-contained in the co-eternal, undetermined quantity.

As I interpret it, Piccolomini's doctrine postulates that prime matter pre-contains every logically conceivable quantitative form (figure, shape, or number). In that sense, Piccolomini accepts Averroes' point that prime matter is also the undetermined potency for all quantities. Some of these inherent quantities (such as some shapes) will be actualised and determined by natural

184 Alessandro Piccolomini, *De certitudine*, f. CVI: *Quamvis enim quantitas, si termino quodam illam constringimus, id ex formae natura sortiatur, quia aliae rerum formae, alios quantitatis terminos sibi vendicant, tamen secluso omni termino, nulli addicta est formae substantiali. Et iccirco dicit Proclus, Super 1 Elementorum, pagina 28, quod materia, ut materia potest augeri in infinitum, sed impeditur a formis ipsis*. Piccolomini most likely refers (under the assumption that he is correctly quoting the folium in Grynaeus' edition) to Proclus, *In Euclidem*, p. 101, l. 16–20: […] προϊὼν δ'ἐκεῖθεν αὐτὸς ὁ πρώτιστος ἑαυτὸν διαστήσας καὶ κινούμενος καὶ ῥέων ἐπ'ἄπειρον καὶ τὴν ἀόριστον δυάδα μιμούμενος κρατεῖται μὲν ὑπὸ τῆς ἰδίας ἀρχῆς, ἐνίζεται δὲ ὑπ'αὐτῆς καὶ περιλαμβάνεται πανταχόθεν. In Morrow's translation: "As it goes forth from that region, this very first of all ideas expands itself, moves, and flows towards infinity and, imitating the indefinite dyad, is mastered by its own principle, unified by it, and constrained on all sides". Piccolomini's idiosyncratic reading of this passage seems to be that, unless limited in some manner, the point, as much as every phantastic quantity composed of points, can be augmented *ad infinitum* without doing any violence to their nature. Another passage of Proclus' *In Euclidem* which might have suggested the second part of the argument, sc., that sensible entities are limited in their quantity by their own nature is found in pp. 86, l. 26–87, l. 6: […] ἐν δὲ τοῖς ἀχωρίστοις ὕλης ἐπιδεδώκασιν ἑαυτοὺς τοῖς περατουμένοις καὶ ἐν ἐκείνοις ἱδρύνθησαν καὶ οἷον μέρη γεγόνασιν ἐκείνων καὶ ἀνεπλήσθησαν τῶν χειρόνων, ὅθεν δὴ καὶ τὸ ἀμερὲς ἐνταύθα τῆς μεριστῆς οὐσίας καὶ τὸ ἀπλατὲς πλάτους μετέσχεν καὶ τὴν ἑαυτῶν ἁπλότητα καὶ τὸ εἰλικρινὲς οὐκ ἔτι τὰ περατοῦντα φυλάξαι δεδύνηται. In Morrow's translation: "[…] But in the forms inseparable from matter the limits surrender themselves to the things they limit; they establish themselves in them, becoming, as it were, parts of them and being filled with their inferior characters. This is why in this region the partless partakes of divisible existence and the breadthless of breadth; and the limiting elements are no longer able to preserve their simplicity and purity […]".

substances. But every one of them? The answer is negative. Together with the potential shapes that will be at some point realised by natural bodies, also mathematical shapes, such as triangles, circles, lines and even points, are eternally inherent to naked matter. Nevertheless, these potencies are never, and will never, be actualised as sensible items. Why? The answer here is straightforward: because no real, natural substance has the bodily shape of, say, a perfect triangle or a line.

Natural substances may approximately resemble geometrical figures, as far as the three-dimensionality and mutability of informed sensible matter allow it. This means that this sphere of wood is only a sphere to the degree that the natural matter that works as its substratum or "subject" (which really belongs to other substances) can embody the properties of a sphere. In other words, this wooden sphere can be described as an instance of a sphere in a hunk of matter: wood. But this matter is a part of the former body of a tree. Thus, the material substrate of this sphere was already informed by the shape and quantity of the wood (the body of the substance tree, with its porousness and irregularities preventing the sphere from being fully realised). As a matter of fact, no sensible substance instantiates a perfect sphere or icosahedron. Since geometrical entities are not substances, they do not possess a matter of their own in the way a tree possesses wood.

How does Piccolomini's theory of co-eternal quantity relate to phantasy's capacity to produce mathematical items?

We have been told that phantasy is the soul-power related to quantitative traits in themselves and that it produces separated phantastic quantities. Now we know that the claim "phantasy produces mathematical items" really stands for "phantasy enacts some of the quantities potentially inherent and co-eternal to prime matter". An immediate objection swiftly arises: if prime matter is something different from phantasy, how can one be the potency of the other? Unless something is added, it might appear that Piccolomini is claiming that the potentiality for enacting quantities is something external, rather than internal to phantasy itself. How does this exactly work?

I shall propose what I find to be the most likely explanation for making sense of Piccolomini's words. As we saw, Piccolomini follows Proclus in conceiving ὕλη νοητή as the actual extension and multiplicity of quantity in itself given in phantasy (*quantitas ipsa in phantasia*). All this reveals that Piccolomini entertains, much like Proclus and, perhaps Aristotle, a two-matter theory, but denies that both matters are identical in the traits they can enact. This stands for what I call "quantitative isomorphism" of phantasy.

This theory entails that both prime matter and phantasy are cut by the same quantitative patterns and endowed with the very same potential to produce every logically conceivable quantity. In other words, phantasy is from the very beginning equipped in such a way that it can grant access to any quantity that the epistemic agent may find in sensible matter. Therefore, every quantity which is actualisable by prime matter, is also phantasisable.

Nonetheless, phantasy is handicapped in some sense vis-à-vis its extramental twin: whereas prime matter can instantiate substances, phantasy is endowed *only* with the capacity for enacting images. Phantasy can produce the image of fire, but this phantastic image will not burn; it can produce the image of a horse, but the phantastic horse will not be alive. Thus, the isomorphy of phantasy vis-à-vis prime matter concerns solely some traits, amongst which it is quantity (and also colour or even sound). This hypothesis also explains why Piccolomini does not consider *materia intelligibilis* as proper matter: to rank as real matter, a substrate should be able to enact substances, a capacity alien to phantasy.

But being unable to hold substances allows phantasy some manoeuvres of its own. Exact mathematical entities will never be fully instantiated in sensible matter. Phantasy, though, does not play according to the rules of sensible matter and its substances: freed from the restrictions of being informed by active principles and alien to any internal form of causality, phantasy can display, in its boundless plasticity, all the logically conceivable, yet naturally impossible, quantities. This allows phantasy to generate geometrical objects with utter *certitudo* (objectual precision).

In sum, by proposing that (1) the *de facto* power to enact every possible quantity belongs to phantasy rather than to real matter, and (2) that phantastic quantities are not copy-paste replicas of sensible quantities stripped out of non-quantitative traits and put in phantasy, Piccolomini accepts that mathematical items (all of them) pre-exist as potentially phantasisable within phantasy.

Postulating the quantitative isomorphism between phantasy and bare prime matter renders phantasy the sole source of mathematical objects. Thus, Piccolomini's syntagm *phantasia efficiens* can be interpreted straightforwardly, as I believe he intended: it is the activity of phantasia, and this power alone, that generates *quantum phantasiatum* from its inherent indeterminate quantity, without being acted upon by sensed natural quantity or the need for any inborn *rationes substantiales*.

3.5.3.2 Three-Fold *abstractio* and the Origin of Mathematical *certitudo*

What, then, does the term *abstractio* stand for in Piccolomini's lexicon?

As mentioned in section 3.5.1, at least since Alexander of Aphrodisias, the term ἀφαίρεσις (and its Latin correlate *abstractio*) fuse two different meanings.[185]

185 De Libera, *L'art des généralités*, pp. 28–32.

One of the meanings (α1) comes from Aristotle's account of the obtention of mathematical objects through the operation of abstraction: it is intended to explain how an epistemic agent gains access and deals with mathematical objects.[186]

The other sense of *abstractio* (α2) has its origin in Aristotle's notion of ἐπαγωγή or induction, defined in *Analytica posteriora* and *Topica*:[187] this is the "way of ascending from particulars (ἀπὸ τῶν καθ'ἕκαστα) to the universal (τὸ καθόλου.)"[188] That is to say, it is the epistemic or logical manoeuvre for obtaining a universal definition from particulars sharing a kind.

As expounded, those two operations were assimilated under the one name of *abstractio* in the Latin tradition. In addition, as we saw in exploring Piccolomini's chapter VII above, the adjective *abstractum* can also mean in the philosophical Latin jargon (α3) "separated". This means that the Greek technical terms ἀφαιρεθέν – the result of abstraction in the sense (α1) – and the perfect participle κεχωρισμένον or that which is separated (either naturally or logically) are also included under the umbrella-term of *abstractio*. As shown in section 3.5.1, the assimilation of abstraction and separation is also found in Alexander's *De intellectu*.

The language of separation (χωρίζειν) is also adopted by Proclus as the triggering condition for projection to happen: the geometer "attempts to separate" (χωρίζειν γὰρ [...] ἐπιχειρεῖ) the circle from its sensible tokens and, as result of this, as he projects the phantastic circle, the geometer simultaneously deals with three circles: the sensible circle used as a stimulus for phantasy, the phantastic circle obtained thereby, upon which demonstrations are made, and the indivisible reason of the circle, which he ultimately investigates.[189]

186 *Metaphysica*, VI, 1, 1026a, l. 5; VII, 3, 1029a, l. 10 and forth; XI, 3, 1061a, l. 28 and forth; XIII, 2, 1076b l. 1–1077b, l. 17; 3, 1077b, l. 17–30.

187 *Analytica posteriora*, I, 18, 81a, l. 40; II, 19, 100a, l. 14–b5. See also *Topica*, I, 12, 105a.

188 *Topica*, I, 12, 105a, l. 13–14. Regarding the ambiguity of the terms καθ'ὅλου in Aristotle's writings and the different (and exclusive) interpretations that thereby were suggested to later commentators, see De Libera, *La querelle des universaux. De Platon à la fin du Moyen Âge* (Paris: Éditions du Seuil, 1996), pp. 29–32.

189 Proclus, *In Euclidem*, p. 54, l. 14–26: τὴν τοίνυν γεωμετρίαν ὅταν περὶ κύκλου τι λέγῃ καὶ διαμέτρου καὶ τῶν περὶ τὸν κύκλον παθημάτων, οἷον ἀφ'ὧν διαιρέσεων τῶν τοιούτων, μήτε περὶ τῶν αἰσθητῶν λέγωμεν ἀναδιδάσκειν – χωρίζειν γὰρ ἀπὸ τούτων ἐπιχειρεῖ – μήτε τοῦ ἐν διανοίᾳ εἴδους. εἷς γὰρ ὁ κύκλος, ἡ δὲ περὶ πολλῶν ποιεῖται τοὺς λόγους καθ' ἕνα ἕκαστον προβάλλουσα καὶ περὶ ἁπάντων τὰ αὐτὰ θεωροῦσα. καὶ ἀδιαίρετος μὲν ἐκεῖνος, διαιρετὸς δὲ ὁ ἐν γεωμετρίᾳ κύκλος. ἀλλὰ τὸ καθόλου μὲν αὐτῷ συγχωρῶμεν ἐπισκοπεῖν, τοῦτο δ' ἄρα ἐκείνῳ τὸ κατατεταγμένον ἐν τοῖς φανταστοῖς κύκλοις, καὶ ἄλλον μὲν ὁρᾶν καὶ κατ' ἄλλον θεωρεῖν τὸν ἐν διὰ νοῦ κύκλον, περὶ ἄλλον δὲ ποιεῖσθαι τὰς ἀποδείξεις. In Morrow's translation (modified): "When, therefore, geometry says something about the circle or its diameter, or about its accidental

It is therefore unsurprising that Piccolomini took χωρίζειν in Proclus' text as describing the production of separated phantastic quantity and put it also as *abstrahere*.

With these three meanings of abstraction in mind, we can now address Piccolomini's theory.

Piccolomini's use of *abstractio* when applied to mathematical items involves (α1) and (α3): it refers to the process of obtaining a phantastic quantity (ἀφαιρεθέν, sense α1) which has been approximately suggested by a natural quantity and indicates that mathematical items exist separated (κεχωρισμένον, sense α2) in phantasy, i.e., as beings wholly discrete from the stimulus-quantity that produced it.

As said, in Piccolomini's ontology of mathematical objects, perceived quantity is not just cloned in phantasy as it is stripped of non-quantitative traits. Rather sensible quantities work like indications for navigating through the phantastic realm of separated quantity: they point towards a shape without really instantiating it, as a sort of *innuendo*. As an example, let us say that a geometry teacher puts before his students three wooden icosahedrons of different sizes and then indicates: "An icosahedron is like these twenty-sided solid figures but with perfect angles, perfect straight lines, absolutely plane surfaces, and not of this or that size, but of whatever size, and it is not made of wood, but of whatever stuff, for the icosahedron is just the shape". By using the stimuli provided by the sense input and trying to understand what the icosahedron as such looks like, the students will be able to map and produce the authentic icosahedron in the abstracted (i.e., separated, α3) region of phantasy: the figure there produced does not causally derive or fully correspond to the tokens brought by the teacher; it has been just suggested by them. Thus, the phantastic icosahedron appears as if it were a cleaned-up, i.e., abstracted (α1) version of the wooden icosahedron – although it is a production of discrete phantasy alone.

But then, why does Piccolomini claim that quantities are more *abstrahibiles*? We have seen that, in his scattered considerations on the epistemic

characteristics, such as tangents to it or segments of it and the like, let us not say that it is instructing us either about the circles in the sense world – for it attempts to abstract from them – or about the form in the understanding. For the circle is one, yet geometry speaks of many circles, projecting them individually and studying the identical features in all of them; and that circle is indivisible, yet the circle in geometry is divisible. Nevertheless, we must grant the geometer that he is investigating the universal, only this universal is obviously the universal present in the phantastic circles. Thus, while he sees one circle, he is studying another, the circle in the understanding, yet he makes his demonstrations about another one".

transparency or *certitudo* of mathematics,[190] he has asserted that undetermined quantity or mathematical items are "separated/abstracted in the easiest way" (*facillime abstrahitur*), that quantity is endowed with an "easiness of separation/abstraction" (*abstractionis facilitas*), and that quantities are "maximally separable/abstractable" (*abstrahibiles maxime*).

To understand the precise meaning of this claim, we must first ask: in comparison to what other things are quantities considered more abstractable? Piccolomini's answer lies in contrasting the epistemic transparency of mathematical objects with the epistemic opacity of natural entities.

> Natural things, though they present their operations to our senses, conceal their ultimate differences, i.e., the very forms and substances from which passions and, ultimately, actions flow. These are deeply immersed in the profound and obscure recesses of nature so that hardly after prolonged and diligent observation and experience of effects they fleetingly become clear to our intellect.[191]

Piccolomini assumes here the doctrine of degrees of abstraction pointed out in section 3.5.1. This means that some natures can be conceived without being instantiated in a particular kind of matter (such as concavity) whereas others need to be instantiated in a given substrate (such as snub-ness, which is only found in a nose). Attending to such a difference, the two meanings of *abstractio* (α1) or obtaining a mathematical object and (α2) or grasping a universal form are unified in Piccolomini's understanding, resulting in the view that more abstractability in a given object implies more knowability of the form or definition of that object (and the other way around). A coherent account of this fused view could be put as follows:

Let us take a natural being (say, a human being) and a quantity (an icosahedron) and let us suppose that an epistemic agent wants to learn the essences of such objects. After observing some human beings, the epistemic agent will obtain a phantastic representation of what a human being looks like for the most part. The epistemic agent uses these conflated images to recognise human beings in everyday life: they are bipedal, they speak, they laugh, they

190 Alessandro Piccolomini, *De certitudine*, f. CVIIIv and f. CIX.

191 Alessandro Piccolomini, *De certitudine*, f. CVIII: *Res vero naturales, quamvis operationes sensui nostro offerant, differentias tamen ultimas, hoc est, formas ipsas ac substantias a quibus passiones et tandem actiones fluunt, occultissime in profundo et obscuro naturae sinu immersas habent, nec vix longa et assidua effectuum observatione et experientia aliquantisper intellectui nostro elucescunt.*

use telephones and cars, they drink coffee, and read books. Nevertheless, this phantastic human being (which may differ according to each one's experience) does not provide the epistemic agent with the specific difference which would reveal what a human being is, the essence of human beings. Much deeper research is required to isolate what is, amongst all its activities and traits, the thing that makes it, amongst other animals, a human being. This trait (rationality) cannot even be represented by phantasy as such, for phantasy is only able to represent images.

Things are quite different when it comes to *quantum phantasiatum*.

After observing some icosahedron-like items and trying to conceive them in isolation, our epistemic agent produces the exact phantastic icosahedron. Unlike the phantastic human being, the true icosahedron is, in fact, the icosahedron obtained in phantasy. In other words, finding the phantastic separated icosahedron (abstraction α1 and α3) and gaining access to what an icosahedron really is are simultaneous events.[192] Whereas the phantastic human being is not an authentic human being (human beings are essentially natural, non-phantastic items), being an authentic icosahedron is nothing else than being an icosahedron in phantasy: for mathematical entities, "what it means to be a real and objective X" coincides with "the subjective, phantastic X".[193] Hence, the intellect abstracts (α2, ἐπαγωγή) the specific difference and the essence (i.e., the form) of the icosahedron(s) without effort, just as the icosahedron is phantasised. Moreover, whereas the phantastic human being may differ in the case of each epistemic agent, all the epistemic agents will have access to exactly the same traits in their respective phantastic icosahedra, so accounting for what we would call the objectivity of mathematics.

So, what is Piccolomini's "easiness for abstraction"? It can be defined as the ease with which an item is formed as a distinct phantastic entity (a *phantasma*) that, as such, fully and immediately reveals its nature to the intellect. This is attainable only with mathematical objects. For this reason, an icosahedron is more abstractable than a human being and, therefore, more certain.

From where does this ease of abstraction in mathematics arise? In other words, what is Piccolomini's mathematical akribology that accounts for the supreme degree of certainty (epistemic transparency) inherent in mathematical

---

192 Because of this aspect of Piccolomini's philosophy, Ulrich Felgner has labelled it as a "constructivist" one: we are able to perfectly know mathematics because we internally fabricate it. See Felgner, *Philosophy of Mathematics in Antiquity and in Modern Times*, pp. 97–99.

193 I want to thank Professor Therese Scarpelli Cory for her illuminating conversation on this particular point.

objects? Piccolomini offers two intertwined reasons underpinning his akribology of mathematics. Herein lies his ultimate explanation for the *certitudo mathematicarum*:[194]

The first reason (1) referred to is an epistemological one. It is already known to us, for it was also adduced by Boccadiferro (section 2.4.3) in explaining Aristotle's claim in *Ethica Nicomachea* (VI, 9, 1142a, l. 6–19) that children can become mathematicians but not natural philosophers.[195] Piccolomini refers to Aristotle's *locus* and repeats the rationale which we already found in Boccadiferro's writings regarding the *sensibile commune*: quantity is the most common accident amongst sensible things; it is perceived all around and through many senses. Being ubiquitous, quantity is the most knowable accident.[196] As we discussed in section 2.3.3, Priscian's commentary (402b, l. 11) also furnished Piccolomini with the idea that quantity is "more perceptible" (αἰσθητικώτερα, *sensatior*) than any other accident or property. This entire argument suggests that, because we constantly engage with surfaces, volumes, and discrete quantities in our everyday lives and receive quantitative inputs through various senses, quantity is *ad nos* easier to grasp than any other accident.

The second reason (2) is an ontological one. Quantities considered in themselves, as co-eternally pre-contained in prime matter and isomorphically present in phantasy, are undetermined regarding their matter (*minor ad materiam determinatio et limitatio*).[197] Therefore, they are in themselves far simpler than natural entities and their accidents. The icosahedron is only its shape and

---

194 See Alessandro Piccolomini, *De certitudine*, f. CVIIIv and f. CIX.

195 Ludovico Boccadiferro, *Explanatio libri Physicorum*, lec. 1, f. 3: *Dico quod sensibile commune licet apprehendatur obscurius a singulo sensu particulariter quam proprium sensibile, tamen ab homine ipso certius apprehenditur sensibile commune quam proprium, quia sensibile commune pluribus sensibus percipitur*; Ludovico Boccadiferro, *Explanatio libri Physicorum*, lec. 1, f. 3: *Sensibilia enim communia eodem modo a naturali et mathematico percipiuntur, tantum ratione different, quia naturalis considerat ea ut termini corporis naturalis quanti, mathematicus ut quantus simpliciter*; Ludovico Boccadiferro, *Explanatio libri Physicorum*, lec. 1, f. 2v: *Ideo* VI *Ethicorum ait Aristoteles: puer potest esse mathematicus, naturalis non, quia res certior est mathematica quam naturalis.*

196 Alessandro Piccolomini, *De certitudine*, f. CVIIv: *Quid amplius quaerimus pro hac sententia, praesertim cum Aristoteles in Ethicis et Eustratius ibidem, assignantes causam cur mathematicae facilem et manifestum darent aditum, etiam et pueris, id non ex demonstrationis vi, sed ex subiecta materia declaraverunt.*

197 See also Alessandro Piccolomini, *De certitudine*, f. CIXv: *Quantitas autem, quia non hanc sibi ascribit materiam magis quam aliam, sed nulla limitata dispositione materiae indiget, nil mirum si abstrahibilis est valde. Et iccirco certiorem et sensatiorem generat cognitionem.*

its spatial relations, not its wood, or its intelligible matter, or any other activity (it has no activities), or any particular size. The icosahedron is completely idle and evacuated of causal powers and the restrictions of real essence and, hence, extremely easy to know. Thus, mathematical items are also *per se* easier to know.[198]

Therefore, quantity is (epistemically, to us) so familiar, and (ontologically, in itself) so inherently simple that the science about it, mathematics, must be the most certain form of cognition in the three senses of *certitudo*, namely (a) irrefutability and necessity, (b) objectual precision, and (c) epistemic transparency.

A last point should be made regarding Piccolomini's confusing hermeneutics of *abstractio*. As we have seen, he adopts this term for describing the process of producing mathematical objects (*imaginatione abstrahentes, ex abstractione*) in a manner which may *prima facie* appear incoherent with his theory of productive phantasy and the occasional intervention of sensible quantity in chapter VII. In other words, the fact that Piccolomini mainly uses the term *abstractio* without further specification may mislead the reader to believe that he is adopting a copy-paste full-correspondence view in the "Aristotelian" manner.

However, things become clearer if we recall that an interpretation of *abstractio* (ἀφαίρεσις) was received by the *Simpliciani* Averroists not only as compatible with but even as integrated into what we call "phantastic projection". As we saw in section 2.1.5, Priscian nuancedly presents Aristotle's positions as follows:

> Aristotle denies the substantiality of the impressions [in phantasy] and says that they arise within (ἐγγίνεσθαι) from abstraction (ἐξ ἀφαιρέσεως) as the soul receives them in accordance with an imitation in respect of sensible objects. It is not because the soul does not **project these too by its own activity**, that is shown by their turning one to truth and precision (ἀκριβές) […], but, because of the very great similarity of the things that are phantasised to the sense-objects they have the place of abstraction (τῆς ἀφαιρέσεως) specifically amongst the phantasised things, **as these are as nearly as possible the same as the sense objects**, without the physical substance, since the impression in phantasy is not like the form but like what is informed (τὸ εἰδοπεποιημένον).[199]

198 See also Biard, "La certitude des mathématiques et ses fondements selon Piccolomini".

199 Translation by Steel (slightly modified) in 'Simplicius', *On Aristotle's On the Soul 3.6–13*, p. 98. Simplicius, *In De anima*, com in III, 429b, l. 18, p. 233, l. 13–25: ὁ δὲ Ἀριστοτέλης τῶν τύπων ἀποφάσκει τὴν οὐσίαν καὶ ἐξ ἀφαιρέσεως ἐγγίνεσθαι λέγει τοὺς τύπους, ὡς κατὰ τὴν

Another token is:

> I mean that mathematics which is wholly concerned with the projected things themselves (περὶ αὐτὰ τὰ προβαλλόμενα), such as shapes, sizes and motions, which Aristotle is accustomed to call "from abstraction" (ἐξ ἀφαιρέσεως εἴωθεν ὁ Ἀριστοτέλης καλεῖν).[200]

In both passages, Priscian refers to Aristotle's ἀφαίρεσις and intends (i) to explain why Aristotle considered phantastic mathematicals to be generated by this process, and (ii) to show that, if rightly understood, Aristotle's doctrine of ἀφαίρεσις is not incompatible with the Neoplatonic putting forward of mathematical entities by phantasy as a generation of its own.

Regarding (i) Priscian teaches that Aristotle named this process "abstraction" to emphasise the resemblance or imitation (μίμησιν) of phantastic quantities to sensible quantities. However, Priscian immediately points out that this "imitation" is not to be understood as denying that the discrete activity of phantasy is the producer of mathematical objects. Thus, he concludes that when Aristotle used the term ἀφαίρεσις he intended to emphasise that there exists a great similarity (ὁμοιότητα) between the two kinds of quantities, the perceived one and the phantastic one. But phantastic quantities are bestowed with a superior objectual precision (ἀκριβές). The reason for this (just apparent) imitative rapport is not that mathematical objects are simplified copies of sensible quantity impressed in phantasy but that both phantastic and natural quantities are expressions of the same mathematical Form (εἶδος). Thus, it follows (ii) that according to Priscian, Aristotle's account, if well understood, is not opposed to that of Plato, but a description of one of its aspects, namely the kinship of phantastic and natural quantities, not as the resemblance of father and son, but as that of siblings of the same ensouled mathematical Form.

The parallels and divergences of Aristotle's, Priscian's, and Piccolomini's models of ἀφαίρεσις/*abstractio* can be illustrated in the following scheme:

πρὸς τὰ αἰσθητὰ μίμησιν προσδεχομένης αὐτοὺς τῆς ψυχῆς. οὐ διότι μὴ κατ᾽οἰκείαν ἐνέργειαν καὶ τούτους ἡ ψυχὴ προβάλλεται (δηλοῖ δὲ ἡ εἰς τὸ ἀληθὲς καὶ ἀκριβὲς περιαγωγή. […]). ἀλλὰ διὰ τὴν τῶν φανταστῶν πρὸς τὰ αἰσθητὰ πλείστην ὁμοιότητα τὸ τῆς ἀφαιρέσεως χώραν ἔσχεν ἰδίως ἐπὶ τῶν φανταστῶν, ὡς οἷον τῶν αὐτῶν ὄντων τοῖς αἰσθητοῖς χωρὶς τῆς φυσικῆς οὐσίας, ἐπεὶ καὶ ὁ ἐν φαντασίᾳ τύπος οὐχ ὡς τὸ εἶδος ἀλλ᾽ὡς τὸ εἰδοπεποιημένον.

200 Translation by Steel in 'Simplicius', *On Aristotle On the Soul 3.6–13*, p. 77. See Simplicius, *In De anima*, comm. in III, 431b, l. 12, p. 277, l. 1–6: λέγω δὲ νῦν μαθηματικὴν […] τὴν ὅλην περὶ αὐτὰ τὰ προβαλλόμενα, οἷον σχήματα καὶ μεγέθη καὶ κινήσεις, στρεφομένην, ἅπερ ἐξ ἀφαιρέσεως εἴωθεν ὁ Ἀριστοτέλης καλεῖν.

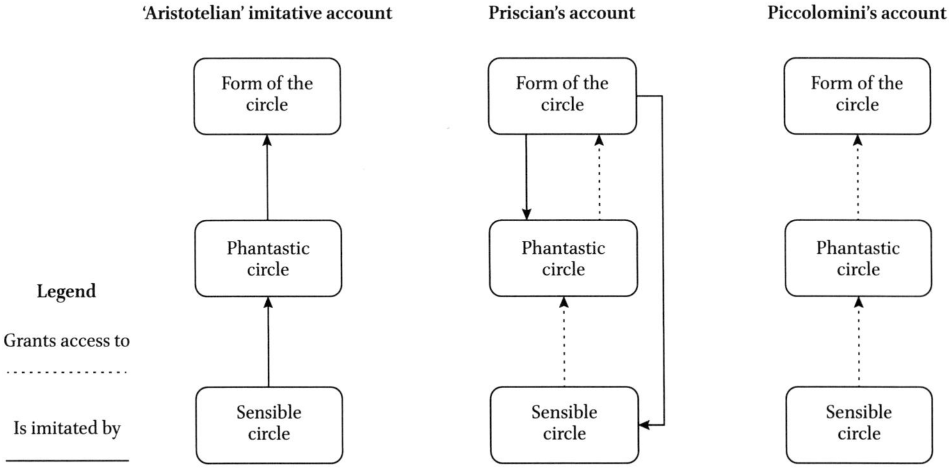

FIGURE 3 Models of abstraction

From this, it is clear that, much like Aristotle, Piccolomini does not consider that there is a single, active paradigmatic cause for both the sensible and the phantastic circle, whereas he takes from Priscian and Proclus that the phantastic circle is not a simplified imitation of the sensible circle. In sum, Piccolomini's model can be understood as Priscian's and Proclus' models but cut off from exemplary causes and ensouled reasons.

Therefore, as for Piccolomini's hermeneutics of *abstractio* (and related terms) in *De certitudine*, he seemingly followed Priscian's conciliatory reading of Aristotle in interpreting Proclus and designing his theory. This ultimately means that *abstractio* is but an Aristotelian name for describing the sensibly-relatable aspect of Proclus' and Priscian's phantastic projection.

3.5.3.3 The Akribological Paradox: Too Certain to Be a Science

The view that the definitions of mathematical items are revealed as they are phantasised closes the logical-metaphysical circle of *De certitudine*. This statement grants us the core reason why *demonstratio potissima* cannot be instantiated in mathematics.

As posited by Piccolomini, the middle-term in the highest sort of demonstration is P's definition.[201] This is so because, if S's definition were already known to us from the very beginning, then the demonstration would be an inane operation, for scientific research seeks to reveal the definitions of subjects.

201 Alessandro Piccolomini, *De certitudine*, f. LXXXXI.

But is this possible in mathematics?

The answer is no, because of the plain reason that, according to Piccolomini, in producing a phantastic mathematical object, we gain immediate access to its definition. i.e., to S's definition:

> Mathematical entities, on their part, being from abstraction, completely and thoroughly present themselves to our sense, and completely open themselves, and they deliver most evidently to our sense not only their properties but also their subjects and their very forms, as all these are quantities.[202]

Therefore, S's definition is always known by the epistemic agent when the object of study is a phantastic quantity. It follows that all mathematical demonstrations will attribute a P-theorem to S through S's definition (as shown by the syllogistic-analytic formalisation of Euclid's I.1). Nevertheless, Piccolomini has already established that the point of doing science is to obtain the definitions of the subjects. This was his argument (much as that of Giles) in claiming that the middle-term in scientific demonstration cannot be S's definition. Otherwise, the scientific enquiry would be redundant.[203] It follows that mathematical demonstrations are either scientifically idle exercises in which passions (theorems) are attributed to S without increasing our (already complete) knowledge of S's definition, or that they are just mere tautological statements (if one intends to demonstrate S's definition though S's definition).

From the point of view of mathematical scientific practice, this might be a puzzling conclusion. How is it that proving new theorems is scientifically irrelevant? The answer is that, in the scientific framework of Piccolomini, genuine knowledge is knowledge about subjects or substances. Thus, knowing what the triangle is according to its definition happens to be epistemically more important than knowing whatever of its theorems (although a triangle is not even a substance). Much like axiomatics, this reveals, yet again, a core tension

---

202 Alessandro Piccolomini, *De certitudine*, f. CVIII: *Res autem mathematicae, cum ex abstractione sint, se ipsas penitus et medullitus sensui nostro praebent seque totas patefaciunt nec solum passiones suas, sed subiecta etiam ipsorumque formas sensui nostro manifestissimas tradunt, cum haec omnia quantitates sint.*

203 Alessandro Piccolomini, *De certitudine*, f. LXXXXII–LXXXXIIv: *Si diffinitio subiecti esset medium in demonstratione potissima, sequeretur quod supervacanea esset omnis demonstratio, quia tunc ad nota dirigeretur. Hoc est absurdum, cum demonstratio per nota ad ignotum procedere debeat.* On this topic, see Longeway, "The Place of *Demonstratio Potissima*".

between the real practice of mathematics and the formal requirements of scientificity of the Aristotelian theory of science.

In sum, the proverbial *dictum* that mathematical demonstration is the most scientific one because it is the most certain (*certissimae ergo potissimae*), and that it is the most certain because its object of study is better known to us (*notior nobis*), is turned by Piccolomini against its long Latin tradition of promoters, from Averroes to Zimara. Piccolomini concludes that phantastic quantities are *so* certain and well-known to us from the very beginning that, paradoxically enough, demonstrations about them cannot teach us anything. Thus, they cannot be classified as really scientific demonstrations.

3.5.3.4 Phantasy: a Kingdom without Bridges

Piccolomini's theory of *phantasia efficiens* and the chance provided by sensible matter to obtain mathematical objects very closely follows Proclus' ontology of mathematics. Three of the four traits proposed in section 2.1.3, namely (i) in-betweenness, (iii) phantastic projectionism, and (iv) insubstantiality, are accepted by Piccolomini in some way. He accepts the pre-existence of all quantities in phantasy instead of (ii) innatism.

Piccolomini accepts (i) in-betweenness by proposing that mathematical objects belong to the ontological realm of *phantasia*, which partakes in both intellectual and sensible traits.

He proposes an interpretation of his own (iii) phantastic projectionism by claiming that *phantasia*, a power of the soul, is efficient of mathematical objects.

Lastly, the Sienese explicitly and repeatedly endorses the (iv) insubstantiality of mathematical objects in his argument against mathematical causality (arguably backed also by his knowledge of Priscian's *In De anima*).

Regarding (ii) innatism, if my interpretation is correct, Piccolomini postulates instead what I have called "quantitative isomorphism:" that mathematical items inherently and co-eternally pre-exist in phantasy as much as they exist in prime matter. Although Piccolomini never applies the terms "potency" or "act" to phantasy or mathematical items, his exposition heavily implies that phantasy is the potency of every quantity.

It is noteworthy that Piccolomini does not accept one of the most characteristic of Proclus' metaphysical tenets regarding the reality of mathematics: that sensible reality ontologically depends on the Soul and, thus, that the ensouled reasons in the World Soul are the causes of sensible realities.[204] Such a view is

204 J. F. Finamore, E. Kutash, "Proclus on the Psychê World Soul and the Individual Soul", in P. D'Hoine, M. Martijn (eds), *All From One*, pp. 122–138; Opsomer, "The Natural World",

not only never suggested by Piccolomini but, moreover, it is wholly incompatible with *De Certitudine*'s metaphysics and alien to its theory of mathematical *phantasia* and quantity.

The reason for this incompatibility is that, in Piccolomini's theory, the intellectual circle, i.e., the definition that the intellect universally grasps from the phantastic circle, is not the cause of the phantastic circle. Whereas in Proclus' view, the ensouled image of the circle (extrinsic to phantasy) is unfolded in phantasy, Piccolomini holds that phantasy *alone* is the producer of *quantum phantasiatum*. Piccolomini's undetermined quantities are inherently present in phantasy and in naked prime matter and nowhere else. In other words, whereas for Proclus (and Priscian) the seeds out of which mathematical items sprout in phantasy are external to it, for Piccolomini mathematical items have their only source in the powers of phantasy, which obtains *quantum phantasiatum* out of itself alone.

In this way, Piccolomini's *phantasia* emerges as a wholly discrete region with laws of its own. *Phantasia* cannot be the cause of anything other than images or phantastic quantity. Conversely, no form of causation stemming from sensible objects is a real cause in *phantasia*. As put above, Piccolomini is extremely cautious in wording how sensible quantities are a part of the process of what he calls *abstractio*: they provide *phantasia* just with the occasion to produce a phantastic quantity within itself, but in no way does the sensory input act upon a passive mathematical phantasy.

The conclusion is that Piccolomini's phantasy is a kingdom without bridges, a realm inhabited by inert items which are excised from any causal chains. The only activity within phantasy is the efficient power of phantasy itself, restricted to its own region, alien to external causation and, hence, also itself cut off from external causal structures.

This sets Piccolomini's wider metaphysical views apart from those of Proclus and Priscian.

---

in P. D'Hoine, M. Martijn (eds), *All From One*, pp. 139–166; Martijn, *Proclus on Nature*; O'Meara, "Geometry and the Divine in Proclus"; O'Meara, *Pythagoras Revived*; Cleary, "Proclus' Philosophy of Mathematics"; MacIsaac, "The Soul and Discursive Reason in the Philosophy of Proclus"; MacIsaac, "Projection and Time in Proclus", pp. 83–105; MacIsaac, "Geometrical First Principles in Proclus' Commentary"; MacIsaac, "Νόησις, dialectique et mathématique chez Proclus"; Helmig, "Iamblichus, Proclus, and Philoponus on Parts, Faculties and *ousiai* of Soul and the Notion of Life", pp. 149–178; Steel, "Breathing Thought"; Mueller, "Foreword".

3.6 Mathematising the Unmathematical: Back to the Mertonian Challenge

In chapter XII, Piccolomini again addresses the problem he treated in the 1st question of the paraphrase six years before, namely the applicability of mathematics to sensible objects (see section 2.5.3). At this point, some of the pivotal issues that were dealt with back then re-emerge. After developing his ontology of mathematics, Piccolomini is now able to restate his former argument against a would-be quantified science able to deal with natural phenomena.

In this section, Piccolomini presents himself with a hypothetical adversary who defends that a mathematical science of natural movement is possible. The adversary is, thus, implying a full-correspondence reading of quantified propositions. By using his now-clarified philosophy of mathematics, Piccolomini answers this challenge:

> [1] However, if one were to argue further, stating that "motion is also a kind of common sensible, just like magnitude; and motion has its own properties and causes, as evident from *Physica* IX and VI, and thus it will be a certain science about motion, that is to say, a natural [science], as a science about quantity, which is mathematical", we can respond to this by stating that if we consider motion in general abstracted from matter, as some continuum, as Aristotle considers it in *Physica* VI, then the consideration will indeed be mathematical, and there is nothing against us. [2] However, if we consider motion insofar as it exists in a celestial body, or in animals, or limited in any other manner, then, since it arises and flows from its own principles and confines itself to its proper matter, it will render knowledge hard and difficult.[205]

205 Alessandro Piccolomini, *De certitudine*, ff. CVIIIv–CIX: [1] *Si vero adhuc replicaretur, quod motus etiam est sensibile quoddam commune, sicut magnitudo; habet autem motus suas passiones et suas causas, ut patet IX et VI Physicorum, ergo ita erit certa de motu scientia, naturalis scilicet, sicut scientia de quantitate, quae mathematica est ad hoc respondere possumus quod si motum consyderabimus in communi abstractum a materia, quatenus continuum quoddam est, sicut consyderat de illo Aristoteles VI Physicorum, tunc consyderatio erit mathematica, et nihil contra nos.* [2] *Si autem quatenus vel est motus in corpore caelesti, vel in animali, vel alio quovis modo limitato, tunc, cum oriatur et fluat ex principiis propriis et materiam sibi propriam limitet, cognitionem reddet arduam et difficilem.* The division into paragraphs and numeration is mine. I will refer to these paragraphs by the assigned numbers.

The argument of the fictional adversary holds in paragraph [1] that, much like quantity, movement is also a common and proper accident to sensible items. Therefore, according to Piccolomini's standards, movement should be also greatly abstractable in the three senses specified, namely (α1) as instantiable in phantasy, (α1) easily known, and (α3) separated.

The adversary also calls upon Aristotle's authority: he refers to the treatment of the continuum in book VI of the *Physica*, where the Stagirite dealt with the problem of continuity in movement, time, and qualitative change by using quantification (points and lines).[206] Book IX of the *Physica* is also invoked. Most probably, Piccolomini here refers either to the last book of *Physica* (namely, book VIII), where Aristotle treats Zeno's paradoxes recurring to mathematical means,[207] or to book I of *De caelo* (following book VIII of the *Physica* in the *corpus Aristotelicum*). The adversary concludes that a natural science of movement which is at the same time mathematical is possible.

Answering this challenge, Piccolomini shows in paragraph [1] how his theory of *phantasia* can defeat a full-correspondence reading of the adduced *loci* from Aristotle's *Physica*. The Sienese follows a treatment akin to that posed regarding the balance in the *quaestio prima* of the paraphrase, where, almost verbatim, the same Aristotelian *locus* was adduced as one of the common places of the Mertonian approach.

Piccolomini first restates his former view on Aristotle's treatment of continuum in *Physica* VI, claiming that in these books Aristotle conducts his inquiry mathematically, not physically.[208] Piccolomini's improved rationale is now given: if phantastic movement is obtained as a continuum and considered as a quantity, then the discourse about it has not a natural entity as its object of study anymore, but a phantastic one. In this case, the entity would be the general quantitative traits of motion as staged in phantasy. It follows that a discipline which quantitatively analyses phantastic movement is another *scientia media* with a separate mathematical object as its subject-matter.

Hence, the very notion of a science of quantity that deals with a natural object is, according to Piccolomini's ontology of scientific objects, a self-defeating one: sciences are either mathematical or natural, but such labels

---

206 B. M. Sattler, *The Concept of Motion in Ancient Greek Thought, Foundations in Logic, Method, and Mathematics*, chap. 7, especially section 7.3; D. Bostock, *Time, Matter, and Form. Essays on Aristotle's Physics*, chap. 10.

207 Aristotle, *Physica*, VIII. 8, 263a, l. 4–263b, l. 9.

208 Alessandro Piccolomini, *In mechanicas quaestiones*, ff. XIXv–XX: *Et hoc praesertim, ut dicunt, occasione sumpta ab iis quae scribitur ab Aristotele in* VI *de Principiis naturalibus, non animadvertentes Aristotelem ipsum in illo libro, disputantem rationibus mathematicis, non naturalibus.*

are totally and irreconcilably exclusive, for they indicate not only some specific demonstrative means (as he has defended in the 1540 paraphrase) but also different modes of being of the objects of study. The conclusion is that a mathematical science of an unmathematical object is just impossible, the "quantification of nature" but an improper way of speaking. This conclusion is even less surprising if we recollect that, according to Piccolomini, all movement in mathematics is metaphorical, for nothing factually moves in a phantastic item.[209]

Finally, Piccolomini details in [2] the conditions under which a discourse concerning sensible movement would be a genuine natural science: movement must be considered as flowing from the proper principles of a substance. In other words, the movement of a planet must be explained through the nature of the heavenly bodies, the colour change of an apple from green to red through the apple's nature, and the movement of an animal through its soul. But, Piccolomini swiftly points out, in doing natural philosophy the nature of a movement considered as an effect of the metaphysical nature of the moved body is linked to the elusive essence of such a substance. Because of that, Piccolomini concludes that the study of real movement is harder and obscurer, i.e., less certain (epistemic transparency) than that of phantastic movement.

Piccolomini returns to the very same question in the text he wrote immediately after *De certitudine*, namely the *Praefatio* to the Latin paraphrase:

> When we say that a human being is generated or that something becomes white, because both generation and the emergence of whiteness indicate motion, we will boldly assert that such statements are natural. However, if we say what a human being is what it is, or [that he is a] whole, or [that he is a] substance, or something related to a genus, then, as the things we say to inhere to the human being are the most common ones, such statements can be called metaphysical. Similarly, by asserting that a human being can be infinitely divided, we establish a mathematical statement. Since Aristotle, in book VI of *Natural Principles*, assuming the mathematical habit (*habitum mathematicum accipiens*), discusses motion.[210]

---

209 Alessandro Piccolomini, *De certitudine*, f. CIIv. See section 3.3.3 above.

210 Alessandro Piccolomini, *In mechanicas quaestiones*, f. IIv: *Cum enim dicimus hominem generari, aut album fieri, quia tam generatio, quam eductio albedinis motum indicat, huiusmodi enuntiationes naturales esse audacter pronuntiabimus. Si vero hominem esse id quod est, vel totum, vel substantiam, vel aliud ad genus, affirmabimus tunc, quando haec quae dicimus inesse homini communissima sunt, huiusmodi enuntiationes metaphysicae dici poterunt. Eodem pacto hominem ad infinitum dividi posse, asseverantes, enuntiationem*

Here, the Sienese reaffirms that: (i) natural philosophy explains natural motion; (ii) metaphysics provides substantial explanations about the most common traits; and (iii) Aristotle's treatment of the continuum and motion in *Physica* VI, as it concludes through mathematical demonstration, is a mathematical discussion, not a piece of natural philosophy. Thus, Piccolomini reasserts, in coherence with his philosophy of mathematics, that the statement "a human being can be divided infinitely" is in no way a physical assertion about the sensible human being. It is a mathematical statement proved through mathematical speculation. Consequently, as we know from Chapter VII of *De certitudine*, it has *quantum phantasiatum* as its exclusive object.

Piccolomini's argument is even more potent given that, in introducing mechanics to the reader of the paraphrase, he blurs the distinction between natural and violent motion to the point of making it almost irrelevant. In his introductory analysis of the expression *praeter naturam* (παρὰ φύσιν)[211] when referred to mechanics, Piccolomini offers the following view:

> Art, although it imitates nature and assists it, so that it may be of benefit to us, operates many things differently from how nature operates. Nevertheless, it [art] should not be considered less an imitator of nature, as it accomplishes whatever it does in a manner that nature would achieve in that same way if it were to do it. Although nature, if not hindered, always observes the same ways in each of its works, art, on its behalf, due to our various utilities, uses, and conveniences, advances through various paths, yet in every path it [art] still follows nature itself.[212]

From the Aristotelian definition of art as an imitation and completion (μιμεῖται, ἐπιτελεῖ) of nature in *Physica* II, 8,[213] Piccolomini concludes that mechanical

---

mathematicam constituimus. Aristoteles enim in VI *libro Naturalium principiorum, habitum accipiens mathematicum, de motu disserit.*

211 On this expression in Pseudo-Aristotle's *Quaestiones mechanicae* see M. J. Schiefsky, "Art and nature in ancient mechanics", in B. Bensaude-Vincent and W. R. Newman (eds), *The Artificial and the Natural: An Evolving Polarity* (Cambridge: Massachusetts Institute of Technology Press, 2007), pp. 67–108.

212 Alessandro Piccolomini, *In mechanicas quaestiones*, f. Vv: *Ars enim quamvis naturam imitetur, ac illam adiuvet, multa tamen, ut usui nobis esse possint, aliter operatur ac illa facit. Neque minus ob id naturae imitatrix existimari debet, quippe quae ex eo modo efficit quidquid facit quo illamet efficeret, si illud faceret. Nam tametsi natura eundem semper in quovis opere suo, si non impeditur, servet modum, ars vero propterea quod nostra utilitas, usus, atque commoditas multiplex est, multiplici etiam et illa itinere progrediatur, in quovis tamen itinere naturam et ipsa sequitur.*

213 Aristotle, *Physica*, II, 8, 199a, l. 8–20: ἔτι ἐν ὅσοις τέλος ἔστι τι, τούτου ἕνεκα πράττεται τὸ πρότερον καὶ τὸ ἐφεξῆς. οὐκοῦν ὡς πράττεται, οὕτω πέφυκε, καὶ ὡς πέφυκεν, οὕτω πράττεται

practice also depends upon and imitates (follows) and completes (goes beyond) the principles of nature.

What mechanics does is unfold new ways in which nature could (but does not) behave according to its own principles. Although a catapult does not exist by nature, its functioning depends on the natural principles ruling, say, the behaviour of weights, elasticity, and the fall of bodies. The Sienese hence concludes that the expression *praeter naturam* when applied to mechanics means the additional powers or capacities that this discipline bestows upon human beings (such as shooting a great rock with easiness by using a catapult) that they would be naturally unable to perform if they did not have mechanics. In this context, when the principles of nature are applied (or forced) in a way that goes beyond its inherent order and simplicity for the sake of human benefit and usefulness, mechanics is considered to operate "outside of nature".[214] Perhaps another of the ancient sources behind Piccolomini's claim that mechanical movement is natural is Proclus, who claims in his account of mathematical sciences that the products of Hero's mechanics operate "through ropes and cables, which imitate the motions and impulses of animated beings".[215]

Ultimately, Piccolomini's deliberate weakening of the distinction between motion *secundum naturam* and *praeter naturam* in the use of classical *loci*

ἕκαστον, ἂν μή τι ἐμποδίζῃ. πράττεται δ' ἕνεκά του· καὶ πέφυκεν ἄρα ἕνεκά του. οἷον εἰ οἰκία τῶν φύσει γιγνομένων ἦν, οὕτως ἂν ἐγίγνετο ὡς νῦν ὑπὸ τῆς τέχνης· εἰ δὲ τὰ φύσει μὴ μόνον φύσει ἀλλὰ καὶ τέχνῃ γίγνοιτο, ὡσαύτως ἂν γίγνοιτο ᾗ πέφυκεν. ἕνεκα ἄρα θατέρου θάτερον. ὅλως δὲ ἡ τέχνη τὰ μὲν ἐπιτελεῖ ἃ ἡ φύσις ἀδυνατεῖ ἀπεργάσασθαι, τὰ δὲ μιμεῖται. εἰ οὖν τὰ κατὰ τέχνην ἕνεκά του, δῆλον ὅτι καὶ τὰ κατὰ φύσιν· ὁμοίως γὰρ ἔχει πρὸς ἄλληλα ἐν τοῖς κατὰ τέχνην καὶ ἐν τοῖς κατὰ φύσιν τὰ ὕστερα πρὸς τὰ πρότερα. Translation by J. Barnes at Aristotle, *Complete Works of Aristotle*, Vol. 1: "Further, where there is an end, all the preceding steps are for the sake of that. Now surely as in action, so in nature; and as in nature, so it is in each action, if nothing interferes. Now action is for the sake of an end; therefore, the nature of things also is so. Thus, if a house, e.g., had been a thing made by nature, it would have been made in the same way as it is now by art; and if things made by nature were made also by art, they would come to be in the same way as by nature. The one, then, is for the sake of the other; and, generally, art in some cases completes what nature cannot bring to a finish, and in others imitates her. If, therefore, artificial products are for the sake of an end, so clearly also are natural products. The relation of the later to the earlier items is the same in both".

214 Alessandro Piccolomini, *In mechanicas quaestiones*, f. vv: *Natura igitur ex sua simplicitate nequaquam multiformis, si quando evenit ut eius simplicitate nos propter multiplices nostros usus non contenti adversus ipsam quippiam moliamur, tunc illa contra nitens ac contendens nostris conatibus difficultatem imponit. Quam difficultatem seu potius haesitationem quandam ut superemus arte quadam indigemus, qua tandem naturae conatus vel penitus confringentes vel, aliqua saltem ex parte, demolientes, nostrum opus perficiamus. Illud itaque artis genus, quo huic difficultati mederi possumus, mechanicum appellamus.*

215 Proclus, *In Euclidem* p. 41, l. 13–15: διὰ νεύρων καὶ σπαρτῶν ἔμψυχος ὁλκὰς καὶ κινήσεις ἀπομιμουμένων.

implies that scientific explanation of the movement of sensible items (including the movement of machines) will always belong to natural philosophy. It is highly significant that, after having devoted one hundred and forty folia to paraphrasing and discussing the *Quaestiones mechanicae*, Piccolomini states both within the 1540 paraphrase and again in the last chapter of *De certitudine* six years later that the object of study of mechanics is not real movement and, consequently, that mechanics cannot provide scientific accounts of the natural principles behind the movements of the machines.

My conclusion in this respect is that *De certitudine* is not just an idle, if curious, addition to Piccolomini's *In mechanicas quaestiones* that happened to have great fortune. On the contrary, it is a work which, through an elaborated philosophy of mathematics, integrally underpins the *prima facie* paradoxical scientific status attributed to mechanics in the 1456 *Praefatio* to the paraphrase and offers philosophical foundations for the anti-Mertonian views there expounded some years before.

The resulting theoretical framework holds that the discipline of mechanics, as far as it is a middle, mathematical science grounded in *quantum phantasiatum*, is unable to explain the causes of any kind of sensible movement. Contrarily, as far as mechanical speculations approximately match sensible items in practical affairs, mechanics is not a mathematical science, much like the approximate, applied geometry of the carpenter is not a science, but part of his art.

Piccolomini's position can be analysed in three different levels of argumentation, offering a multi-layered account of why a mathematised science of nature is impossible. Those three intertwined arguments are:

(a) ***Non-correspondence view***: Quantified propositions approximate, but do not completely fit or correspond to, the sensible entities they resemble. The abyss separating phantastic quantity from sensible items is unsurmountable. Ex. gr., mathematical truths concluded about quantified general motion will never hold exactly true in the motion of sensible items. Phantastic quantities only approximatively and accidentally resemble unmathematical items.

(b) ***Discretion of the object of study***: Even if, *ad impossibile*, a perfect isomorphy existed linking up phantastic quantities to their sensible correlates, a quantified science would nonetheless retain phantastic quantity as its proper and unique object. Such a science would only accidentally and obliquely reveal something about motion, i.e., as far as motion is quantity. Therefore, a science which is at the same time mathematical and natural remains absurd.

(c) ***Lack of causal explanation***: Even if, again *ad impossibile*, a mathematical-and-natural science were possible, the quantifications of such science would only describe sensible movement, but quantifications would remain

> silent about the real, substantial causes of movement. Thus, even if perfectly isomorphic quantifications of sensible items were possible, such a quantified theory of movement would only describe "how" natural things move but would have no say about "why" natural things move. The "why" of natural motion necessarily requires engaging in a science about substances.

This theory reveals that Piccolomini's scope in *De certitudine* goes clearly beyond the widely acknowledged challenge to mathematical demonstrations. In his attempt to make mechanics a discrete mathematical science innocuous to natural philosophy, Piccolomini also launched an arrow to the heart of the so-called new science and the 16th- to 18th-century attempts to mathematise natural philosophy. Nevertheless, the reception and consequences of this momentous aspect of Piccolomini's theory have received, to my knowledge, no attention in scholarship.

3.7 Common Mathematics and *De certitudine*'s Consistency: a Hypothesis

One of the most puzzling aspects of Piccolomini's argumentation in *De certitudine* could be portrayed in this manner: all the examples of mathematical objects and mathematical demonstrations proposed in his philosophising are geometrical. Even his philosophical theories, using the Averroistic *dimensiones* co-eternal to prime matter, Proclus' elaborations on intelligible matter, or his portrait of phantastic motion as a mathematical continuum suggest that, when Piccolomini speaks of quantity in his philosophical argumentation, he exclusively refers to continuous quantity. Nevertheless, the Sienese extracts from these arguments conclusions that he applies to all mathematical disciplines.

I may suggest that hints can be found in Piccolomini's text of an argument for solving this perplexing theoretical hindrance, of which he seems to be aware. To achieve this, he would require a method to unify geometry and arithmetic, enabling him to draw conclusions in both disciplines by relying on arguments from just one. This could be done if there were a superior genus of which arithmetic and geometry are species, and this is exactly what Piccolomini, following Proclus, postulates in his only two mentions of arithmetic in the opusculum (both in the same section). The crucial passage is included after the exposition of his philosophy of mathematics in chapter VII:

> And here, incidentally, one crucial point must be noted: when we have indicated that phantasised quantity (*quantum phantasiatum*) is the subject-matter of mathematics, this is indeed not said of geometry or arithmetic,

> which are the two primary genera of mathematics, but it is declared to be a certain faculty common to both geometry and arithmetic ([…] *cuiusdam facultatis communis ad geometriam et arithmeticam*). For Proclus, most explicitly in books I and II [of *In Euclidem*], shows everywhere the existence of a common science encompassing these two [geometry and arithmetic], claiming its own subject and distinct properties, principles specific to itself, and subordinating both geometry and arithmetic to itself.[216]

Piccolomini establishes that *quantum phantasiatum*, the key concept of his philosophy he has just tackled, is not the subject-matter of geometry or arithmetic. We are told that both mathematical sciences share phantastic objects of study by virtue of belonging to a higher science to which subalternate both continuous and discrete quantity, a science which has *quantum phantasiatum* as such as its object of scrutiny. In introducing this notion, he once more invokes Proclus' *In Euclidem*, pointing to the notion of ὅλη or κοινὴ μαθηματική there delivered.[217] Following the guide of Proclus, Piccolomini calls this

216 Alessandro Piccolomini, *De certitudine*, f. LXXXXVIIv: *Et hic obiter unum est notandum magni ponderis, quod cum quantum phantasiatum ostenderimus esse mathematicorum materiam sive subiectum, hoc quidem non geometriae vel arithmeticae, quae duae sunt mathematicae prima genera, subiectum esse dicitur, sed cuiusdam facultatis communis ad geometriam et arithmeticam. Nam manifestissime Proclus in I et II libro passim ostendit dari quandam scientiam communem ad illas duas, quae proprium subiectum et proprias passiones, propria principia sibi vindicat, et illas duas sibi subalternat.*

217 Proclus, *In Euclidem*, p. 7, l. 13–22: ἀλλ'ὅτι μὲν ἀρχαὶ καὶ τῶν μαθημάτων αὗται προεστήκασιν, αἳ καὶ τῶν ὄντων ἁπάντων, φανερόν. ὥσπερ δὲ τὰς κοινὰς ἀρχὰς αὐτῶν τεθεωρήκαμεν καὶ διὰ πάντων διηκούσας τῶν μαθηματικῶν γενῶν, οὕτω δὴ καὶ τὰ κοινὰ αὐτῶν θεωρήματα καὶ ἁπλᾶ καὶ τῆς μιᾶς ἐπιστήμης ἔγγονα τῆς πάσας ὁμοῦ τὰς μαθηματικὰς γνώσεις ἐν ἑνὶ ἐπεχούσης ἀναλογισώμεθα, καὶ ὅπως ἐφαρμόττει πάσαις καὶ δύναται καὶ ἐν ἀριθμοῖς καὶ ἐν μεγέθεσι καὶ ἐν κινήσεσι θεωρεῖσθαι σκοπήσωμεν. In Morrow's translation: "It is clear, then, that the principles primary in mathematics are those that preside over all things. Just as we have noted these common principles and seen that they pervade all classes of mathematical objects, so let us enumerate the simple theorems that are common to them all, that is, the theorems generated by the single science that embraces alike all forms of mathematical knowledge; and let us see how they fit into all these sciences and can be observed alike in numbers, magnitudes, and motions". Proclus, *In Euclidem*, p. 9, l. 11–19: οὐ γὰρ δήπου τῶν μὲν μεριστῶν εἰσὶν ἐπιστῆμαι καὶ γνώσεις, τῶν δὲ ἀΰλων καὶ τῆς νοερᾶς θεωρίας ἐγγυτέρω τεταγμένων οὐδὲ μίαν ἔχομεν ἐπιστήμην. ἀλλὰ πολλῷ πρότερον ἡ ἐκείνων γνῶσίς ἐστιν ἐπιστήμη καὶ ἀπ'ἐκείνης αἱ πολλαὶ τοὺς κοινοὺς ὑποδέχονται λόγους καὶ μέχρι τοσούτου γνώσεων ἄνοδος ἀπὸ τῶν μερικωτέρων ἐπὶ τὰς ὁλικωτέρας, ἕως ἂν ἐπ'αὐτὴν ἀναδράμωμεν τὴν τοῦ ὄντος, ᾗ ὄν ἐστιν, ἐπιστήμην. In Morrow's translation: "It cannot be that we have sciences of particular areas of being and knowledge of them but have no single science of the immaterial objects that stand much closer to intellectual inspection. Knowledge of those objects is by far the prior science, and from it the several sciences get their common propositions,

superior mathematics "*communis*" and proposes then that some general passions and principles can be explored and established in it. The principles and truths of such common science of quantity would be shared by every mathematical discipline.

Piccolomini briefly hints at two technical cases to make evident the existence of this maximally general science about quantity. Following Proclus yet again, he claims that some of the theorems from Euclid's books IV and VI, where magnitude or μέγεθος is treated, are repeated in the arithmetical book VII and VIII, but applied to number or ἀριθμός.[218] Presumably, Piccolomini has given the incorrect book number and refers to the theory of proportion presented in books V (and IV) of Euclid's *Elementa*, traditionally attributed to Eudoxus.[219] Two cases of the same theorem applied to different kinds of quantity (i.e., to magnitude and number) are theorems 12 of the V and VII books, respectively.[220]

---

our knowledge ascending from the more partial to the more general until at last we reach the science of being as being". Proclus, *In Euclidem*, p. 48, l. 1–5: τὰ μὲν δὴ κοινὰ καὶ ἐπὶ πᾶσαν διατείνοντα τὴν μαθηματικὴν ἐπιστήμην ἐν τοῖς προειρημένοις λόγοις τεθεάμεθα τῷ τε Πλάτωνι συμπορευόμενοι καὶ παρὰ τῶν ἄλλων ἀναλεγόμενοι τὰ πρὸς τὴν παροῦσαν πραγματείαν ἡμῖν συντείνοντα νοήματα. In Morrow's translation: "In the preceding discourse we have examined the common characters pervading all mathematical science, following Plato's lead and also using thoughts collected from other sources that are relevant to the present study". See Rabouin, *Mathesis universalis*, chap. I, II, and III; Napolitano Valditara, *Le idee, i numeri, l'ordine*; Crapulli, *Mathesis universalis*; O'Meara, "Mathematics and the Sciences"; Bechtle, "Sur la mathématique générale".

218 Proclus, *In Euclidem*, pp. 7, l. 22–8, l. 4: τοιαῦτα δέ ἐστι τά τε τῶν ἀναλογιῶν καὶ τὰ τῶν συνθέσεων καὶ διαιρέσεων καὶ τῶν ἀναστροφῶν καὶ ἐναλλαγῶν, ἔτι δὲ τὰ τῶν λόγων πάντων, οἷον πολλαπλασίων καὶ ἐπιμορίων [καὶ] ἐπιμερῶν, καὶ τῶν τούτοις ἀντικειμένων, καὶ ἁπλῶς τὰ περὶ τὸ ἴσον καὶ ἄνισον καθόλου θεωρούμενα καὶ κοινῶς, οὐ καθόσον ἔστιν ἐν σχήμασιν ἢ ἀριθμοῖς ἢ κινήσεσιν, ἀλλ' αὐτὸ καθ'αὑτὸ τούτων ἑκάτερον φύσιν τινὰ ἔχον κοινὴν καὶ γνῶσιν ἑαυτοῦ παρεχόμενον ἁπλουστέραν. In Morrow's translation: "Such are the theorems governing proportion, namely, the rules of compounding, dividing, converting, and alternating; likewise the theorems concerning ratios of all kinds, multiple, superparticular, superpartient, and their counterparts; and the theorems about equality and inequality in their most general and universal aspects, not equality or inequality of figures, numbers, or motions, but each of the two by itself as having a nature common to all its forms and capable of more simple apprehension". Piccolomini's example from Euclid is likely motivated by Proclus' observation in the second prologue that everything which is "speakable" (i.e., rational) and knowable in geometry is susceptible to be arithmetically treated, p. 49, l. 12–13: πᾶν γάρ, ὅσον ἔστι ῥητὸν ἐν αὐτῇ [γεωμετρίᾳ] καὶ γνωστόν, ἐκ τῶν ἀριθμητικῶν ἀφορίζεται λόγων. In Morrow's translation: "[…] for everything that is expressible and knowable in geometry is determined by arithmetical ratios".

219 Heath's introduction to book V in Euclid, *The Thirteen Books of the Elements*, vol. 2, book V, p. 112 and forth.

220 Compare proposition V, 12: "If any number of magnitudes are proportional, then one of the antecedents is to one of the consequents as the sum of the antecedents is to the sum of the consequents' with proposition VII, 12: 'If any number of numbers are proportional,

The second source supporting the existence of this common mathematical science is Aristotle's chapter 5 in the first book of *Analytica posteriora*.[221] There, Aristotle expounds on a theory of permutability of the proportions (A : B = C : D) and explains how this property holds in numbers, lines, solid geometry, and even time if considered as a continuum.[222]

That there are some principles and propositions (theorems) which regard quantity as such is neither explicitly affirmed nor denied by Proclus. Nevertheless, this claim fits Iamblichus' explicit views on the matter. Amongst the objects of study of common mathematics presented by Iamblichus in *De communi mathematica scientia*, ποσὸν ἁπλῶς or unqualified quantity is indexed.[223] Although, as Robichaud has shown, Iamblichus' work was extant at least in manuscript form in the Latin translation of Marsilio Ficino,[224] and therefore it was available also in Greek manuscripts, it seems highly unlikely that Piccolomini was familiar with this text. Iamblichus' name is never mentioned in *De certitudine*, and there is no evidence pointing to the Sienese being acquainted with this source. It is more likely that Piccolomini interpreted Proclus as proposing a science about general quantity. In Piccolomini's outlook, a science about unqualified quantity stands as a science about unqualified phantastic quantity – as he has indicated in the quotation above.

So, what would be Piccolomini's use of this common science? Why is he even introducing it at the end of his chapter on the ontology of mathematics? This leads us, I believe, back to our initial puzzle, namely understanding whether, by his theoretical standards, Piccolomini can legitimately draw conclusions which apply to mathematics as a whole by using exclusively geometrical cases. I find that Piccolomini's inclusion of common mathematics as a final remark to chapter VII may be meant to shield him against this sort of objection (whether his move is effective or not is a different question). If the conclusions drawn from his meta-mathematical argumentation in *De certitudine* would be taken

---

then one of the antecedents is to one of the consequents as the sum of the antecedents is to the sum of the consequents". The only difference in the phrasing of both theorems is that the term "magnitude" in book V is substituted by the term "number" in book VII. See Heath's introduction to book V in Euclid, *The Thirteen Books of the Elements*, vol. 2, book V, pp. 159–160 and book VII, pp. 312–313.

221 Aristotle, *Analytica posteriora*, I, 5,74a, l. 4–13; l. 17–25.

222 Alessandro Piccolomini, *De certitudine*, f. LXXXXVIII: *Sed quam plurima theoremata in IV, VI et dehinc repetita in VII et VIII Elementorum apertissime hoc ostendunt, veluti de quattuor quantitatibus proportionalibus, et earum permutata proportione, et de aliis patere potest. Quod etiam aperte innuere videtur Aristoteles I Posteriorum analyticorum, cap. 5.*

223 Iamblichus, *De communi*, p. 19, l. 2.

224 Robichaud, *Plato's Persona*, p. 171 and forth.

as concerning unqualified phantastic quantity, as Piccolomini intends, they would be theorems of common mathematics.

It should not be surprising to find that philosophical propositions about the being of unqualified quantity could also be part of common mathematics: Proclus' take on κοινὴ μαθηματική portrays it as a discipline bridging mathematics and metaphysics. Common mathematics speculate on mathematical properties considered in themselves, such as, as put by Proclus, unqualifiedly universal equality and inequality ([…] ἁπλῶς τὰ περὶ τὸ ἴσον καὶ ἄνισον καθόλου […]).[225] Therefore, the reflection on meta-mathematical questions and definitions leads, according to the Diadochus' account, to a speculation bridging to metaphysics on the nature of the most general quantitative traits. Proclus' reasoning would also apply (as it factually does in Iamblichus' account) to the concept of quantity: asking questions about unqualified quantity necessarily leads the practitioner of mathematics to speculation on the characteristics of unqualified quantity *qua* quantity or, in Piccolomini's lexicon, about *quantum phantasiatum* as such.

Thus, Piccolomini's doctrine on the nature of quantity, if validly concluded of unqualified phantastic quantity, will hold true in every branch of mathematics. More exhaustively, the rationale hypothetically doing the leap to common mathematics and preserving the coherence of the opusculum could be expressed as follows:

(1) There is a common mathematical science whose object of study is unqualified phantastic quantity.

(2) Every truth demonstrated about the object of common mathematics holds true in every science dealing with phantastic quantity.

(3) Every truth demonstrated of unqualified phantastic quantity is demonstrated of the object of common mathematics.

(4) The geometrical arguments contained in *De certitudine* (1.17 and 1.32) reveal meta-mathematical theorems, not of phantastic quantity *qua* continuous, but of unqualified phantastic quantity (including meta-mathematical propositions such as "phantastic quantities are causally idle").

(5) Therefore, (concluded from 3 and 4) the truths demonstrated from the geometrical arguments in *De certitudine* hold true about the object of common mathematics.

(6) Therefore, (concluded from 2 and 5) the truths demonstrated from the geometrical arguments in *De certitudine* hold in every science dealing with phantastic quantity.

225 Proclus, *In Euclidem*, p. 7, l. 26–27.

(7) All mathematical sciences deal with phantastic quantity and, conversely, every science dealing with phantastic quantity is mathematical.
(8) Therefore, (concluded from 6 and 7) the truths demonstrated from the geometrical arguments in *De certitudine* hold in all mathematical sciences.

As both Crapulli and Rabouin have pointed out,[226] Piccolomini's theory of *communis scientia* in his reception of Proclus' *In Euclidem* may be considered one of the most relevant 16th-century iterations of what is today known as *mathesis universalis*. It is perhaps one of the earliest statements in the Latin West of a superior mathematical science linking arithmetic and geometry.[227]

To this, it should be added that, if my interpretation is right, brief and gratuitous as it might appear, Piccolomini's inclusion of common mathematics in chapter VII should not be understood just as a note in passing: its presence in the opusculum is not a neutral matter of curiosity or an erudite *fioritura* of a humanist. For, as shown, without a logical manoeuvre able to unify arithmetic and geometry as belonging to one single genus on which the treatise would be focused, the logical and doctrinal assemblage of *De certitudine* would collapse.

226 Crapulli, *Mathesis universalis*, pp. 9–62; Rabouin, *Mathesis universalis*, p. 201 and forth.
227 Crapulli, *Mathesis universalis*, p. 32 and forth.

PART 4

Conclusions

In this book, I have tried to place Alessandro Piccolomini's opusculum *De certitudine mathematicarum disciplinarum* in its historical and intellectual context and to account for his philosophy of certainty or akribology. I have highlighted five contextual aspects which have remained obscure in modern scholarship. These hermeneutical keys allow us to better understand Piccolomini's aims and the nuances of his ontology of mathematics. These have been: (i) the Italian anti-Mertonian tradition, (ii) the philosophy of mechanics as cultivated in the 16th-century Italian context, (iii) the tradition of the "Paduan" theory of demonstrative regress, (iv) the "anti-Galenic" theory of analysis, and (v) Piccolomini's agreements and disagreements with the so-called "Simplician" Averroism of Genua De' Passeri as well as the decisive influence of Priscian of Lydia's *In De anima* in the Sienese's reading of Proclus' *In Euclidem*.

Regarding the philosophy of mechanics, I have shown that *De certitudine* should be read in light of the controversies regarding the Pseudo-Aristotelian *Quaestiones mechanicae*, the problems raised by the quantification of motion, and the relation between mathematical statements and natural objects. Discussions including the certainty of mathematics vis-à-vis that of natural philosophy and the place of mechanics amongst the sciences were promulgated in the circles of erudite technicians such as Tartaglia and both his and Piccolomini's patron, Diego de Hurtado y Mendoza. I have made the case that Piccolomini's opusculum intended to deliver a philosophy of mathematics that would make mechanics a friendly discipline to an erudite readership most likely hostile to quantification in the explanation of movement. Therefore, to fully account for *De certitudine*'s philosophy and aims, it should be taken into consideration together with Piccolomini's 1540 Latin paraphrase to the *Quaestiones mechanicae* and its 1546 *Praefatio*, to which *De certitudine* was added as an attachment in the 1547 publication.

The contents of the *Praefatio* and the paraphrase are revealing. The first question of the paraphrase contains some pieces on the philosophy of mathematics which can be deemed as an early, piecemeal version of Piccolomini's later theories. It portrays some of the intuitions that would be fully developed six years later in *De certitudine*. In the paraphrase, Piccolomini opposes what I have called a "full-correspondence" reading of quantification, a view that he acknowledges as the implicit undercurrent in the philosophy of the Italian epigones of the Mertonian *calculatores*.

 | DOI:10.1163/9789004724143_005

The full-correspondence approach to quantification can be portrayed in two summary points:

(1) Quantified statements about natural realities correspond to real or possible extra-mathematical realities of which they are formalisations. In other words, there is an isomorphism between quantified statements and (actual or possible) quantified realities.
(2) It follows from the isomorphy granted in (1) that quantified statements are also epistemically valuable, i.e., that they (can) convey or reveal something about the isomorphically quantified natural realities.

Ever since a full-correspondence reading of quantification became the default philosophy of many of the Italian Calculators, arguably through the influence of Cajetan of Thiene, an increasingly hostile reaction took place. I have analysed the views of some representative voices of the anti-Mertonian reaction as developed in the 15th and early 16th century, such as Pietro Pomponazzi and Agostino Nifo. I have also shown how Ludovico Boccadiferro, Piccolomini's teacher, developed a philosophy of mathematics designed to exalt natural philosophy over mathematics, some of whose themes are echoed in *De certitudine*.

Piccolomini's piecemeal philosophy in the first question of his paraphrase belongs to this tradition of anti-Mertonian thought. He defends both that (a) mathematical principles do not hold when applied to sensible entities and, (b) that all quantified discourse is discretely mathematical. As shown by the reiteration of this theme in the last chapter of *De certitudine* and in the *Praefatio*, his opposition to a full-correspondence reading of quantification remained a main part of his outlook to the time he wrote the opusculum. In these sections, Piccolomini defends through his by then mature philosophy of mathematics that a mathematical science of natural motion is simply impossible, as it is an oxymoron.

I have emphasised the Averroistic pedigree of the starting points of the controversy. Two passages from Averroes dictated the views of Latin scholars on the issues of certainty and demonstration in mathematics vis-à-vis natural philosophy, namely (1) his comment on *Metaphysica* II, 3, and (2) his *Prooemium* to the *Magnum commentarium* to the *Phyisca*. In the former, Averroes had conceded the highest certainty to mathematics as its objects are deprived of matter, thus surpassing natural philosophy and its certainty. In the same passage, he also related diverse degrees of certainty to the most scientific demonstration (*demonstratio simpliciter*, *absoluta*, or *potissima*). In the latter passage, the Commentator had sustained the existence of this superior, most scientific logical procedure, and declared that it is found mostly in mathematics.

From these two *loci* of Averroes the Latin tradition held (with some variations) the position which I have summarised with the lemma *certissimae ergo potissimae*. This was the shared opinion that mathematical demonstrations were *potissimae* because of the superior certainty degree allowed by the special ontological status of their objects of study. This position was sustained by many of the 13th- to 16th-century Latin authorities whom Piccolomini explicitly opposes in *De certitudine*, to wit (the Parisian sources) Albert the Great, Thomas Aquinas, Giles of Rome, Marsilius of Inghen, (and the Italian sources) Donato Acciaioli, Nifo, and Marco Antonio Zimara.

This leads us to the idiosyncratic traits of Piccolomini's position. Piccolomini's *De certitudine* seeks to break the long-held link *certissimae ergo potissimae* by putting forward an argument that he initially presents as somewhat paradoxical, namely that, in truth, mathematical demonstrations cannot be of the most scientific kind or *potissimae* precisely because they are *certissimae*. For him to defend and clarify this puzzling statement, Piccolomini needs to develop a full ontology of mathematics. It is then that Proclus' *In Euclidem* appears as the primary authority in his writing, which Piccolomini reads as a precursor to his position. The two main sections from Proclus' *In Euclidem* as an authority backing his thesis are (1) Proclus' criticisms of the three flaws of proposition I.32 as a demonstration "by signs" and his recognition that mathematical necessity stems from the special nature of the mathematical object of study (p. 206, l. 12–23), and (2) Proclus' testimony that Amphinomus and Aristotle did not admit any form of causation or scientific explanative account in geometry (p. 202, l. 9–12). As another central argumentative piece of *De certitudine*, Piccolomini would paraphrase and adapt several sections of Proclus' second prologue.

In explaining Piccolomini's reception of Proclus' *In Euclidem*, I have brought attention to the overlooked fact that Piccolomini, as someone educated at Padua at the peak of the so-called "Simplician" Averroism, and who had had the greatest exponent of this form of Averroism, Genua De' Passeri, as his professor, was heavily influenced by the doctrines of Priscian's *In De anima* (quoted twice in *De certitudine*). As shown, Piccolomini was seemingly attributing to Proclus some of the views that were familiar to him from Priscian's text and reading Proclus' lexicon through the somewhat similar lexicon of his better-known Priscian. In doing that, he would produce what I have called a "Simplician" Proclus, i.e., a Proclus who, much like Priscian, sustains the causal idleness of phantastic mathematical objects – the cardinal doctrine in Piccolomini's argumentation. One likely explanation for Piccolomini's deliberate attribution of such doctrines to Proclus might be the convenience of showing that his argument was backed by the mathematical and philosophical authority of no less of a figure than the only surviving ancient Greek commentator on Euclid. Gaining Proclus to his side was an almost certain victory.

Thus, Piccolomini's denial of mathematical causation has as its origin Priscian's commentary (mostly in p. 210, l. 5–9; p. 233, l. 22–30; p. 276, l. 27–35), whose views were spread and known through the influential teaching of Genua De' Passeri (*In De anima*, III, 4, p. 174v). To accomplish his denial of mathematical causality, Piccolomini would use other arguments which he collected from various sources beyond Proclus and Priscian, including Averroes, Alexander of Aphrodisias, and, perhaps, his Bolognese master Boccadiferro (see Table 9 in the Appendix).

Additionally, Priscian also provides the key for grasping Piccolomini's theory of abstraction as, somewhat confusingly, presented in his opusculum. The precise meaning of Piccolomini's *abstractio* risks being understood (and inaccurately so, in my view) as suggesting that phantastic mathematical objects are derivative from and, hence, straightforward simplified imitations of physical objects. One section from Priscian's commentary (p. 233, l. 13–25), explicitly invoked by Piccolomini in his discussion of abstractability and certainty, provides the key for cracking the meaning of *abstractio* in *De certitudine*.

In Priscian's account, Aristotle referred to the process of obtaining mathematical objects in phantasy as "abstraction" to highlight the resemblance of phantastic quantities to sensible quantities. However, Priscian clarifies that this "imitation" does not negate the fact that phantasy, in its distinct activity, is the only generator of mathematical objects. Consequently, when Aristotle used the term "abstraction", he aimed to emphasise the similarity between the two species of quantities. The apparent imitative connection lies in that both quantitative entities are expressions of the same mathematical reason in two different substrates: the circle can be instantiated (with diverse objectual precision) both in sensible matter and in phantasy. Because of this, Priscian held, Aristotle was referring to the phantastic projection of ensouled reasons – in agreement with the Platonic tradition. In sum, "abstraction" is but another name for mathematical projection, stressing the resemblance amongst phantastic and sensible participating entities.

This account will be partially accepted by Piccolomini. He borrows from Proclus and Priscian the idea that the phantastic circle is not a simplified imitation of the sensible circle, but a *sui generis*, discrete generation of phantasy. This is explained by positing that, although the production of the phantastic circle is triggered by the perception of the sensible circle, sensible quantity is just the stimulus affording the occasion of the phantastic generation of mathematical quantities. I have shown that Piccolomini expresses this crucial idea by adopting a characteristic Latin phrase taken from Genua and the Paduan *Simpliciani*: *habita occasione*. However, unlike Proclus and Priscian, the Sienese does not posit a single, paradigmatic cause for both the sensible and phantastic mathematical items: ensouled mathematical reasons and forms participated in by various strata of reality are absent from *De certitudine*'s worldview.

Therefore, drawing from these two Neoplatonic sources and trimming from them the exemplary causes and representative relations, Piccolomini composed an ontology of mathematics which is at the heart of his 1546 work. This philosophy could be expressed in these ten points:

(i) Mathematical items exist as abstracted (separated) quantitative beings in phantasy (*quantum phantasiatum*).
(ii) Phantastic quantitative entities are insubstantial and, thus, causally idle.
(iii) Mathematical items are not totally freed from sensible-like traits, but they manifest ontological and epistemological traits unattainable in sensible matter.
(iv) Mathematical items, therefore, exist in a middle way (*modus medius*) between sensible matter and intellect, i.e., phantasy.
(v) Mathematical items are produced by phantasy alone (*phantasia efficiens*).
(vi) The occasional affordance (*habita occasione*) provided by sensible quantity triggers phantasy's production of mathematical objects.
(vii) The intellect gathers universal mathematical concepts from phantastic quantities.
(viii) The object of study of middle or subalternate sciences (*scientiae mediae*) such as mechanics is phantastic quantity – and nothing else.
(ix) Phantasy is a power of the soul between perception and intellect (*medium locum tenet*) – but tied to the body.
(x) Therefore, phantasy is improperly called "intelligible matter" and also "passive intellect".

In addition to my exposition and exegesis of Piccolomini's ontology of mathematics in chapter VII, I have also highlighted one of the theoretical problems which this theory confronts. If phantastic quantities are neither expressions of superior principles nor imitations of sensible quantities, where do they come from? Piccolomini needs to posit a source of phantastic quantity in phantasy itself. I have provided the hypothesis of "quantitative isomorphy" between sensible matter and phantasy to address this hindrance, a conjecture which is formally coherent and that finds textual support. It is founded in (i) Averroes' theory of the co-eternal quantity to prime matter (referred to by Piccolomini in explaining abstractability), (ii) the explicit claim that phantastic quantities are found (*reperiuntur*) in phantasy, and (iii) that, following Proclus, Piccolomini refers to phantasy as "intelligible matter" in an equivocal sense. Piccolomini assumes a sort of commonality between both substrates, namely that both real and intelligible matter inherently pre-contain every possible quantity.

I have also posited a conjecture closing Piccolomini's philosophy of mathematics: as he was relying solely on geometrical cases, Piccolomini strategically fortified his argumentation by finding a way to include every mathematical branch in its scope. This safeguard is achieved by incorporating the notion of a *communis mathematica* (later known as *mathesis universalis*) from Proclus' *In Euclidem* as a final addition to his philosophy of mathematics. According to Piccolomini, "common mathematics" deals with unqualified phantastic quantity or *quantum phantasiatum*. Consequently, all his theorems regarding the insubstantiality and causal idleness of quantity, if considered applicable to unqualified quantity, extend their validity to discrete and continuous quantity. Therefore, they hold across every conceivable mathematical branch.

Regarding Piccolomini's akribology of mathematics, I have shown that this is thoroughly explained by a close consideration of his ontology and sources. Piccolomini teaches that, lacking the complexity of substances and being causally idle, the definitions of mathematical items are revealed to epistemic agents as they phantasise mathematical objects. This epistemological advantage is granted by the unparalleled "easiness of abstraction" (*facilitas abstrahendi*) with which quantity is endowed. This means that, in mathematics, the real object is the one produced by the epistemic agent, and in such an epistemically transparent way that, in the very act of production, the essence of the subject is revealed. Therefore, the three senses of *certitudo*, as I have defined it in Piccolomini's text, are maximally found in mathematics: (a) the rank of objectual precision, where matter and accidental features are not obstacles to manifesting and apprehending the definitions; (b) the epistemic transparency following the objectual precision, allowing immediate knowledge of the definitions and no chances of error; and (c) the necessary and irrefutable demonstrations thereby produced.

However, from the fact that mathematics possesses the greatest degree of certainty, severe consequences follow. This is so because, to Piccolomini's standards, mathematical certainty and its ontological conditions prevent unqualifiedly scientific demonstrations from being arranged in this discipline. Piccolomini shows this through two different arguments, one regarding the content of the demonstration, and another concerning its form:

Firstly, (1) according to the Aristotelian demonstrative theory, the middle-term of a syllogism must express the cause accounting for the fact of which scientific knowledge is sought. Nevertheless, Piccolomini's ontology of mathematics denies any causation stemming from mathematical items. Thus, it necessarily follows that the content of the middle-term of a mathematical proof put in syllogistic form will express none of the four causes and, hence, that it is not a real demonstration.

Secondly, (2) regarding its formal aspects, Piccolomini engages in a lengthy discussion on the arrangement of *demonstratio potissima*. I have provided formal and textual evidence that Piccolomini's model of unqualified demonstration is a form of the long-discussed Paduan demonstrative regress which lacks step 2 and states that the middle-term must be the definition of the predicate or passion that is attributed to its subject. Piccolomini argues this point by signalling (arguably following Giles of Rome) that the goal of scientific inquiry is to obtain the definition of the subjects. Therefore, the definition of the subject cannot be a condition, but a consequence of the most scientific demonstration. Nevertheless, in the case of mathematical disciplines, as said above, the epistemic transparency of its objects allows for the immediate grasping of their definitions. It follows (paradoxically so) that mathematical objects are so certain and easily known that real scientific inquiry about them is just impossible. Mathematics appears in this account as a discipline in which theorems taken as demonstrable attributes (i.e., *per se* accidents) are idly demonstrated of a subject whose definition is known from the beginning of the enquiry.

The formal differences between mathematical demonstration and *demonstratio potissima* are also evident in Piccolomini's resolution of Euclid I.1. I have argued that *De certitudine* presents an anti-Galenic interpretation of resolution and composition, a debate that persisted since Peter of Abano and was addressed in the 16th century. The Galenian view that Piccolomini challenges equates the logical processes of resolution and composition with the demonstrations *quia* and *propter quid*, respectively. I have suggested that Piccolomini's motivation appears to stem from the significant emphasis Proclus placed on mathematical analysis and his formal description of it. Indeed, if Proclus' discussions on mathematical analysis are interpreted in the Galenic manner, it would lead to the conclusion that a demonstrative regress composed of two syllogisms, *quia* and *propter quid* (thus constituting *demonstratio potissima*), can be applied within mathematics. As this reading is deleterious to Piccolomini's intentions, he was bound to counter it. He did so in two ways: firstly (i) he offered a five-fold classification of analysis which deprived analysis in general (and mathematical analysis in particular) of any heuristic and demonstrative powers, depicting it as a logical validity checking tool. Secondly (ii), Piccolomini resolved, in a syllogistic manner, Euclid's proposition I.1. This evinces the unsurmountable discrepancy between the syllogistic structure of the mathematical proof and Piccolomini's model of demonstrative regress. This formalisation also reveals that the definition of the subject needs to be taken as the middle-term in the syllogistic form of the mathematical demonstration.

The conclusion is that formally speaking, mathematical demonstrations, in which the middle-term is always the definition of the subject, are neither of the highest scientific rank, nor scientific at all.

Through both of these arguments, Piccolomini subverts the traditional Latin lemma *certissimae ergo potissimae*: mathematical demonstrations are neither in content nor in form *potissimae* precisely because they are *certissimae* – something that, according to Piccolomini, was known to the Greek commentators, but ignored by the Latin ones. This ultimately stands as a highly innovative epistemological outlook, namely that the certainty in the demonstrations and methods of a given discipline tells us nothing about its scientific status and, contrarily, that science does not necessarily involve certainty.

What was Piccolomini's point in his argumentation in *De certitudine*?

This leads us again to the question of mechanics. I have shown that, if we read the *Praefatio* to the Latin paraphrase (written after *De certitudine*), some of the fruits of Piccolomini's philosophy of mathematics, particularly regarding mechanics, are revealed. There we learn that mechanics, like every mathematical discipline, is a form of contemplative philosophy. This follows from two unconventional tenets of Piccolomini:

Firstly, (1) Piccolomini's theory of the middle sciences postulates that, regardless of the appearances, all quantified discourses invariably have separated phantastic quantity as their object of study. Thus, interpreting Proclus' in-betweenness of mathematics, Piccolomini proposes that subordinated disciplines such as harmonics, astronomy, and mechanics are called "middle" because their object of study, i.e., phantasy, ontologically lies in-between intellect and perception. From this, it also follows that geometry and arithmetic are middle sciences.

Secondly, (2) Piccolomini entertains that, for a discipline to be contemplative, dealing with substances or real causation is not required. The whole contemplation just lies in grasping universal principles ruling over a certain type of being and the theorems following from such principles. He even conceives of a sort of speculation exercised through the abstract consideration of mathematical principles with practical ends, such as the mathematical invention or the design of machines. Arguably, Piccolomini's statement that mechanics (and generally all middle sciences) is a part of contemplative philosophy targets the overly anti-mathematical views of De' Passeri and his followers.

Piccolomini's outlook regarding mechanics can be summarised in the following eight points:

(i) Every quantified or mathematical discipline has phantastic quantities as its object of study.
(ii) Therefore, every quantified or mathematical discipline is not concerned with natural objects. This means that the object of study of mechanics is not sensible motion.
(iii) Phantastic quantity is deprived of substantiality and causation.
(iv) Because of this, demonstration in mathematics does not provide any causal explanations. It features demonstrations "from signs" and in no way *potissimae*.
(v) As a consequence, mechanics (or quantified kinetics, generally) cannot provide, either in its form or content, causal explanations, neither of phantastic (i.e., metaphorical) motion, nor of real motions.
(vi) Nevertheless, the investigation of the universal principles and theorems ruling phantastic quantity is a form of contemplative philosophy.
(vii) It follows that the exercise of mechanical research and invention, regardless of whether it concerns insubstantial and causally idle objects, and ultimately directed to practice, is a form of contemplative philosophy.
(viii) As a corollary, the quantification of nature is an oxymoron.

As I have explained, Piccolomini delivers a philosophy of mathematics able to dignify the trade of mechanical technicians as the practitioners of a kind of lofty contemplative philosophy. At the same time, it opens such a methodological and ontological chasm between the two disciplines that any emulation between them is prevented. Hence, the early 1540 anti-Mertonian philosophy as he put it in the paraphrase is restated in 1546 in the last chapter of *De certitudine* where, on the eve of the so-called scientific revolution, Piccolomini firmly concludes that a mathematical science of natural motion is impossible.

What was the outcome of Piccolomini's mechanical project?

In 1564, Pietro Catena, doubtlessly acquainted with Piccolomini's work, delivered a series of lectures on the *Quaestiones mechanicae* at Padua marking the pioneering step in European education of introducing Pseudo-Aristotle's work into the mathematical curriculum. These lectures were attended by Guidobaldo del Monte, whose elaborations on mechanics in his 1577 *Mechanicorum liber* paved the way for Galileo Galilei's later achievements. In this respect, I conclude that Piccolomini's wider project to provide a dignified and uncontroversial place for mechanics within the disciplinary distribution of his time can be considered a critical chapter in the early modern history of science and technology. However, the impact (or even the existence) of

Piccolomini's philosophy of mechanics has been obscured by his most challenging and resonant positions in *De certitudine*. For, as widely acknowledged, because of its lingering echoes and imposing challenge, haunting Europe's mathematical culture for more than one century, Piccolomini's interpretation of Proclus in *De certitudine mathematicarum* can be deemed as one of the most influential elaborations in the early modern history of philosophy and science.

To close, we return to the question that initiated our journey:

Is mathematics certain?

It truly is but at the cost of its reality. Indeed, mathematical objects are not real things; if they were, their cognition would be uncertain.

Piccolomini's audacious inquiry reminds us that what we often take for granted is not so easily pinned down. He invites us to the following thought: when we ponder the mathematical precision of fields such as theoretical physics or aeronautics, we are actually thinking about mental constructs rather than about the things themselves.

To "mathematise nature" is but to cloak the world's bewildering complexity with an illusory, if comforting, veil of mathematical trustworthiness; a veil that we weave and project within our minds and that holds us within the realm of the most certain imagination. As we turn to ourselves in unfolding mathematics, any certainty that we may perceive in mathematically doing physics or aeronautics pertains, in the end, to the inner mental constructs we interpose between us and the real things outside us.

In other words, to taste certainty, we retreat from reality into mathematical fiction. But if unreality is the price of certainty and precision, uncertainty and approximation are the costs we must bear to engage with real things. Dealing with reality is, perhaps, dealing with uncertainty.

Appendix

TABLE 7 Mentions of Proclus in Piccolomini's *Paraphrasis* and *De certitudine*

Alessandro Piccolomini, *In mechanicas quaestiones Aristotelis paraphrasis*; *De certitudine mathematicarum scientiarum* (Rome: Antonio Blado, 1547)	Mentions of Proclus (34 in total)
In mechanicas quaestiones Aristotelis paraphrasis, *Praefatio*, f. III.	Verumenimvero, et nos alias disputantes sustentavimus, et nunc constanter asserimus, demonstrationes quibus utuntur mathematici non esse potissimas illas praecipuas, quas Aristoteles in libris *De demonstratione* omni diligentia inquirit. Quo fit ut alia sit causa qua in primo certitudinis gradu mathematicae scientiae reponantur, sicut nos abunde in libello quem fecimus *De certitudine mathematicarum scientiarum* opinionem hanc nostram manifestissimis rationibus declaravimus. Cuius quidem opinionis ansam praebuit Proclus in I *Elementorum*.
In mechanicas quaestiones Aristotelis paraphrasis, q. XXX, f. LVIII–LVIIIv.	Non enim inconveniens est eandem complexionem sive (ut ita dicam) conclusionem ex pluribus assumptionibus pluribusque veris causis in mathematicis haberi posse, ut scribit Proclus, et ut quilibet cognoscere potest, qui mathematicas disciplinas quam modice degustaverit.
De certitudine mathematicarum scientiarum, *Praefatio*, f. LXXII.	Verumenimvero hanc meam sententiam, quamvis mihi constantissime probaretur ac quampluribus rationibus fulciretur, in me ipso tamen, ne quasi παράδοξον a me dictum videretur, eo usque comprimendam duxi, donec Proclum ipsum hoc idem cognoscens, maxima animi laetitia affectus testem tam locupletem nactus, id ipsum dehinc clara voce frequenter asserui [...].
Chap. I, f. LXXIVv.	Illud autem, quod libri alicuius, vel ὑποκείμενον ut Alexander in *Priorum analyticorum* vel ὕλη ut Ammonius, Proclus, et Simplicius vocare soliti sunt, a Latinis "subiectum" circa quod vel adaequatum nuncupatur.

 | DOI:10.1163/9789004724143_006

TABLE 7 (*cont.*)

Chap. V, f. LXXXVI.	Amplius dicit Proclus *In I Elementorum*, libro III, non omnia theoremata vel problemata Euclidis ex dignitatibus actu constare, sed pauca quaedam.
Chap. V, f. LXXXVIV.	Qui omnes per quid deffinitionem quid nominis, non autem quid rei, interpretantur. Et hoc non iniuria, nam deffinitiones quae accipiuntur inter dignitates et positiones scientarum, teste Proclo, Philopono et aliis, sunt deffinitiones quid nominis, non quid rei.
Chap. V, f. LXXXVIIIV.	Hoc idem asserit Proclus I *Elementorum*, libro III, necnon Averroes videtur hoc innuere I *Posteriorum*, 11.
Chap. V, f. LXXXVIIIV.	Responderi potest in mathematicis saltem valere dictam consequentiam propterea quod deffinitiones in illis praecognitae, quamvis sint quid nominis, tantundem tamen possunt ac si essent quid rei, teste Proclo, et Simplicio, et aliis.
Chap. V, f. LXXXIXV.	Isti solent negare hanc consequentiam, nam in geometria, verbi gratia, minores illae erunt tandem ostensae ex constructione, quae est quarta pars theorematis vel problematis (dividuntur enim in sex partes, ut dicit Proclus).
Chap. VII, f. LXXXXVII.	De materia mathematicarum facultatem disputans Proclus *In I Elementorum*, libro II, dicit, primum, entia ipsa mathematica, cum nihil aliud sint, quam quantitates ipsae abstractae, non posse inter ipsa penitus sensibilia connumerari.
Chap. VII, f. LXXXXVII–LXXXXVIIV.	Concludit ergo Proclus ex Platone, quod res ipsae mathematicae, de quibus fiunt demonstrationes, nec omnino in subiecto sensibiles sunt, nec penitus ab ipso liberatae, sed in ipsa phantasia reperiuntur figurae illae mathematicae, habita tamen occasione a quantitatibus in materia sensibili repertis. Intellectus autem, ex iis, quae in phantasia sunt quantitatibus, rationes illas universales colligit. Materia ergo harum scientiarum erit quantum ipsum, hoc modo, ut ita dicam, phantasiatum. Et id plerisque, quamvis non satis proprie, materia intellegibilis nuncupatur.
Chap. VII, f. LXXXXVIIV.	Ex hoc patet quare mathematicae disciplinae a Proclo, a Simplicio *In I De anima*, et a Ptolomeo in praefatione *Almagesti*, mediae scientiae nuncupantur.

TABLE 7 (*cont.*)

Chap. VII, f. LXXXXVIIV.	Nam manifestissime Proclus in I et II libro passim ostendit, dari quandam scientiam communem ad illas duas [geometriam et arithmeticam] quae proprium subiectum et proprias passiones propriaque principia sibi vindicat et illas duas sibi subalternat.
Chap. VIII, LXXXXVIII.	Dico igitur, ex Proclo, quod quamvis forte non possit reperiri demonstratio in illis ex fine, quia figurae illae, quatenus tales, non habent operationem ad quam constitui possit demonstratio.
Chap. VIII, LXXXXVIIIV.	Praeterea demonstrat Proclus mathematicas ad philosophiam de moribus, non solum esse utiles, sed vere necessarias. Ad divinam vero philosophiam utiles sunt, non solum exercendo intellectum nostrum ad abstractionem ipsam, qua maxime indiget theologus: sed multas ob alias causas, quas apud eumdem Proclum videre possumus. Quod tandem ad naturalem philosophum emolumenta afferant et per se patet, et a Proclo affatim exponitur et declaratur.
Chap. VIII, f. LXXXXVIIIV.	Quod ex hoc arguit Proclus, quod cum primum aliquis primis quasi labris mathematicas degustaverit, tali delectatione capitur ut, posthabitis curis omnibus, in hoc totum versans caeteras amplius delectationes non magni faciat.
Chap. IX, f. LXXXXVIIII.	Nam in hoc (teste Proclo) differt problema a theoremate, quod problema illud est in quo quippiam, cum primum non sit, proponitur inveniendum ac construendum.
Chap. IX, f. LXXXXVIIII–LXXXXVIIIIV.	Quod, ut melius intellegatur, sciendum est ex Proclo quodlibet vel theorema vel problema sex partes, ut plurimum, continere, quae dicuntur a Graecis πρόθεσις, ἔκθεσις, διορισμός, κατασκευή, ἀπόδειξις, συμπέρασμα. Nos vero "propositionem", "expositionem", "determinationem", "constructionem", "ostensionem", et "conclusionem" appellare possimus.

TABLE 7 (*cont.*)

Chap. IX, f. C.	Hoc idem in aliis theorematibus et problematibus invenietur, praeterquam quod quandoque a constructione supersedetur, cum scilicet ostensio non indiget constructione, et ex hoc patet veritas eius rei, quod dicit Proclus pagina 46, 97, 58, et 61, quod scilicet theoremata et problemata Euclidis comprehendunt etiam illa, quae dicunt esse Theonis, tanquam suas partes necessarias.
Chap. X, f. CV.	Et hac via procedit frequenter Euclides, dico frequenter quia quandoque deductione utitur ad impossibile, per secundum scilicet modum hypotheticorum syllogismorum, ut dicit Proclus et Philoponus. Quae deductiones etiam resolvi possunt, ut patet in logica, et ego Euclidem exponens etiam resolvebam. Cum vero iudicium facere volumus de huiusmodi compositivo processu, tunc regredientes ab ultima conclusione per praemissas et praemissarum praemissas, tandem ascendentes ad prima principia resolvere dicimur. Qua resolutione dignoscimus an per veras et proprias praemissas composita prius fuerit illa conclusio. Hoc dicit Proclus pluribus in locis, in I et II libro *Super I Elementorum*.
Chap. X, ff. CIIIIV–CV.	Siquis autem diceret quod quamvis plures eiusdem passionis demonstrationes esse possint, una tamen erit sola potissima, per medium immediatum, responderi potest quod hoc est falsum quia Proclus aperte asserit diversas aequo modo perfectas posse fieri in mathematicis passionum demonstrationes.
Chap. XI, f. CV.	Et quod hoc sit verum videmus quod mathematici easdem passiones de eisdem subiectis variis assumptis mediis demonstrant. Aliter enim triangulum habere tres ostendit Theon vel Euclides; aliter Campanus, et aliter etiam Proclus, qui hoc ipsum quod ego dico animadvertit. Et demonstratio quam facit Proclus de habere tres videtur esse ex intimiore medio quam demonstratio Theonis. Non enim accipit quiddam extrinsecum, sed ex perpendiculari intrinseca ostendit propositum.

TABLE 7 (*cont.*)

Chap. XI, f. CV.	Quamvis tamen diversimode (ut dixi) possint demonstrari passiones mathematicae, videtur tamen, sicut innuit Proclus super 32, I, dicendum quod illa ex omnibus demonstrationibus eiusdem passionis erit laudabilior quae breviori tramite et per pauciores syllogismos concluserit intentum.
Chap. XI, f. CVII–CVIIV.	Sed haec sufficiat, praesertim cum authoritates habeamus mirabiles. Et primo Proclum virum illustrem in mathematicis. Proclus enim *Super I Elementorum*, pagina 21, dicit plura elementa, id est, propositiones Euclidis posse esse sibi invicem elementa, quod profecto non esset super veras causas demonstrarent: nihil enim sui ipsius potest esse causa et unius rei tantum est una definitio, teste Aristotele, VI *Topicorum*, capitulo 3. Addit etiam non omnia Euclidis theoremata quasi per catenam sibi ipsis semper ordine quodam inservire, quia nimis arduum esset, ne dicam impossibile, illum ordinem invenire. Et ideo (dicit ipse) diversi diversimode demonstrant, non alius alio melius, nisi quia forte brevius. Dicit etiam pagina 17 quod geometria utitur omnibus dialecticae instrumentis. Alibi vero deducere frequenter asserit ad impossibile, quae deductio est ponenda sub demonstratione signi, teste Averrroe primo *Coeli*, 94. Pagina vero 47 inquit Proclus geometrarum rationes habere vim et necessitatem ex subiecta materia, non autem ex natura demonstrationis. Praeterea pagina 46 ponit haec verba apertissima et manifesta: τήν γε μὲν αἰτίαν καὶ τὸ διὰ τί πολλοῖς ἔδοξεν ἡ γεωμετρία μὴ θεωρεῖν. [Καὶ] ταύτης γάρ ἐστι καὶ ὁ Ἀμφίνομος τῆς δόξης Ἀριστοτέλους κατάρξαντος. Quid amplius quaerimus pro hac sententia?
Chap. XII, f. CIX.	Quod idem asseverat Proclus *In Elementa*, pagina 47, ubi habet haec legiptima [*sic*] verba: πανταχοῦ γὰρ τὸ ἀναγκαῖον ἔχουσιν οἱ γεωμετρικοὶ λόγοι διὰ τὴν ὑποκειμένην ὕλην.

TABLE 8 Translating ἀνάλυσις and σύνθεσις in Proclus' *In Euclidem*

Author	Locus	Text
Proclus	Friedlein f. 8, l. 4–6.	Καὶ μὴν καὶ τὸ κάλλος καὶ ἡ τάξις κοινὰ πάντων ἐστὶ τῶν μαθημάτων καὶ ἡ ἀπὸ τῶν γνωριμωτέρων ὁδός ἐπὶ τὰ ζητούμενα καὶ ἡ ἐκ τούτων ἐπ᾽ἐκεῖνα μετάβασις, ἃς δὴ καλοῦσιν ἀναλύσεις καὶ συνθέσεις.
Valla	*De expetendis et fugiendis rebus*, chap. XV.	Quin pulchritudo quoque et ordo omnibus communia sunt disciplinis a notioribusque ad ea quae in quaestione posita sunt via a quibus ulterius proceditur, quae sane resolutiones et compositiones vocantur.
Zamberti	f. 2v, 27–30 in ff. 1r–154v of München MS Clm 6.	Pulchritudo siquidem et omnium mathematicarum communis ordo est ut ab his quae sunt notiora ad ea quae quaerimus processus fiunt, et ex his ad illas, quae compositiones et resolutiones vocantur.
Gabia	f. 6v, 5–11 in Ambrosiana MS P. 51 sup.	Quin enim pulchritudo atque ordo communes sunt omnium disciplinarum viaque a notioribus ad ea qua quaeruntur, et ab his ad illa reditus, quas sane vocant resolutiones et compositiones.
Barozzi	chap. 3, f. 4.	Atqui pulchritudo quoque et ordo omnibus communia sunt mathematicis disciplinis, et a notioribus ad ea quae quaeruntur via, et ab his ad ea transitus, quae sane resolutiones et compositiones appellantur.
Dasypodius	*Mathematicum complectens praecepta mathematica, astronomica et logistica*, Vol. 2, 1570, pp. 167–168.	Atqui pulchritudo quoque et ordo communis est omnibus scientiis mathematicis et via seu methodus quae fit a notioribus ad ea quae in questione sunt proposita, et ab his ad illa reditus, quae appellantur resolutiones et compositiones.
Mariner	f. 18, 18–22, in Madrid, Biblioteca Nacional de España, MS 9871, ff. 45–299.	Etenim pulchritudo et ordo communia omnium sunt mathematum et ipsa via eorum, quae magis cognoscibilia sunt, in ea quae quaeruntur et transitus et regressus quae ultra ab his efficitur, quas quidem vocant analyses et compositiones.

TABLE 9 Piccolomini's views and sources

	Piccolomini's tenets	Sources (*op. cit.*)
1	Mathematical demonstration adopts multiple middle-terms.	Proclus (*In Euclidem*, p. 379, l. 2–15).
2	There is no causation of any kind in mathematics.	Proclus (*In Euclidem*, p. 202, l. 9–12), Priscian (*In De anima*, p. 210, l. 5–9; p. 233, l. 22–30; pp. 276–277, l. 27–35), Genua De' Passeri (*In De anima*, III, 4, p. 174v).
3	Mathematical objects are not substances.	Aristotle (XIII, 2, 1077a, l. 32), Priscian (*In De anima*, p. 210, l. 5–9; p. 233, l. 22–30; pp. 276–277, l. 27–35).
4	There is no efficient cause in mathematics.	Alexander (*In Metaphysicam* III, 996a, l. 21, p. 181, l. 34–p. 182, l. 4; p. 182, l. 22–28), Proclus (*In Euclidem*, p. 186, l. 9–14).
5	There is no final cause in mathematics.	Aristotle (*Metaphysica* III, 2, 996a), Alexander (*In Metaphysicam* III, 996a, l. 21, pp. 181, l. 34–p. 182, l. 4; p. 182, l. 22–28), Averroes (*Commentaria in libros De physico auditu*, VI, 3, f. 79v).
6	There is no material cause in mathematics.	Boccadiferro (*In De anima*, lec. 34, pp. 57–58).
7	There is no formal (i.e., substantial) cause in mathematics.	Aristotle (*Metaphysica*, XIII, 2, 1077a, l. 32), Priscian (*In De anima*, p. 210, l. 5–9; p. 233, l. 22–30; pp. 276–277, l. 27–35).
8	There is no *demonstratio potissima* in mathematics.	Proclus (*In Euclidem*, p. 206, l. 12–23).
9	Mathematics is not a science in the strict sense.	Priscian (III, 431b, l. 12, pp. 276–277, l. 27–35), Genua De' Passeri (*In De anima,* III, 4, p. 174v).
10	Mathematical certainty stems from the object of study of mathematics.	Aristotle (*Metaphysica*, II, 3), Proclus (*In Euclidem*, p. 206, l. 12–23), Priscian (*In De anima*, III, 429b, l. 18, p. 233, l. 13–25), Averroes (*In Metaphysicam*, II, 3, p. 35v, l. 16–27).

Bibliography

Manuscripts

Arezzo, Biblioteca Città di Arezzo, 390.
Bologna, Biblioteca Universitaria, 2293, fol. 1–149v.
Florence, Biblioteca Nazionale Centrale di Firenze, II, IV, 214.
Florence, Biblioteca Medicea Laurenziana, Pl. XXVIII, 45.
Madrid, Biblioteca Nacional de España, 9871, fol. 45–299.
Milan, Biblioteca Ambrosiana P. 51 sup.
Milan, Biblioteca Ambrosiana R 181 (16) inf. 1.
Modena, Biblioteca Estense, α. T. 9. 11, fol. 67–241v.
Munich, Bayerische Staatsbibliothek, A. gr. b. 954.
Munich, Bayerische Staatsbibliothek, Clm 10674, fol. 337r–340v.
Munich, Bayerische Staatsbibliothek, Clm 6, fol. 1r–154v.
Munich, Bayerische Staatsbibliothek, Cod. Graec. 427, fol. 1–234v.
Naples, Biblioteca Nazionale di Napoli, VIII, 33, fol. 21r–300r.
Oxford, Bodleian Library, Barocci 161.
Oxford, Corpus Christi College Library, 97.
Vatican City, Biblioteca Apostolica Vaticana, Urb. Gr. 71.
Vatican City, Biblioteca Apostolica Vaticana, Vat. Barb. Gr. 101.
Vatican City, Biblioteca Apostolica Vaticana, Vat. Barb. Gr. 145.
Vatican City, Biblioteca Apostolica Vaticana, Vat. Lat. 1905, fol. 5–36v.
Vatican City, Biblioteca Apostolica Vaticana, Vat. Lat. 4575, fol. 1r–19r.
Venice, Biblioteca Marciana, VI. 11 (coll. 1409).
Venice, Biblioteca Marciana, Lat. VIII, 83, f. 4–15v.
Venice, Biblioteca Marciana, Gr. Z. 214.
Venice, Biblioteca Marciana, Gr. Z. 306.

Primary Sources

Agostino Nifo, *Expositio super octo Aristotelis Stagirite libros De physico auditu*, Venice, Hieronymus Scotus, 1559.

Agostino Nifo, *Metaphysicarum disputationum dilucidarium*, Naples, Sigismundus Mayr Alemannus, 1511.

Agostino Nifo, *Super libros De anima*, Venice, Giunti, 1544.

Agostino Nifo, *Super tres libros De Anima*, Venice, Alexander Calcedonius, 1503.

Agustine of Hippo, *Contra Academicos*, eds W. Green, K. Daur, Corpus Christianorum. Series Latina 29, Turnhout, Brepols, 1970.

Alanus Alanus ab Insulis, *The Plaint of nature* ed. J. J. Sheridan, Toronto, Pontifical Institute of mediaeval studies, 1980.

Albert the Great, *Liber primus Posteriorum Analyticorum*, in *Opera omnia*, éd. par Auguste Borgnet, vol. 11, Parisiis, apud Vives, 1890.

Albert the Great, *De animalibus*, ed. A. Borgnet, vol. 11–12, Paris, Ludovicus Vives, 1890.

Albert the Great, *In libros Physicorum Aristotelis*, ed. A. Borgnet, vol. 3, Paris, Ludovicus Vives, 1890.

Albert the Great, *In libros Posteriorum analyticorum*, ed. A. Borgnet, vol. 2, Paris, Ludovicus Vives, 1890.

Albert the Great, *Super Euclidem*, ed. P. M. J. E. Tummers, Münster, Aschendorff, 2014.

Alessandro Piccolomini, *De certitudine mathematicarum scientiarum*, Rome, Antonio Blado, 1547.

Alessandro Piccolomini, *Della grandezza della terra et dell'acqua*, Venice, Ziletti, 1558.

Alessandro Piccolomini, *De la institutione di tutta la vita del huomo nato nobile*, Venice, Hyeronimus Scotus, 1545.

Alessandro Piccolomini, *Della sfera del mondo, Delle stelle fisse,* Venice, Pozzo, 1540.

Alessandro Piccolomini, *La economica tradotta di lingua greca in lingua toscana*, Venice, Pozzo, 1540.

Alessandro Piccolomini, *Il libro della Poetica d'Aristotele*, Siena, Bonetti, 1572.

Alessandro Piccolomini, *In mechanicas quaestiones Aristotelis paraphrasis*, Rome, Antonio Blado, 1547.

Alessandro Piccolomini, *Instrumento della filosofia*, Venice, Giovanmaria Bonelli, 1552.

Alessandro Piccolomini, *Della filosofia naturale parte prima*, Venice, Lorenzini, 1560.

Alessandro Piccolomini, *Libro sesto* in *I sei primi libri dell'Eneide di Virgilio tradotti*, Venice, Niccolò Zoppino, 1540.

Alessandro Piccolomini, *I tre libri della Retorica d'Aristotele tradotti in lingua volgare*, Venice, Francesco de' Franceschi, 1571.

Alessandro Piccolomini, *La prima parte de le theoriche*, Venice, Giordano Ziletti, 1558.

Alessandro Piccolomini, *Lettere al Varchi*, in F. Cerrata, *Alessandro Piccolomini: letterato e filosofo senese del Cinquecento*, Siena, Accademia degli Intronati, 1960, pp. 264–272.

Alessandro Piccolomini, *Prima parte della filosofia naturale*, Venice, Vincenzo Valgrisi, 1551.

Alessandro Piccolomini, *Seconda parte della filosofia naturale*, Venice, Vincenzo Valgrisi, 1554.

Alessandro Piccolomini, *Parafrasi di monsignor Alessandro Piccolomini arciuescouo di Patras, sopra le Mecaniche d'Aristotile, tradotta da Oreste Vannocci Biringucci, gentilomo senese*, transl. Oreste Vannocci Biringucci, Rome, Francesco Zanetti, 1582.

Alexander of Aphrodisias, *De anima*, ed. I. Bruns, Berlin, G. Raimer, 1887.

Alfarabius, *De scientiis*, transl. Gerard of Cremona, eds A. Galnnier, M. Lejbowicz, J. Jolivet, Turnhout, Brebols, 2016.

Anaritius, *Commentary on Euclid. The Latin Translation, I–IV*, ed. P. M. J. E. Tummers, Münster, Aschendorff, 1993.

An-Nayrizi, *The Commentary of al-Nayrizi on Book II–IV of Euclid's Elements of Geometry*, ed. A. Lo Bello, Leiden Boston, Brill, 2009.

Aristotle, *Analytica posteriora*, ed. I. Bekker, vol. 1, Berlin, Accademia regia Borusica, 1831.

Aristotle, *Analytica priora*, ed. I. Bekker, vol. 1, Berlin, Accademia regia Borusica, 1831.

Aristotle, *Categoriae*, ed. I. Bekker, vol. 1, Berlin, Accademia regia Borusica, 1831.

Aristotle, *Complete Works of Aristotle*, ed. J. Barnes, Vol. 1–2, Chichester, Princeton University Press, 1984.

Aristotle, *De anima*, ed. I. Bekker, vol. 1, Berlin, Accademia regia Borusica, 1831.

Aristotle, *De caelo*, ed. I. Bekker, vol. 1, Berlin, Accademia regia Borusica, 1831.

Aristotle, *De memoria*, ed. I. Bekker, vol. 1, Berlin, Accademia regia Borusica, 1831.

Aristotle, *De somno*, ed. I. Bekker, vol. 1, Berlin, Accademia regia Borusica, 1831.

Aristotle, *Ethica Nicomachea*, ed. I. Bekker, vol. 2, Berlin, Accademia regia Borusica, 1831.

Aristotle, *Metaphysica*, ed. I. Bekker, vol. 2, Berlin, Accademia regia Borusica, 1831.

Aristotle, *Physica*, ed. I. Bekker, vol. 1, Berlin, Accademia regia Borusica, 1831.

Aristotle, *Topica*, ed. I. Bekker, vol. 1, Berlin, Accademia regia Borusica, 1831.

Aristotle, *Topica*, transl. Anonymus saec. XII, eds L. Minio-Paluello, B. G. Dod, Turnhout, Brepols, 1969.

Aristotle, *Metaphysica: translatio media*, transl. Anonymus saec. XII vel XIII, ed. G. Vuillemin-Diem, Turnhout, Brepols, 1976.

Aristotle, *Metaphysica: translatio vetustissima*, transl. James of Venice, ed. G. Vuillemin-Diem, Turnhout, Brepols, 1970.

Aristotle, *Aristotelis Stagyritae Ethicorum libri X cum Averrois Cordubensis exactis commentariis*, transl. Leonardo Bruni, Venice, Giunti, 1542.

Aristotle, *Ethica Nicomachea*, transl. Robert Grosseteste, ed. R. A. Gauthier, Turnhout, Brepols, 1973.

Aristotle, *Metaphysica*, recens. William of Moerbeke, ed. G. Vuillemin-Diem, Turnhout, Brepols, 1995.

Averroes, *Aristotelis opera cum Averrois commentariis*, vol. 1–9 (Venice: Giunti, 1562–1574).

Averroes, *Commentaria in libros De physico auditu Aristotelis*, in *Aristotelis opera cum Averrois commentariis*, vol. 4, Venice, Giunti, 1562.

Averroes, *Commentaria in libros Metaphysicorum Aristotelis*, in *Aristotelis opera cum Averrois commentariis*, vol. 8, Venice, Giunti, 1562.

Averroes, *Commentarium magnum in Aristotelis De anima*, ed. F. Stuart Crawford, Cambridge, Massachusetts, The Medieval Academy of America, 1953.

Averroes, *De substantia orbis*, in *Aristotelis opera cum Averrois commentariis*, vol. 9, Venice, Giunti, 1562.

Averroes, *Epitome in primum librum posteriorum*, in *Aristotelis opera cum Averrois commentariis*, vol. 1.2, Venice, Giunti, 1562.

Averroes, *Expositio magna in libros posteriorum resolutoriorum*, in *Aristotelis opera cum Averrois commentariis*, vol. 1.2, Venice, Giunti, 1562.

Averroes, *Expositio media in libros Physicorum*, in *Aristotelis opera cum Averrois commentariis*, vol. 4, Venice, Giunti, 1562.

Averroes, *Sharh al-Burhān li-Aristū wa-Talkhīs al-Burhān, al-Silsila al-turāthiyya*, vol. 12, Al-Kuwayt, Al-Kuwayt: Al-Majlis al-Watanī li-l-Thaqāfa wa-l-Funūn wa-l-Ādāb, 1984.

Avicenna Liber de philosophia prima sive scientia divina, édition critique de la traduction médiévale par Simone van Riet, introduction doctrinale par Gérard Verbeke, Leiden, Brill, 3 vol., 1977, 1980, 1983.

Avicenna, *Liber de philosophia prima sive scientia divina*, transl. Simone van Riet, ed. G. Verbeke, vol. 1–3, Leiden, Brill, 1977–1983.

Benito Pereyra, *De communibus omnium rerum naturalium principiis et affectionibus libri quindecim* (Paris: Sonnius, 1576).

Boethius, *Consolatio philosophiae*, ed H. F. Stewart, E. K. Rand, S. J. Tester, Cambridge MA, Harvard University Press, 1973.

Calcidius, *Timaeus a Calcidio translatus commentarioque instructus*, eds W. J. Verdenius, J. H. Waszink, vol. 4, London Leiden, Warburg Institute, Brill, 1962.

Cicero, *Scripta Quae Manserunt Omnia*, ed. O. Plasberg, Leipzig, Teubner, 1922.

Cicero, *De finibus bonorum et malorum*, ed. H. Hackham, Cambridge MA, Harvard University Press, 2015.

Claudius Ptolemy, J. L. Heiberg (ed.), *Syntaxis astronomica* in *Claudii Ptolemaei opera quae exstant omnia* (Leipzig: Teubner, 1898) Vol. I.

Conradus Dasypodius, Christianus Herlinus, *Analysis geometriae sex librorum Euclidis*, Strassburg, J. Richelius, 1566.

Dominicus Gundissalinus, *De anima*, ed. J. T. Murckle, in *Mediaeval Studies* 2/1, pp. 23–103 (1940).

Donato Acciaoli, *Ethicorum ad Nicomachum: libri decem, Ioanne Argyropylo Byzantio interprete, ad Graecum exemplar diligentissime recogniti, cum Donati Acciaioli Florentini, viri doctissimi, commentariis*, Lyon, Antonius Vincentius, 1544.

Duns Scotus, *Quaestiones in librum Porphyrii Isagoge*, eds R. Andrews, G. Etzkorn, G. Gál, R. Green, T. Noone, R. Wood, vol. 1, St. Bonaventure NY, The Franciscan Institute Press, 1999.

Euclid, *The Thirteen Books of the Elements*, ed. Th. Heath, vol. 1–3, New York, Dover Publications, 1956.

Euclid, Niccolò Tartaglia, *Euclide Megarense philosopho, solo introduttore delle Scientie Mathematice*, Venice, Venturino Ruffinelli, 1543.

Euclid, Proclus, *Στοιχείων βιβλία ιε΄ ἐκ τοῦ Θέωνος συνουσιῶν. εἰς τοῦ αὐτοῦ τὸ πρῶτον ἐξηγημάτων Πρόκλου βιβλίον*, ed. Simon Grynaeus, Basel, I. Hervagius, 1533.

Eustratius, *In Analyticorum posteriorum librum secundum commentarium*, ed. M. Hayduck, Berlin, Academia Litterarum Regiae Borrusicae, 1907.

Francesco Barozzi, *Opusculum in quo una oratio, et duae quaestiones: altera de certitudine, et altera de medietate mathematicarum continentur*, Padua, Grazioso Percacino, 1560.

Francesco Barozzi, *Procli Diadochi Lycii philosophi platonici ac matematici probatissimi in primum Euclidis Elementorum librum commentariorum aduniuersam mathematicam disciplinam principium eruditionis tradentium libri IV*, Padua, Gratiosus Perchacinus, 1560.

Francis Petrarch, *Familiarium rerum libri* eds V. Rossi, U. Bosco, Turin, Aragno, 2004.

Francisco Suárez, *Metaphysical Disputation I: On the Nature of First Philosophy or Metaphysics*, ed. Duarte, S., Catholic University of America Press, 2021.

Galen, *Ars medica quae et ars parva dicitur*, Venice, Guillielmus Rovillius, 1548.

Galen, I. Jonhson (ed.), *The Art of Medicine*, (Cambridge: Loeb Classical Library, 2016).

Gellius, *Noctes Atticae*, ed. K. Marshall, Leipzig, Teubner, 1968.

Giles of Rome, *Super libros duos Posteriorum analyticorum Aristotelis*, Venice, Bonetus Locatellus, 1488.

Giles of Rome, *Super libros duos Posteriorum analyticorum Aristotelis*, eds Joanne Jacobo de Puteo, Petro de Nardis, Padua, Maufer, 1478.

Giles of Rome, *Super libros Posteriorum analyticorum*, Venice, Bonetus Locatellus, 1488 (réimp. anast. Frankfurt, Minerva, 1967).

Giovanni Francesco Pico della Mirandola, *Liber de imaginatione*, Rome, Aldus, 1501.

Historia Augusta, ed. D. Magie, Cambridge MA, Harvard University Press, 2015.

Iamblichus, *De communi mathematica scientia*, ed. N. Festa, Leipzig, Teubner, 1891.

Iamblichus, *On the General Science of Mathematics*, eds. J. M. Dillon, J. O. Urmson, S. Gertz, London: Bloomsbury Academic, 2020.

Isiodore of Seville, *Etymologiarum sive originum libri XX*, ed. W. M. Lindsay, Oxford, Clarendon, 1911.

Isiodore of Seville, *Liber differentiarum*, ed. M. A. Andrés Sanz, Corpus Christianorum. Series Latina 111A, Turnhout, Brepols 2006.

Jacopo Forlì, *Super Tegni Galeni*, Padua, Giunti, 1475.

Jacopo Zabarella, *De methodis* in *Opera logica*, Venice, Paulus Meietus, 1578.

Jacopo Zabarella, *De methodis libri quattuor: Liber de Regressu*, ed. C. Vasoli, Bologna, Cooperativa Libraria Universitaria Editrice Bologna, 1985.

Jean de Jandun, *Quaestiones perspicacissimi peripatetici Ioannis de Ianduno in duodecim libros Metaphysicae*, Venice, Hyeronimus Scotus, 1553.

John Buridan, *In Metaphysicen Aristotelis questiones argutissimae*, Paris, Iodicus Badius Ascensius, 1518, repr. in *Kommentar zur Aristotelischen Metaphysik*, Frankfurt am Maine, Minerva, 1964.

John Buridan, *Quaestiones in duos libros Aristotelis Posteriorum analyticorum*, ed. H. Hubien, unpublished typescript available at www.logicmuseum.com.

John Scot Eriugena, *Annotationes in Marcianum*, ed. I. Ramelli, Il pensiero occidentale, Milan, Bompiani, 2006.

John Scot Eriugena, *Expositiones super hierarchiam caelestem S. Dionysii*, ed. J. Barbet, Turnhout, Brepols, 1975.

Ludovico Boccadiferro, *Explanatio libri Physicorum Aristotelis*, Venice, Academia Veneta, 1558.

Ludovico Boccadiferro, *In duos libros Aristotelis de generatione et corruptione: doctissima commentaria*, Venice, Franciscus de Franciscis Senensis, 1571.

Ludovico Boccadiferro, *Lectiones in Aristotelis Stagiritae libros, quos vocant Parva naturalia, De sensu et sensili*, Venice, Hyeronimus Scotus, 1560.

Ludovico Boccadiferro, *Lectiones super tres libros De anima Aristotelis*, Venice, Iohannes Baptiasta Somaschus et fratres, 1566.

Marco Antonio (Genua) De' Passeri, *In tres libros Aristotelis De anima*, Venice, Damiano Zenaro, 1576.

Marco Antonio Zimara, *Theoremata seu memorabilium propositionum limitationes*, Venice, Hyeronimus Scotus, 1543.

Marsilio Ficino, ed R. Marcel, *Théologie Platonicienne de l'immortalaté des ames*, Paris, Le Belles Lettres, 2007.

Marsilius of Inghen, *Quaestiones proutiles in Aristotelis posteriora* in *Parvorum logicalium liber*, Vienna, Singrenius, 1516.

Michael Psellus, *Compendium mathematicum*, Leyden: Elzevirii, 1647.

Niccolò Leoniceno, *De tribus doctrinis ordinatis secundum Galeni sententiam liber*, ed. Andreas Leenius, *Opuscula*, Basel, Andrea Cratandrus et Ioannes Bebelius, 1532.

Niccolò Leonico Tomeo, *Opuscula nuper in lucem edita*, Venice, Vitalis, 1525.

Paul of Venice, *Summa philosophiae naturalis*, Venice, Iodocus Badius Ascensius, 1503.

Peter of Abano, *Conciliator differentiarum, quae inter philosophos et medicos versantur*, Venice, Giunti, 1496.

Philoponus, *Aristotelis Analytica Posteriora Commentaria*, Berlin, De Gruyter, 1962.

Pico della Mirandola, *Conclusiones nongentae*, ed. S. A. Farmer, *Syncretism in the West: Pico's 900 Theses (1486): The Evolution of Traditional Religious and Philosophical Systems*, Tempe AZ: Arizona Center for Medieval and Renaissance Studies, 1998.

Pietro Catena, *Oratio pro idea methodi ad nobilem Federicum Sforciam*, Padua, Gratiosus Perchacinus, 1563.

Pietro Pomponazzi, *De intensione et remissione formarum ac parvitate et magnitudine*, eds Raimondi, F. P, García Valverde, J. M., Milan, Bompiani, 2013.

Pietro Pomponazzi, *De reactione*, eds Raimondi, F. P, García Valverde, J. M., Milan, Bompiani, 2013.

Pietro Pomponazzi, *Quaestio de regressu*, ed. A. Poppi, *Corsi inediti dell'insegnamento padovano*, vol. 1, Padua, Antenore, 1970.

Plato, *Epinomis* ed. J. Burnet, vol. 5, Oxford, Clarendon, 1907.

Plato, *Respublica* ed. J. Burnet, vol. 2, Oxford, Clarendon, 1992.

Plato, *Timaeus* ed. J. Burnet, vol. 3, Oxford, Clarendon, 1989.

Plotinus, *Enneades*, eds P. Henry, H. R. Schwyzer, vol. 1–2 (Oxford: Clarendon, 1989–1990).

Proclus, *A Commentary on the First Book of Euclid's Elements*, ed. G. R. Morrow, New Jersey, Princeton University Press, 1970.

Proclus, *Commentary on the First Book Of Euclid's Elements*, ed. Th. Taylor, London, Printed for the Author, 1792.

Proclus, *Commento al I libro degli «Elementi» di Euclide*, ed. M. Timpanaro Cardini, Pisa, Giardini Editori, 1978.

Proclus, in E. R. Dodds (ed.) *Elements of Theology*, Oxford, Oxford University Press, 1963.

Proclus, *In Alcibiadem*, ed. V. Cousin, Paris, Minerva, 1864.

Proclus, *In Platonis Cratylum commentaria,* ed. G. Pasquali, Leipzig, Teubner, 1908.

Proclus, *In Platonis Parmenidem commentaria*, ed. C. Steel, vol. 1–3, Oxford, Oxford University Press, 2007–2009.

Proclus, *In Platonis Rem Publicam commentarii*, ed. W. Kroll, Leipzig, Teubner, 1899.

Proclus, *In Platonis Timaeum commentaria*, ed. E. Diehl, vol. 1–2, Leipzig, Teubner, 1903–1904.

Proclus, *In primum Euclidis Elementorum librum commentarii*, ed. G. Friedlein, Leipzig, Teubner, 1874.

Proclus, *The Elements of Theology: A Revised Text with Translation, Introduction, and Commentary*, ed. E. R. Dodds, Oxford, Clarendon Press, 1992.

Proclus, *Théologie platonicenne*, eds H. D. Saffrey, L. G. Westerink, vol. 1, Paris, Belles Lettres, 1968.

Pseudo-Aristotle, *Mechanica*, ed. Vittore Fausto, Paris, Josse Bade, 1517.

Pseudo-Aristotle, *Quaestiones mechanicae*, ed. J. P. van Capelle, Amsterdam, Petrus Den Hengst et filius, 1812.

Quintilianus, *Institutio Oratoria*, ed. M. Winterbottom, Leipzig, Teubner, 1970.

Raffaele Maffei, *Commentariorum urbanorum 38 libri*, Rome, J. Besicken, 1502.

Richard Swineshead, *Calculationes noviter emendatae atque revisae*, Venice, Octavianus Scotus, 1520.

Robert Grosseteste, *Commentarius in Posteriorum analyticorum libros*, eds. L. S. Olschki, P. Rossi (Florence: Testi e Studi 1981).

Roger Bacon, *Summulae dialectices*, ed. R. Steele, Oxford, Clarendon, 1940.

Rudolf Agricola, *De inventione dialectica libri omnes et integri et recogniti*, Amsterdam, Alardus, 1539.

Simon of Faversham, *Questions on the Posterior analytics*, ed. J. Longeway, Ph.D., Ithaca NY, Cornell University, 2016.

Simplicius, *Commentaria Simplicii profundissimi, et acutissimi philosophi in treis* [sic] *libros De anima Aristotelis*, ed Giovanni Faseolo, Venice, Hieronymus Scotus, 1543.

Simplicius, *On Aristotle's On the Soul* 1.1–2.4, Simplicius, ed. J. O. Urmson, London: Bloomsbury Academic, 1994.

Simplicius, *On Aristotle's On the Soul 3.1–5*, ed. H. J. Blumenthal, London, Bloomsbury Academic, 2014.

'Simplicius', *On Aristotle's On the Soul 3.6–13*, ed. C. Steel, London, Bloomsbury Academic, 2014.

Simplicius, *In libros Aristotelis De anima commentaria*, ed. M. Hayduck, Reimer, Königlich Preussische Akademie der Wissenschaften zu Berlin, 1882.

Syrianus, *In Metaphysica commentaria*, ed. W. Kroll, Berlin, Reimer, 1902.

Syrianus, *On Aristotle Metaphysics 3–4*, eds. Dillon, J. M., O'Meara, D. J., London, Bloomsbury Academic, 2008.

Syrianus, *On Aristotle Metaphysics 13–14*, eds. Dillon, J. M., O'Meara, D. J., London, Bloomsbury Academic, 2006.

Tacitus, *Annales*, ed. C. D. Fisher, Leipzig, Teubner, 1906.

Themistius, *In libros Aristotelis De anima paraphrasis*, ed. R. Heinze, Berlin, Reimer, 1899.

Themistius, *Analyticorum posteriorum paraphrasis*, ed. M. Wallies, Berlin, Academiae litterarum regiae Borusicae, 1900.

Themistius, ed R. B. Todd, *Themistius: On Aristotle On the Soul* (London: Bloomsbury Academic, 1996).

Thomas Aquinas, *Sententia libri De anima*, ed. Leonina, vol. 45.1, Paris, Commissio Leonina, 1984.

Thomas Aquinas, *Expositio in libros Posteriorum analyticorum*, ed. Leonina, vol. 1.2, Paris, Commissio Leonina, 1989.

Thomas Aquinas, *Phyiscorum Aristotelis expositio*, ed. Leonina, vol. 2, Paris, Commissio Leonina, 1884.

Thomas Aquinas, *Sententia libri Metaphysicae*, ed. M. R. Cathala, R. M. Spazzi, Turin, Marietti, 1950.

Ugo Benzi, *Super libros Tegni Galeni*, Venice, Octavianus Scotus, 1518.

Walter Burley, *Quaestiones super librum Posteriorum analyticorum*, ed. M. C. Sommers, Toronto, Pontifical Institute of Mediaeval Studies, 2000.

William Wheatley, *Expositio in Boethii De scholarium disciplina*, in Thomas Aquinas, *Opera omnia*, vol. 24, Parma, Pietro Fiaccadori, 1869.

William of Ockham, *Summa logicae*, eds E. A. Moody, Ph. Boehner, G. Gál, S. Brown, A. Gambatese, St. Bonaventure NY, The Franciscan Institute Press, 1999.

Secondary Sources

Adams, M. P., "Demarcating Aristotelian Rhetoric: Rhetoric, the Subalternate Sciences, and Boundary Crossing", in *Apeiron* 48 (2015), pp. 99–122.

Adamson, P., "Yahyá ibn 'Adi and Averroes on «Metaphysics» Alpha Elatton", in *Documenti E Studi Sulla Tradizione Filosofica Medievale* (2010), 21, pp. 343–374.

Adamson, P., Karfík, F., "Proclus' Legacy", in P. D'Hoine, M. Martijn (eds), *All From One: A Guide to Proclus*, Oxford, Oxford University Press, 2016, pp. 290–332.

Aertsen, J. A., "Ontology and Henology in Medieval Philosophy (Thomas Aquinas, Master Eckhart and Berthold of Moosburg)", in E. P. Bos, P. A. Meijer (eds), *On Proclus and His Influence in Medieval Philosophy*, Leiden, New York, and Cologne, 1992, pp. 120–140.

Allen, J., *Inference from Signs: Ancient Debates about the Nature of Evidence* (Oxford: Clarendon Press, 2001).

Allen, M. J. B., "Marsilio Ficino", in S. Gersh (ed.), *Interpreting Proclus: From Antiquity to the Renaissance*, Cambridge, Cambridge University Press, 2014, pp. 353–379.

Axworthy, A., *Motion and Genetic Definitions in the Sixteenth-Century Euclidean Tradition*, Birkhäuser Cham, Springer, 2021.

Baldi, B., *Le vite de' Matematici. Edizione annotata e commentata della parte medievale e rinascimentale a cura di Elio Nenci*, Milan, Franco Angeli, 1998.

Bechtle, G., "Sur la mathématique générale comme σκοπός (sujet) du *De communi mathematica scientia* de Jamblique et sur la composition de ce traité", in A. Lernould (ed.), *Études sur le Commentaire de Proclus au premier livre des Éléments d'Euclide*, Saint-Paul Fribourg, Presses Universitaires du Septentrion, 2007, pp. 31–44.

Beekes, R., *Etymological Dictionary of Greek*, Leiden, Brill, 2010.

Berger, H., "On the Reception of English Logic at Universities of Central Europe: Helmoldus de Zoltwedel (Prague, Leipzig) on the Liar-Paradox", in D. A. Di Liscia, E. D. Sylla (eds), *Quantifying Aristotle: The Impact, Spread and Decline of the Calculatores Tradition*, Leiden Boston, Brill, 2022, pp. 215–231.

Bernard, A., "L'arrière-plan rhétorique de la theórie de l'activité mathématique chez Proclus", A. Lernould (ed.), in *Études sur le Commentaire de Proclus au premier livre des Éléments d'Euclide*, Saint-Paul Fribourg, Presses Universitaires du Septentrion, 2007, pp. 67–88.

Bertagna, M., "Giles of Rome and Paul of Venice on *Demonstratio Potissima*", in *Documenti e studi sulla Tradizione filosofica medievale* 20 (2009), pp. 375–394.

Bertato, F., "Sobre as Formalizações Silogísticas dos Elementos, efetuadas por Herlinus, Dasypodius, Clavius e Hérigone", in *Actas dos Encontros Luso-Brasileiros de História da Matemática* (2014), pp. 923–957.

Beumberg-Chaumont, J., "Découverte, analyse et démonstration chez les premiers commentateurs médiévaux des Seconds Analytiques", in J. Biard (ed.), *Raison et démonstration: Les commentaires médiévaux sur le Seconds analytiques*, Turnhout, Brepols, 2015, pp. 71–96.

Bianchi, L., "Fra Ermolao Barbaro e Ludovico Boccadiferro: qualche considerazione sulle trasformazioni della "fisica medievale" nel Rinascimento italiano", in *Medioevo* 29 (2004), pp. 341–378.

Bianchi, L., "Per una storia dell'aristotelismo 'volgare' nel Rinascimento: problemi e prospettive di ricerca", in *Bruniana & Campanelliana* 15/2 (2009), pp. 367–385.

Biard, J., "Blasius of Parma on the Calculation of the Variation of Qualities and Aristotelian Physics", in D. A. Di Liscia, E. D. Sylla (eds), *Quantifying Aristotle: The Impact, Spread and Decline of the Calculatores Tradition*, Leiden Boston, Brill, 2022, pp. 232–254.

Biard, J., "Certitudo", in I. Atucha, D. Calma, C. Koenig-Pralong, I. Zavattero (eds), *Mots médiévaux offerts à Ruedi Imbach*, Porto, Fédération Internationale des Études Médiévales, 2011, pp. 153–162.

Biard, L., "La certitude des mathématiques et ses fondements selon Piccolomini" in M. F. Piéjus, M. Residori (eds) *Alessandro Piccolomini (1508–1579), un siennois à la croisée des genres et des savoirs* (Paris: Sorbonne PSN, 2012), pp. 247–58.

Biard, J., "Quante tipologie di dimostrazione? Origini incrociate nella tradizione degli Analitici Secondi", in A. Musco, G. Musotto (eds), *Traduzioni e Tradizioni: Il pensiero medievale nell'incontro tra le culture mediterranee. Siracusa, 26–29 Settembre 2011*, Palermo, Officina Studi Medievali, 2014, pp. 61–78.

Black, D. L., "Knowledge ('ilm) and certitude (yaqin) in al-Farabi's epistemology", in *Arabic Sciences and Philosophy* 16/1 (2006), pp. 11–45.

Boer, S., "The (human) soul" H. Lagerlund, B. Hill (eds), *The Routledge Companion to Sixteenth Century Philosophy*, New York, Routledge, 2017, pp. 411–435.

Bohle, B., "Proclus on the Pneumatic Ochema", in S. Coughlin, D. Leith, O. Lewis (eds) *The Concept of Pneuma after Aristotle*, Berlin, Edition Topoi, 2020, pp. 343–364.

Bostock, D., *Time, Matter, and Form Essays on Aristotle's Physics*, Oxford, Oxford University Press, 2006.

Bronstein, D., *Aristotle on Knowledge and Learning: The Posterior Analytics*, Oxford, Oxford University Press, 2016.

Bundy, M. W., *The theory of imagination in classical and mediaeval thought*, Illinois, Folcroft library, 1927.

Burnet, J. (ed.), *Platonis opera*, Vol. 1–5 (Oxford: Clarendon, 1989–1995).

Burnett, C., Mendelsohn, A., "Aristotle and Averroes on Method in the Middle Ages and Renaissance: The "Oxford Gloss" to the Physics and Pietro D'Afeltro's Expositio Proemii Averroys", in D. Di Liscia, A. R. Kessler, Ch. Methien (eds), *Method and Order in the Renaissance Philosophy of Nature: The Aristotle Commentary Tradition*, Aldershot, Hants, Ashgate, 1997, pp. 53–111.

Busa SI, R., "De phantasia et imaginatione iuxta S. Thomam", in *Phantasia – imaginatio, V. Colloqui internazionale Roma 9–11 gennaio 1986*, Rome, Edizioni dell'Ateneo Roma, 1988, pp. 136–152.

Caiazzo, I., "La première réception du *Liber de causis* en Occident (XIIe–XIIIe siècles)", in D. Calma (ed.), *Reading Proclus and the Book of Causes: Western Scholarly Networks and Debates*, Leiden Boston, Brill, 2019, vol. 1, pp. 18–46.

Calma, D., Imbach, R., "Heymeric de Campo, auteur d'un traité de métaphysique. Étude et édition partielle du Colliget principiorum", in *Archives d'Histoire Doctrinale et Littéraire du Moyen Age* 80 (2013), pp. 277–412.

Calma, D., "Du néoplatonisme au réalisme et retour, parcours latins du *Liber de causis* aux XIIIe–XVIe siècles", in *Bulletin de philosophie médiévale* 54 (2012), pp. 217–276.

Calma, D., "Introductio to Reading Proclus and the Book of Causes: Notes on the Western Scholarly Networks and Debates", in D. Calma (ed.), *Reading Proclus and the Book of Causes: Western Scholarly Networks and Debates*, Leiden Boston, Brill, 2019, pp. 18–46.

Calma, D., "Metaphysics as a Way of Life: Heymericus de Campo on Universals and the "Inner Man"", in *Vivarium* 58/4 (2020), pp. 305–334.

Calma, D., "The Exegetical Tradition of Medieval Neoplatonism. Considerations on a Recently Discovered Corpus of Texts", in D. Calma (ed.), *Neoplatonism in the Middle Ages, Vol. 1: New Commentaries on* Liber de causis *(ca. 1250–1350)*, Turnhout, Brepolis, 2016, pp. 11–52.

Campillo Bo, Á. J., "From Poetics to Mathematics: Vicente Mariner's Latin Translation of Proclus' *In Euclidem*", in *Noctua* 11/2 (2024), pp. 258–294.

Campillo Bo, Á. J., "Jerónimo Muñoz's Reception of Proclus' *In Euclidem*: Philosophy of Mathematics and an Attempt to Prove the Parallel Postulate", in *Early Science and Medicine* (2023) pp. 631–658.

Campillo Bo, Á. J., "The Forgotten Gifts of Hermes: The Latin Reception of Proclus' *Commentary on Euclid's Elements*", in *Mediterranea, International Journal on the Transfer of Knowledge* 8 (2023), pp. 193–278.

Carolino, L. M., "Mathematics and the late aristotelian theory of science: The '*Quaestio de Certitudine Mathematicarum*' in seventeenth-century Portuguese universities", in V. Navarro, W. Eamon (eds), *Beyond the black legend: Spain and the Scientific Revolution*, Valencia, Universidad de Valencia, 2007, pp. 399–412.

Casteigt, J., "Le chapitre xi[xii] du Liber de causis commenté par Albert le Grand: l'unité de la cause et de son effet comme immanence réciproque", in D. Calma (ed.), *Reading Proclus and the Book of Causes: Western Scholarly Networks and Debates*, Leiden Boston, Brill, 2019, pp. 476–502.

Cerrata, F., *Alessandro Piccolomini: letterato e filosofo senese del Cinquecento*, Siena, Accademia degli Intronati, 1960.

Cerreta, F., "An Account of the Early Life of the Accademia degli Infiammati in the letters of Alessandro Piccolomini", in *The Romanic Review* 48 (1957), pp. 249–264.

Charles, D., *The Undivided Self. Aristotle and the Mind-Body Problem*, Oxford, Oxford University Press, 2021.

Chlup, R., *Proclus, An Introduction*, Cambridge, Cambridge University Press, 2012.

Claessens, G., "Clavius, Proclus, and the Limits of Interpretation: Snapshot-idealization versus Projectionism", in *History of Science* 47/3 (2009), pp. 317–336.

Claessens, G., "Het denken verbeeld. De vroegmoderne receptie (1533–1650) van Proclus' *Commentaar op het eerste boek van Euclides' Elementen*", Ph.D., Leuven, Katholieke Universiteit Leuven, 2011.

Claessens, G., "Kepler's Platonism in the *Harmonice mundi*", in *De mundi recentioribus phænomenis: Cosmologie et science dans l'Europe des Temps modernes, xv*[e]*–xvii*[e] *siècles*. Essais en l'honneur de Miguel Ángel Granada, Brepols, 2022, pp. 303–317.

Claessens, G., "Proclus in the Renaissance", in M. Sgarbi (ed.), *Encyclopedia of Renaissance Philosophy*, Cham, Springer International Publishing Switzerland, 2022.

Claessens, G., "Reception and the Textuality of History: Ramus and Kepler on Proclus' History and Philosophy of Geometry", *Texts, Transmissions, Receptions: Modern Approaches to Narratives* (Boston: Brill, 2015), pp. 281–294.

Claessens, G., "Sixteenth- and Seventeenth-Century Views on Mathematical Imagination", in Ch. Lüthy, C. Swan, p. J. J. M. Bakker, C. Zittel (eds), *Image, Imagination, and Cognition: Medieval and Early Modern Theory and Practice*, Leiden Boston, Brill, 2018, pp. 208–226.

Claessens, G., "The Drawing Board of Imagination: Federico Commandino and John Philoponus", *Journal of the History of Ideas*, 76/4 (2015), pp. 499–515.

Clagget, M., *The Science of Mechanics in the Middle Ages*, Madison WI, University of Wisconsin, 1959.

Cleary, J. J., *Aristotle and Mathematics: Aporetic Method in Cosmology and Metaphysics*, Leiden New York Köln, Brill, 1995.

Cleary, J. L., "Proclus' Philosophy of Mathematics", in G. Bechtle and D. J. O'Meara (eds), *La philosophie des mathématiques de l'Antiquité tardive Actes du colloque international Fribourg, Suisse, 24–26 septembre*, Saint-Paul Fribourg, Editions Universitaires Friboug Suisse, 1998, pp. 85–102.

Conti, A. D., "Opinions on Universals and Predication in Late Middle Ages: Sharpe's and Paul of Venice's Theories Compared", in *Documenti e Studi sulla Tradizione Filosofica Medievale* 18 (2007), pp. 483–500.

Corbini, A., *La teoria della scienza nel XIII secolo. I commenti agli Analitici Secondi*, Florence, Testi e Studi, 2006.

Corbini, A., "Jean Buridan et Marsile d'Inghen sur la predication par soi", in J. Biard (ed.), *Raison et démonstration: Les commentaires médiévaux sur le Seconds analytiques*, Turnhout, Brepols, 2015, pp. 169–184.

Cosci, M., "Francesco Barozzi", in M. Sgarbi (ed.), *Encyclopedia of Renaissance Philosophy*, Cham: Springer International Publishing Switzerland, 2022.

Cotugno, A., "Alessandro Piccolomini", in M. Sgarbi (ed.), *Encyclopedia of Renaissance Philosophy*, Cham: Springer International Publishing Switzerland, 2022.

Counet, J. M., "Duns Scot et le *Liber de causis*", in D. Calma (ed.), *Reading Proclus and the Book of Causes: Western Scholarly Networks and Debates*, Leiden Boston, Brill, 2019, vol. 1, pp. 210–251.

Courtenay, W. J., "The Early Stages in the Introduction of Oxford Logic into Italy", in A. Maierù (ed.), *English Logic in Italy in the 14th and 15th Centuries*, Naples, Bibliopolis, 1982, pp. 13–32.

Cozzoli, D., "Alessandro Piccolomini and the certitude of mathematics", in *History and Philosophy of Logic* 28/2 (2007), pp. 151–171.

Crapulli, G., *Mathesis universalis: Genesi di una idea nel XVI secolo*, Rome, Edizioni dell'Ateneo Roma, 1969.

D'Addario, A., "Donato Acciaioli", in *Dizionario biografico degli Italiani*, vol. 1, Roma, Istituto della enciclopedia italiana, 1960, pp. 80–82.

D'Ancona, C., "The *Liber de causis*", in S. Gersh (ed.), *Interpreting Proclus: From Antiquity to the Renaissance*, Cambridge, Cambridge University Press, 2014, pp. 137–161.

D'Hoine, P., "Syrianus and Proclus on Aristotle", in A. Falcon (ed.) *Brill's Companion to the Reception of Aristotle in Antiquity*, Leiden, Brill, 2016, pp. 374–393.

De Bellis, E., *Il pensiero logico di Agostino Nifo*, Galatina, Congedo, 1998.

De Bellis, E., "La trattazione della teoria aristotelica della demonstratio nel Ms. Gal. 27 di Galileo Galilei", in *Rivista di Filosofia Neo-Scolastica* 109/3 (2017), pp. 725–736.

De Garay, J., "Proclus' Reception in the Sixteenth Century: Commentary on the First Book of Euclid's Elements Jesús de Garay", in E. Anagnostou, K. Parry, *Later Platonists and their Heirs among Christians, Jews, and Muslims*, Leiden Boston, Brill, 2023, pp. 459–479.

De Groot, J., "*Dunamis* and the Science of Mechanics: Aristotle on Animal Motion", in *Journal of the History of Philosophy* 46/1 (2008), pp. 43–67.

De Haas, F. A. J., "Priscian of Lydia and Pseudo-Simplicius on the soul" in L. P. Gerson (ed.) *The Cambridge History of Philosophy in Late Antiquity*, Vol. 2 Cambridge, Cambridge University Press, 2010, pp. 756–764.

De Libera, A., *La querelle des universaux. De Platon à la fin du Moyen Âge*, Paris, Éditions du Seuil, 1996.

De Libera, A., *L'art des généralités: Théories de l'abstraction*, Paris, Aubier, 1999.

Demange, D., *Jean Duns Scotus, La théorie du savoir*, Paris, Sic et Non Librairie Philosophique, 2007.

De Pace, A., *Le matematiche e il mondo: Ricerche su un dibattito in Italia nella seconda meta` del Cinquecento*, Milan, Francoangeli, 1993.

De Risi, V., *Leibniz on the Parallel Postulate and the Foundations of Geometry*, Cham, Springer, 2010.

De Risi, V., "The development of Euclidean axiomatics", in *Archive for History of Exact Sciences* 70 (2016), pp. 591–676.

De Vaan, M., *Etymological Dictionary of Latin*, Leiden, Brill, 2008.

Deslauriers, M., *Aristotle on Definition*, Leiden, Brill, 2007.

Di Liscia, D. A., "Biagio Pelacani da Parma's Geometrisation of Latitudes and the Problems of the Mean Degree Theorem", in J. Biard, A. Robert (eds), *La philosophie de Blaise de Parme: Physique, psychologie, éthique*, Florence, Micrologus Library, 2019, pp. 129–152.

Di Liscia, D. A., "Introduction", in D. A. Di Liscia, E. D. Sylla (eds), *Quantifying Aristotle: The Impact, Spread and Decline of the Calculatores Tradition*, Leiden Boston, Brill, 2022, pp. 1–2.

Di Liscia, D. A., "Perfections and Latitudes: The Development of the Calculators' Tradition and the Geometrisation of Metaphysics and Theology", in D. A. Di Liscia, E. D. Sylla (eds), *Quantifying Aristotle: The Impact, Spread and Decline of the Calculatores Tradition*, Leiden Boston, Brill, 2022, pp. 278–328.

Di Mare, S., "Vincenzo Maggi", in M. Sgarbi (ed.), *Encyclopedia of Renaissance Philosophy*, Cham, Springer International Publishing Switzerland, 2022.

Di Napoli, G., *L'immortalità dell'anima nel Rinascimento*, Turin, Società Editrice Internazionale, 1963.

Dijksterhuis, E. J., *The Mechanization Of The World Picture*, Oxford, Oxford University Press, 1961.

Dillon, J. M., "Dionysius the Areopagite", in S. Gersh (ed.), *Interpreting Proclus: From Antiquity to the Renaissance*, Cambridge, Cambridge University Press, 2014, pp. 111–124.

Di Pasquale Barbanti, M., *Ochema-Pneuma e phantasia nel neoplatonismo. Aspetti psicologici e prospettive religiose*, Catania, Symbolon, 1998.

Distelzweig, P. M., "The intersection of the mathematical and natural sciences: The subordinate sciences in Aristotle", in *Apeiron* 46/2 (2013), pp. 85–105.

Dubouclez, O., *Descartes et la voie de l'analyse*, Paris, Presses Universitaires de France, 2013.

Ebbesen, S., "The Posterior Analytics 1100–1400 in East and West", in J. Biard (ed.), *Raison et démonstration: Les commentaires médiévaux sur le Seconds analytiques*, Turnhout, Brepols, 2015, pp. 11–21.

Évrard, E., "FANTASIA chez Proclus", in M. Fattori, M. Bianchi (ed.) *Phantasia – imaginatio, V. Colloqui internazionale Roma 9–11 gennaio 1986*, Rome, Edizioni dell'Ateneo Roma, 1988, pp. 57–68.

Facciolati, J., *Fasti Gymnasii Patavini*, Sala Bolognese, Forni, 1978.

Felgner, U., *Philosophy of Mathematics in Antiquity and in Modern Times* (Springer: Birkhäuser Cham, 2023).

Fellina, S., *Modelli di episteme neoplatonica nella Firenze del '400. Le gnoseologie di Giovanni Pico della Mirandola e di Marsilio Ficino*, Florence, Olschki, 2014.

Fellina, S., *Platone allo Studium Fiorentino-Pisano (1576–1635). L'insegnamento di Francesco de' Vieri, Jacopo Mazzoni, Carlo Tomasi, Cosimo Boscagli, Girolamo Bardi* (Mantova: Scripta, 2019).

Ferrari, G., "Dimostrazione Matematiche e Conoscenza Scientifica in Alessandro Piccolomini", in H. Burns, F. P. Di Teodoro, G. Bacci (eds), *Saggi di Letteratura architettonica da Vitruvio a Winckelmann III*, Firenze, Olschki, 2010, pp. 215–233.

Finamore, J. F., Kutash, E., "Proclus on the Psychê World Soul and the Individual Soul", in P. D'Hoine, M. Martijn (eds), *All From One: A Guide to Proclus*, Oxford, Oxford University Press, 2016, pp. 122–138.

Forcellini, E., *Lexicon Totius Latinitatis*, Padua, J. Manfrè, 1771.

Frede, D., "The Cognitive Role of Phantasia in Aristotle", in (eds) M. C. Nussbaum, A. O. Rorty, *Essays on Aristotle's De Anima*, Oxford: Oxford University Press, 1995, pp. 279–296.

Freguglia, P., *Ars analytica. Matematica e methodus nella seconda metà del Cinquecento*, Busto Arsizio, Bramante Editore, 1988.

Führer, M., Gersh, S., "Dietrich of Freiberg and Berthold of Moosburg", in S. Gersh (ed.), *Interpreting Proclus: From Antiquity to the Renaissance*, Cambridge, Cambridge University Press, 2014, pp. 299–317.

Gabbe, M., "Themistius as a Commentator on Aristotle: Understanding and Appreciating his Conception of Nous Pathetikos and Phantasia", *Dionysius* 26 (2008), pp. 73–92.

Gabbey, A., "Between Ars and Philosophia Naturalis: Reflections on the Historiography of Early Modern Mechanics", in J. V. Field, F. A. J. L. James (eds), *Renaissance and Revolution: Humanists, Scholars, Craftsman, and Natural Philosophers in Early Modern Europe*, Cambridge New York, Cambridge University Press, 1993, pp. 133–145.

Gabor, G., *Colloquium 1 The Authorship of the Pseudo-Simplician Neoplatonic Commentary on the De Anima. Proceedings of the Boston Area Colloquium of Ancient Philosophy*, Leiden, Brill, 2020.

Galluzzo, G., *The Medieval Reception of Book Zeta of Aristotle's Metaphysics* (*Studien Und Texte Zur Geistesgeschichte Des Mittelalters*), Leiden Boston, Brill, 2001.

Gamba, E., "Leonico Tomeo, Niccolò", in M. Sgarbi (ed.), *Encyclopedia of Renaissance Philosophy*, Cham, Springer International Publishing Switzerland, 2022.

Garin, E., "La giovinezza di Donato Acciaiuoli", in *Rinascimento*, 1/1 (1950), pp. 43–70.

Gersh, S., "Nicholas of Cusa", in S. Gersh (ed.), *Interpreting Proclus: From Antiquity to the Renaissance*, Cambridge, Cambridge University Press, 2014, pp. 318–352.

Gersh, S., "One Thousand Years of Proclus", in S. Gersh (ed.), *Interpreting Proclus: From Antiquity to the Renaissance*, Cambridge, Cambridge University Press, 2014, pp. 4–5.

Giacobbe, G. C., "Alcune cinquecentine riguardanti il processo di rivalutazione epistemologica della matematica nell'ambito della rivoluzione scientifica rinascimentale" *La Berio* 13 (1973), pp. 7–44.

Giacobbe, G. C., "Epigoni nel seicento della "Questio de certitudine mathematicarum": Giuseppe Biancani", *Physis*, 18 (1976), pp. 5–40.

Giacobbe, G. C., "Francesco Barozzi e la *quaestio de certitudine mathematicarum*", in *Physis* 14/4 (1972), pp. 162–193.

Giacobbe, G. C., "Il *Commentarium De Certitudine Mathematicarum Disciplinarum* di Alessandro Piccolomini", in *Physis*, 14/2, (1972), pp. 162–193.

Giacobbe, G. C., "La riflessione metamatematica di Pietro Catena", in *Physis* 15/2 (1973), pp. 178–196.

Giacobbe, G. C., "La Quaestio de certitudine mathematicarum all'interno della scuola padovana", *Atti del convegno di storia della logica* (1974), pp. 95–112.

Giacobbe, G. C., "Un gesuita progressista nella "Quaestio de certitudine mathematicarum" rinascimentale: Benito Pereyra", *Physis*, 19 (1977), pp. 51–86.

Gilbert, N. W., *Renaissance Concepts of Method*, New York Chichester, Columbia University Press, 1960.

Girardi, M. T., "Accademia degli Infiammati", in M. Sgarbi (ed.), *Encyclopedia of Renaissance Philosophy*, Cham, Springer International Publishing Switzerland, 2022.

Girardi, M. T., "Bernardino Tomitano", in M. Sgarbi (ed.), *Encyclopedia of Renaissance Philosophy*, Cham: Springer International Publishing Switzerland, 2022.

Gómez-Lobo, A., "Definitions in Aristotle's *Posterior Analytics*", in D. J. O'Meara (ed.), *Studies in Aristotle*, Washington DC, Catholic University of America Press, 1981, pp. 25–46.

Goulding, R., *Defending Hypatia: Ramus, Savile, and the Renaissance Rediscovery of Mathematical History*, Dordrecht, Springer, 2010.

Granger, H., "The *Differentia* and the *Per Se* Accident in Aristotle", in *Archiv für Geschichte der Philosophie* 63 (1981), pp. 118–129.

Grellard, Ch., "Certainty", in H. Lagerlund (ed.), *Encyclopedia of Medieval Philosophy*, Dordrecht, Springer, 2011, pp. 206–208.

Grendler, P. F., *The Universities of the Italian Renaissance*, Baltimore, Johns Hopkins University Press, 2004.

Guicciardini, N., *Isaac Newton on Mathematical Certainty and Method*, Cambridge MA, Massachusetts Institute of Technology Press, 2011.

Hamesse, J., "Imaginatio et phantasia chez les auteurs philosophiques du 12e et du 13e siècle", in *Phantasia – imaginatio, V. Colloqui internazionale Roma 9–11 gennaio 1986*, Rome, Edizioni dell'Ateneo Roma, 1988, pp. 153–184.

Hangai, A., *Alexander of Aphrodisias on Phantasia: An Aristotelian Account of Mental Representation in 2nd–3rd Centuries CE*, Ph.D., Budapest, Central European University, 2016.

Hangai, A., "Alexander of Aphrodisias' Criticism of the Stoic Theory of Perception: typos and typōsis", in *Elenchos*, 43/2 (2022), pp. 339–362.

Hankinson, R. J. and M. Havrda (eds.), *Galen's Epistemology: Experience, Reason, and Method in Ancient Medicine* (Cambridge: Cambridge University Press, 2021).

Harari, O., "Les preuves démonstratives dans le Commentaire de Proclus au premier livre des *Éléments* d'Euclide", in A. Lernould (ed.), *Études sur le Commentaire de Proclus au premier livre des Éléments d'Euclide*, Saint-Paul Fribourg, Presses Universitaires du Septentrion, 2007, pp. 197–216.

Harari, O., "Methexis and Geometrical Reasoning in Proclus' *Commentary on Euclid's Elements*", in *Oxford Studies in Ancient Philosophy* 30 (2006), pp. 361–389.

Harari, O., "Proclus' Account of Explanatory Demonstrations in Mathematics and its Context", in *Archiv für Geschichte der Philosophie* 90 (2008), pp. 137–164.

Harris, A., Taub, L., "Quantification and precision: a brief look at some ancient accounts", in *Annals of Science* 81 (2023), pp. 10–29.

Hasse, D. N., "Three Double Translations from Arabic into Latin by Gerard of Cremona and Dominicus Gundisalvi", in D. Calma (ed.), *Reading Proclus and the Book of Causes: Western Scholarly Networks and Debates*, Leiden Boston, Brill, 2019, vol. 2, pp. 247–276.

Helbing, M. O., "La fortune des commentaires de Proclus sur le premier livre des Éléments d'Euclide à l'époque Galilée", in G. Bechtle and D. J. O'Meara (eds), *La philosophie des mathématiques de l'Antiquité tardive Actes du colloque international Fribourg, Suisse, 24–26 septembre*, Saint-Paul Fribourg, Editions Universitaires Fribourg Suisse, 1998, pp. 173–193.

Hellyer, M., *Catholic Physics: Jesuit Natural Philosophy in Early Modern Germany*, Notre Dame, University of Notre Dame, 2005.

Helmig, C., *Forms and Concepts: Concept Formation in the Platonic Tradition*, Berlin, De Gruyter, 2012.

Helmig, C., "Iamblichus, Proclus, and Philoponus on Parts, Faculties and *ousiai* of Soul and the Notion of Life", in K. Corcilius, D. Perler (eds), *Partition of the Soul in Ancient, Medieval and Early Modern Philosophy*, Berlin, De Gruyter, 2014, pp. 149–178.

Helmig, C., "Proclus' Criticism of Aristotle's Theory of Concept Formation in *Analytica Posteriora* II.19", in F. A. J. de Haas, M. Leunissen, M. Martijn (eds), *Interpreting Aristotle's Posterior Analytics in Late Antiquity and Beyond*, Leiden, Brill, 2010, pp. 27–55.

Helmig, C., "The Truth Can Never Be Refuted: Syrianus' View(s) on Aristotle Reconsidered", in Longo, A., Corti, L., D'Andrès, N., Del Forno, D., Maffi, E., Schmidhauser, A. (eds) *Syrianus et la métaphysique de l'antiquité tardive. Actes du colloque international, université de Genève, 29 septembre–1er octobre 2006*, Elenchos 51 (Napoli: Bibliopolis, 2009), pp. 347–380.

Higashi, S., *Penser les mathématiques au XVIe siècle*, Paris, Editions Classiques Garnier, 2018.

Higashi, S., "Giorgio Valla et Alessandro Piccolomini. Quelques aspects de la réception de Proclus ê la Renaissance", in *Bulletin of Liberal Arts Education Center* 27 (2007), pp. 31–52.

Hintikka, J., Remes, U., *The Method of Analysis: Its Geometrical Origin and Its General Significance*, Dordrecht Boston, Springer, 1974.

Hintikka, J., "Method of Analysis: A Paradigm of Mathematical Reasoning?", in *History and Philosophy of Logic* 33/1 (2012), pp. 49–67.

Hobson, A., *Renaissance Book Collecting: Jean Grolier and Diego Hurtado de Mendoza, their Books and Bindings*, Cambridge, Cambridge University Press, 1999.

Imbach, R., "Au-delà de la métaphysique: Notule sur l'importance du Commentaire de Berthold de Moosburg OP sur les *Éléments de théologie*", in D. Calma (ed.), *Reading Proclus and the Book of Causes: Western Scholarly Networks and Debates*, Leiden Boston, Brill, 2019, vol. 1, pp. 393–376.

Imbach, R., "Le (néo-)platonisme médiéval: Proclus latin et l'école dominicaine allemande", in *Revue de théologie et de philosophie* 110 (1978), pp. 427–448.

Janseens, J., "Doubles traductions et omissions: une approche critique en vue d'une édition de la traduction latine du *Liber de causis*", in D. Calma (ed.), *Reading Proclus and the Book of Causes: Western Scholarly Networks and Debates*, Leiden Boston, Brill, 2019, vol. 2, pp. 275–316.

Jardine, N., "Epistemology of the sciences", in C. B. Schmitt, Q. Skinner, E. Kessler (eds), *The Cambridge History of Renaissance Philosophy*, Cambridge, Cambridge University Press, 1988, pp. 685–711.

Jardine, N., "Galileo's road to truth and the demonstrative regress", in *Studies in History and Philosophy of Science* 7/A (1976), pp. 277–318.

Johansen, T. K., *The Powers of Aristotle's Soul*, Oxford, Oxford University Press, 2012.

Judson, L., "Aristotle and Crossing the Boundaries between the Sciences", in *Archiv für Geschichte der Philosophie* 28 (2019), pp. 177–204.

Keller, A. G., "Machines as Mathematical Instruments", in L. B. Cormack, S. A. Walton, J. A. Schuster (eds), *Mathematical Practitioners and the Transformation of Natural Knowledge in Early Modern Europe*, Cham, Springer, 2017, pp. 115–130.

Keller, A. G., "Mathematics, Mechanics and the Origins of the Culture of Mechanical Invention", in *Minerva* 23/3 (1985), pp. 348–361.

King, E., *Supersapientia: Berthold of Moosburg and the Divine Science of the Platonists*, Leiden Boston, Brill, 2021.

King, E., "*Sapiens modernus*: The Reception of Dietrich of Freiberg in Berthold of Moosburg", in D. Calma, E. King (eds), *The Renewal of Medieval Metaphysics Berthold of Moosburg's Expositio on Proclus' Elements of Theology*, Leiden Boston, Brill, 2021, pp. 243–281.

King, E., "Eriugenism in Berthold of Moosburg's *Expositio super Elementationem theologicam Procli*", in D. Calma (ed.), *Reading Proclus and the Book of Causes: Western Scholarly Networks and Debates*, Leiden Boston, Brill, 2019, vol. 1, pp. 394–437.

Kiosoglou, S. A., "Proclus' Theory of Seemingly "Useless" Theorems", in *Méthexis*, 34/1 (2022), pp. 148–165.

Klitenic Wear, S., *The Teachings of Syrianus on Plato's Timaeus and Parmenides*, Leiden Boston, Brill, 2011.

Knöll, P. (ed.), *Contra Academicos libri tres*, Vienna Leipzig, Corpus Scriptorum Ecclesiasticorum Latinorum, 1922.

Krause, K., Anzulewicz, H., "From Content to Method: the *Liber de causis* in Albert the Great", in D. Calma (ed.), *Reading Proclus and the Book of Causes: Western Scholarly Networks and Debates*, Leiden Boston, Brill, 2019, vol. 1, pp. 152–180.

Kristeller, P. O., "Proclus as a Reader of Plato and Plotinus, and His Influence in the Middle Ages and in the Renaissance", in J. Pépin, H. D. Saffrey (eds) *Proclus lecteur et interprète des anciens, Actes du colloque international du Centre National de la Recherche Scientifique, 2–4 octobre 1985*, Paris, Centre national de la recherche scientifique, 1978, pp. 191–211.

Laird, W. R., *The Renaissance of Mechanics: Ancient Science in the Age of Humanism*, Cham, Springer, 2024.

Laird, W. R., "The Scope of Renaissance Mechanics", in *Osiris* 2 (1986), pp. 43–68.

Lear, J., "Aristotle's philosophy of mathematics", in *Philosophical Review* 91 (1982), pp. 161–192.

Lefebvre, R., "La phantasia chez aristote: Subliminalité, indistinction et pathologie de la perception", in *Les Etudes Philosophiques. Aristote sur la imagination* (1997), pp. 41–58.

Leijenhorst, C., "The Nature of the Understanding: Intellect, Conception, and Concepts" H. Lagerlund, B. Hill (eds), *The Routledge Companion to Sixteenth Century Philosophy*, New York, Routledge, 2017, pp. 475–492.

Leinkauf, T., *Il neoplatonismo di Francesco Patrizi come presupposto della sua critica ad Aristotele*, Florence, La Nuova Italia Editrice, 1990.

Leinkauf, T., "Francesco Patrizi", in S. Gersh (ed.), *Interpreting Proclus: From Antiquity to the Renaissance*, Cambridge, Cambridge University Press, 2014, pp. 380–402.

Lennox, J. G., "Aristotle, Galileo, and 'Mixed Sciences'", in W. A. Wallace (ed.), *Reinterpreting Galileo*, Washington, D.C., Catholic University of America Press, 1986, pp. 29–51.

Lernould, A., ""Introduction" and "Analyse des Prologus I t II de l'*In Euclidem*"", in A. Lernould (ed.), *Études sur le Commentaire de Proclus au premier livre des Éléments d'Euclide*, Saint-Paul Fribourg, Presses Universitaires du Septentrion, 2007, pp. 11–25.

Levitin, D., *Ancient Wisdom in the Age of the New Science*, Cambridge, Cambridge University Press, 2015.

Levitin, D., *The Kingdom of Darkness: Bayle, Newton, and the Emancipation of the European Mind from Philosophy*, Cambridge New York, Cambridge University Press, 2022.

Lewis, C. T., Short, C., *A Latin Dictionary* (Oxford: Clarendon Press, 1879).

Lines, A. D., "Natural Philosophy in Renaissance Italy: The University of Bologna and the Beginnings of Specialization", in *Early Science and Medicine* 6/4 (2001), pp. 267–323.

Longeway, J., *Demonstration and Scientific Knowledge in William of Ockham: A Translation of Summa Logicae III–II: De Syllogismo Demonstrativo, and Selections from the Prologue to the Ordinatio*, Notre Dame, University of Notre Dame, 2007.

Longeway, J., "Aegidius Romanus and Albertus Magnus vs. Thomas Aquinas on the highest sort of demonstration (*demonstratio potissima*)", in *Documenti e Studi sulla Tradizione Filosofica Medievale* 13 (2002), pp. 373–434.

Longeway, J., "The Place of *Demonstratio Potissima* in some 16th-Century Accounts of Mathematics", in J. Biard (ed.), *Raison et démonstration: Les commentaires médiévaux sur le Seconds analytiques*, Turnhout, Brepols, 2015, pp. 223–251.

Longo, A., "*In Euclidem*, prologue I, chapitre 6: Ce que Proclus doit à son maître Syrianus dans les arguments avancés contre la thèse de l'existence "postérieure" des objets mathématiques", in A. Lernould (ed.), *Études sur le Commentaire de Proclus au premier livre des Éléments d'Euclide*, Saint-Paul Fribourg, Presses Universitaires du Septentrion, 2007, pp. 45–54.

Longo, A., "Réflexions ontologiques et renvois à la *Métaphysique* dans le commentaire de Thomas d'Aquin sur les *Seconds Analytiques* d'Aristote", in J. Biard (ed.), *Raison et démonstration: Les commentaires médiévaux sur le Seconds analytiques*, Turnhout, Brepols, 2015, pp. 97–112.

Machado Mota, B., *O estatuto da Matemática em Portugal nos Séculos XVI e XVII*, Lisboa, Fundação Calouste Gulbenkian, 2011.

MacIsaac, G., "Νόησις, dialectique et mathématique chez Proclus", in A. Lernould (ed.), *Études sur le Commentaire de Proclus au premier livre des Éléments d'Euclide*, Saint-Paul Fribourg, Presses Universitaires du Septentrion, 2007, pp. 125–138.

MacIsaac, G., "Geometrical First Principles in Proclus' Commentary on the First Book of Euclid's Elements", in *Phronesis* 59 (2014), pp. 44–98.

MacIsaac, G., "Phantasia between soul and body in Proclus' Euclid Commentary", in *Dionysius* 19 (2001), pp. 125–136.

MacIsaac, G., "Projection and Time in Proclus", in J. Inglis (ed.), *Medieval Philosophy and the Classical Tradition in Islam, Judaism and Christianity*, London, Curzon Press, 2009, pp. 83–105.

MacIsaac, G., "The Soul and Discursive Reason in the Philosophy of Proclus", Ph.D., Notre Dame, University of Notre Dame, 2001.

Mahoney, E. P., *Two Aristotelians of the Italian Renaissance: Nicoletto Vernia and Agostino Nifo*, Aldershot Burlington, Ashgate, 2000.

Mahoney, M. S., "Barrow's mathematics: between ancients and moderns", in M. Feingold (ed.), *Before Newton: The Life and Times of Isaac Barrow*, Cambridge New York, Cambridge University Press, 2021 (Transformations: Studies in the History of Science and Technology).

Maierù, L., "La diffusione di Proclo, commentatore di Euclide, nel Cinquecento", in *11° Annuario del Liceo Scientifico "B. G. Scorza"*, Soveria Mannelli, Calabria Letteraria Editrice, 1999, pp. 49–68.

Malgieri, M. E., "Citing the Book of Causes, IV: Henry of Ghent and His (?) Questions on the Metaphysics", in D. Calma (ed.), *Reading Proclus and the Book of Causes: Western Scholarly Networks and Debates*, Leiden Boston, Brill, 2019, vol. 1, pp. 181–209.

Malink, M., "Demonstration by reductio ad impossibile in Posterior Analytics 1. 26", in Victor Caston (ed.), *Oxford Studies in Ancient Philosophy*, Vol. LVIII., Oxford, Oxford University Press, 2020, pp. 91–155.

Mancosu, P. "Aristotelian logic and Euclidean mathematics: Seventeenth-century developments of the quaestio de certitudine mathematicarum", in *Studies in History and Philosophy of Science* 23/2 (1992) pp. 241–265.

Mancosu, P., *Philosophy of Mathematics and Mathematical Practice in the Seventeenth Century*, Oxford, Oxford University Press, 1999.

Mancosu, P., Mugnai, M., *Syllogistic Logic and Mathematical Proof*, Oxford, Oxford University Press, 2023.

Martijn, M., *Proclus on Nature: Philosophy of Nature and Its Methods in Proclus' Commentary on Plato's Timaeus*, Leiden Boston, Brill, 2010.

Martijn, M., "Proclus' Geometrical Method", in P. Remes, S. Slaveva-Griffin (eds), *The Routledge Handbook of Neoplatonism*, Abingdon, Routledge, 2014, pp. 145–159.

Martin, C., "Humanism and the Assessment of Averroes in the Renaissance", in A. Akasoy, G. Giglioni (eds), *Renaissance Averroism and Its Aftermath: Arabic Philosophy in Early Modern Europe*, New York, London, Springer, 2012, pp. 65–81.

Megna, P., "Per Ficino e Proclo", in F. Baussi, V. Fera (eds), *Laurentia laurus: Scritti offerti a Mario Martelli*, Messina, Centro Interdipartimentale di Studi Umanistici, 2004, pp. 313–362.

Meliado, M., *Sapienza peripatetica: Eimerico di Campo e i percorsi del tardo albertismo*, Münster, Aschendorff, 2018.

Miolo, L., "Le *Liber de causis* et l'*Elementatio theologica* dans deux bibliothèques anglaises: Merton College (Oxford) et Peterhouse (Cambridge)", in D. Calma (ed.), *Reading Proclus and the Book of Causes: Western Scholarly Networks and Debates*, Leiden Boston, Brill, 2019, vol. 1, pp. 71–120.

Modrak, D., "Nominal Definition in Aristotle", in D. Charles (ed.), *Definition in Greek Philosophy*, Oxford, Oxford University Press, 2010, pp. 252–285.

Molland, A. G., "Colonizing the world for mathematics", in E. Grant, J. E. Murdoch (eds), *Mathematics and its applications to science and natural philosophy in the Middle Ages: essays in honor of Marshall Clagett*, Cambridge New York, Cambridge University Press, 1987, pp. 45–65.

Molland, A. G., "Mathematics in the Thought of Albertus Magnus", in J. A. Weisheipl (ed.), *Albertus Magnus and the Sciences: Commemorative Essays*, Toronto, Pontifical Institute of Mediaeval Studies, 1980, pp. 463–478.

Morison, B., "Logic", R. J. Hankinson (ed.), *The Cambridge Companion to Galen* (Cambridge: Cambridge University Press, 2008) chapter 4, pp. 66–15.

Mueller, I., "Aristotelian Objections and Post-Aristotelian Responses to Plato's Elemental Theory", in C. Horn, J. Wilberding (eds), *Neoplatonism and the Philosophy of Nature*, Oxford, Oxford University Press, 2012, pp. 129–146.

Mueller, I., "Aristotle on geometric objects", in J. Barnes, M. Schofield, Richard Sorabji (eds), *Articles on Aristotle*, vol. 3, London, Duckworth, 1979, pp. 96–107.

Mueller, I., "Aristotle's Doctrine of Abstraction in the commentators", in R. Sorabji (ed.), *Aristotle Transformed: the Ancient Commentators and their Influence*, Ithaca NY, Duckworth, 1990, pp. 463–479.

Mueller, I., "Foreword", in Proclus, G. R. Morrow, *Proclus: A Commentary on the First Book of Euclid's Elements*, New Jersey, Princeton University Press, 1970, pp. ix–xxxi.

Mueller, I., "Mathematics and Philosophy in Proclus' *Commentary on Book I of Euclid's Elements*", in *Proclus, lecteur et interprète des anciens: Actes du Colloque international du CNRS, Paris, 2–4 octobre 1985*, pp. 305–318.

Mueller, I., "Syrianus and the Concept of Mathematical Number", in G. Bechtle and D. J. O'Meara (eds), *La philosophie des mathématiques de l'Antiquité tardive Actes du colloque international Fribourg, Suisse, 24–26 septembre*, Saint-Paul Fribourg, Editions Universitaires Friboug Suisse, 1998, pp. 71–84.

Murdoch, J. E., "Infinity and continuity", in N. Kretzmann, A. Kenny, J. Pinborg, E. Stump (eds), *The Cambridge History of Later Medieval Philosophy*, Cambridge, Cambridge University Press, 1982, from VII – Natural philosophy, pp. 564–592.

Napolitano Valditara, L. M., *Le idee, i numeri, l'ordine: La dottrina della «Mathesis universalis» dall'Accademia antica al Neoplatonismo*, Padua, Bibliopolis, 2010.

Nardi, B., *Saggi sull'Aristotelismo padovano dal secolo XIV al XVI*, Florence, Sansone, 1958.

Navarro-Brotons, V., *Disciplinas, saberes y prácticas: Filosofía natural, matemáticas y astronomía en la sociedad española de la época moderna*, Valencia, Universidad de Valencia, 2014.

Navarro-Brotons, V., *Jerónimo Muñoz: Matemáticas, cosmología y humanismo en la época del Renacimiento*, Valencia, Universidad de Valencia, 2019.

Navarro-Brotons, V., Rodríguez Galdeano, E., "The Cultivation of Astronomy in Spanish Universities in the Latter Half of the 16th Century", in M. Feingold, V. Navarro-Brotons (eds), *Universities and Science in the Early Modern Period* (New York London: Springer), 2006, pp. 83–98.

Nikulin, D., *Matter, Imagination and Geometry: Ontology, Natural Philosophy and Mathematics in Plotinus, Proclus and Descartes*, Ashgate, Aldershot, 2002.

Nikulin, D., "Imagination et mathématiques chez Proclus", in A. Lernould (ed.), *Études sur le Commentaire de Proclus au premier livre des Éléments d'Euclide*, Saint-Paul Fribourg, Presses Universitaires du Septentrion, 2007, pp. 139–160.

Nikulin, D., "Physica more geometrico demonstrata: Natural philosophy in Proclus and Aristotle", in *Proceedings of the Boston Area Colloquium in Ancient Philosophy* 17 (2003), pp. 183–209.

O'Meara, D. J., *Pythagoras Revived: Mathematics and Philosophy in late Antiquity*, Oxford, Oxford University Press, 1989.

O'Meara, D. J., "Geometry and the Divine in Proclus", in T. Koetsier and L. Bergmans (eds), *Mathematics and the Divine* (Amsterdam: Elsevier 2005), pp. 133–145.

O'Meara, D. J., "L'eloge des sciences mathématiques dans la philosophi de l'Antiquité tardive", in A. Lernould (ed.), *Études sur le Commentaire de Proclus au premier livre des Éléments d'Euclide*, Saint-Paul Fribourg, Presses Universitaires du Septentrion, 2007, pp. 57–67.

O'Meara, D. J., "Mathematics and the sciences", in P. D'Hoine, M. Martijn (eds), *All From One: A Guide to Proclus*, Oxford, Oxford University Press, 2016, pp. 167–182.

Ogden, S. O., *Averroes on Intellect. From Aristotelian Origins to Aquinas' Critique*, Oxford, Oxford University Press, 2022.

Opsomer, J., "Was sind irrationale Seelen?", in *Perkams/Piccione* (2006), pp. 136–166.

Opsomer, J., "Mathematical Explanation and the Philosophy of Nature, Late Ancient Philosophy: Astronomy and the Theory of the Elements", in *Documenti e studi sulla tradizione filosofica medievale* 23 (2012), pp. 65–106.

Opsomer, J., "Organiser la philosophie selon ses éléments: Structures argumentatives dans les *Éléments de Théologie*", in G. Aubry, L. Brisson, P. Hoffmann, L. Lavaud (eds), *Relire les Éléments de théologie de Proclus: Réceptions, interprétations antiques et modernes*, Paris, Hermann, 2021, pp. 133–176.

Opsomer, J., "Proclus' *Elements of Physics* and the axiomatization of kinematics", in F. Baghdassarian, I. Papachristou, S. Toulouse (eds), *Relectures néoplatoniciennes de la théologie d'Aristote*, Sankt Augustin, Academia, 2020, pp. 83–102.

Opsomer, J., "Proclus' *Elements of Theology* and Platonic Dialectic", in D. Calma (ed.), *Reading Proclus and the Book of Causes: Western Scholarly Networks and Debates*, Leiden Boston, Brill, 2019, vol. 3, pp. 17–36.

Opsomer, J., "Syrianus, Proclus, and Damascius", in F. Sheffield and J. Warren (eds), *Routledge Companion to Ancient Philosophy*, New York London, Routledge, 2013, pp. 626–642.

Opsomer, J., "The Natural World", in P. D'Hoine, M. Martijn (eds), *All From One: A Guide to Proclus*, Oxford, Oxford University Press, 2016, pp. 139–166.

Palmén, R., *Richard of St. Victor's Theory of Imagination*, Leiden, Brill, 2012.

Palmieri, P., "On *scientia* and *regressus*" H. Lagerlund, B. Hill (eds), *The Routledge Companion to Sixteenth Century Philosophy*, New York, Routledge, 2017, pp. 319–349.

Pasnau, R., *Metaphysical Themes 1274–1671*, Oxford, Oxford University Press, 2013.

Pedrazzi, M., "Sul tentativo di Alessandro Piccolomini di ridurre a sillogismo la I dimostrazione degli Elementi di Euclide", *Cultura e Scuola*, 13 (1974), pp. 221–230.

Pellegrin, P., "Definition in Aristotle's Posterior Analytics", in J. Lennox, R. Bolton (eds) *Being, Nature, and Life in Aristotle: Essays in Honor of Allan Gotthelf*, Cambridge, Cambridge University Press, 2010, pp. 122–146.

Piché, D., "Causa prima et esse dans le *Liber de causis* selon Thomas d'Aquin et Siger de Brabant", in *Dialogue* 38 (1999), pp. 75–97.

Pisano, R., Capecchi, D., *Tartaglia's Science of Weights and Mechanics in the Sixteenth Century, Selections from Quesiti et inventioni diverse: Books VII–VIII*, New York London, Springer, 2016.

Poppi, A., "Pietro Pomponazzi tra Averroismo e Galenismo sul Problema del "Regressus"", in *Rivista Critica di Storia della Filosofia* 24/3 (1969), pp. 243–266.

Premuda, L., "Un discepolo di Leoniceno tra filologia ed empirismo: G. Manardo e il «libero esame» dei classici della medicina in funzione di più spregiudicati orientamenti metodologici", in *Atti del Convegno internazionale per la celebrazione del V Centenario della nascita di Giovanni Manardo, 1462–1536. Ferrara, 8–9 dicembre 1962*, (Ferrara: Università degli studi di Ferrara 1963), pp. 43–56.

Rabouin, D., *Mathesis universalis: L'idée de « mathématique universelle » d'Aristote à Descartes*, Paris, Presses Universitaires de France, 2009.

Rabouin, D., "Le rôle de Proclus dans les débats sur la 'mathématique universelle' à la Renaissance", in A. Lernould (ed.), *Études sur le Commentaire de Proclus au premier livre des Éléments d'Euclide*, Saint-Paul Fribourg, Presses Universitaires du Septentrion, 2007, pp. 217–235.

Rachak, J., "La présence du *Liber de causis* dans l'œuvre d'Ibn Bagga et pseudo-Ibn Bagga: un philosophe péripatéticien du XIe/XIIe siècle de l'Occident islamique", in D. Calma (ed.), *Reading Proclus and the Book of Causes: Western Scholarly Networks and Debates*, Leiden Boston, Brill, 2019, vol. 2, pp. 233–245.

Raimondi, F. P., "Pomponazzi e la polemica anticalculatoria", in *Pietro Pomponazzi, Tutti i trattati peripatetici*, Milan, Bompiani, 2013, pp. 9–86.

Randall, J. H., *The School of Padua and the Emergence of the Modern Science*, Padova, Editrice Antenore, 1961.

Rappe, S., *Reading Neoplatonism: Non-discursive Thinking in the Texts of Plotinus, Proclus, and Damascius*, Cambridge, Cambridge University Press, 2007.

Retucci, F., "Liber de causis in Thomas of York", in D. Calma (ed.), *Reading Proclus and the Book of Causes: Western Scholarly Networks and Debates*, Leiden Boston, Brill, 2019, vol. 1, pp. 47–70.

Richardson, K., "Averroism", in H. Lagerlund, B. Hill (eds), *The Routledge Companion to Sixteenth Century Philosophy*, New York, Routledge, 2017, pp. 137–155.

Robichaud, D. J. J., *Plato's Persona: Marsilio Ficino, Renaissance Humanism, and Platonic Traditions*, Philadelphia, University of Pennsylvania Press, 2018.

Robichaud, D. J. J., "Fragments of Marsilio Ficino's Translations and Use of Proclus' *Elements of Theology* and *Elements of Physics*: Evidence and Study", in *Vivarium* 54/1 (2016), pp. 46–107.

Rose, P. L., *A Venetian Patron and Mathematician of the Sixteenth Century: Francesco Barozzi: 1537–1604*, Pisa, Giardini Editori, 1977.

Rose, P. L., Drake, S., "The Pseudo-Aristotelian Questions of Mechanics in Renaissance Culture", in *Studies in the Renaissance* 18 (1971), pp. 65–104.

Rose, P. L., "Certitudo Mathematicarum from Leonardo to Galileo", in *Leonardo nella scienza e nella tecnica: Atti del Simposio Internazionale di storia della Scienza*, Firenze Vinci, Giunti, 1969, pp. 43–49.

Rose, P. L., "Humanist Culture and Renaissance Mathematics: The Italian Libraries of the Quattrocento", in *Studies in the Renaissance* 20 (1973), pp. 46–105.

Rose, P. L., "Patrons, Collectors and Translators: Humanist Origins of the Mathematical Renaissance", in *The Italian Renaissance of Mathematics: Studies on Humanists and Mathematicians from Petrarch to Galileo*, Genève, Librarie Droz, 1975, pp. 26–75.

Rose, P. L., "The New Science of Tartaglia and Benedetti", in *The Italian Renaissance of Mathematics: Studies on Humanists and Mathematicians from Petrarch to Galileo*, Genève, Librarie Droz, 1975, pp. 151–158.

Rossi, P., *I filosofi e le macchine 1400–1700*, Milan, Feltrinelli, 2017.

Rossi, V., Bosco, U. (eds), *Familiarium rerum libri*, Turin, Aragno, 2004.

Roudaut, S., *La mesure de l'être, Le problème de la quantification des formes au Moyen Âge (ca. 1250–1370)*, Leiden, Brill, 2022.

Saffrey, H. D., "Les Néoplatoniciens et les Oracles chaldaïques", in *Revue des études augustiniennes* 27 (1981), pp. 209–225.

Saffrey, H. D., "Pietro Balbi et la première traduction latine de la Théologie platonicienne de Proclus", in P. Cockshaw, M. C. Garand, and P. Jodogne (eds), *Miscellanea codicologica F. Masai dicata*, Gand, E. Story-Scientia, 1979, pp. 425–437.

Sattler, B. M., *The Concept of Motion in Ancient Greek Thought, Foundations in Logic, Method, and Mathematics*, Cambridge, Cambridge University Press, 2020.

Schiefsky, M. J., "Art and nature in ancient mechanics", in B. Bensaude-Vincent and W. R. Newman (eds) *The Artificial and the Natural: An Evolving Polarity*, Cambridge: Massachusetts Institute of Technology Press, 2007, pp. 67–108.

Schmitt, Ch. B., "Hieronymus Picus, Renaissance Platonism and the Calculator", in *The Aristotelian Tradition and Renaissance Universities*, London, Variorum Reprints, 1984, pp. 57–80.

Schmitz, M., *Euclids Geometrie und ihre mathematiktheoretische Grundlegung in der neuplatonischen Philosophie des Proklos*, Würzburg, Königshausen und Neumann, 1997.

Schöttler, T., "From Causes to Relations: The Emergence of a Non-aristotelian Concept of Geometrical Proof out of the *Quaestio De Certitudine Mathematicarum*", in *Society and Politics* 6/2 (2012), pp. 29–47.

Sgarbi, M., *The Italian Mind: Vernacular Logic in Renaissance Italy (1540–1551)*, Leiden, Brill, 2014.

Sheldon-Williams, I. P., "Henads and Angels: Proclus and the Pseudo-Dionysius", in *Studia Patristica* 11 (1972), pp. 65–71.

Sheppard, A., "Phantasia and Mathematical Projection in Iamblichus", in H. J. Blumenthal, J. F. Finamore (eds), *Iamblichus: The Philosopher*, Iowa City, University of Iowa, 1997, pp. 113–120.

Sheppard, A., "Proclus as exegete", in S. Gersh (ed.), *Interpreting Proclus: From Antiquity to the Renaissance*, Cambridge, Cambridge University Press, 2014, pp. 57–79.

Siasiri, N. G., *Arts and sciences at Padua: the studium of Padua before 1350*, Toronto, Pontifical Institute of Mediaeval Studies, 1973.

Siorvanes, L., *Proclus: Neoplatonic Philosophy and Science*, New Haven, Yale University Press, Edinburgh University Press, 1996.

Siorvanes, L., "Proclus' life, works, and education of the soul", in S. Gersh (ed.), *Interpreting Proclus: From Antiquity to the Renaissance*, Cambridge, Cambridge University Press, 2014, pp. 33–56.

Slaveva-Griffin, S., *Plotinus on Number*, Oxford, Oxford University Press, 2009.

Solère, J. L., "L'ordre axiomatique comme modèle d'écriture philosophique dans l'Antiquité et au Moyen Age", in *Revue d'Histoire des Sciences* 56/2 (2003), pp. 323–345.

Sorabji, R., "Introduction", in R. Sorabji (ed.), *Aristotle Transformed: the Ancient Commentators and their Influence*, Ithaca NY, Duckworth, 1990.

Spinoza, G., "Phantasia e imaginatio nell'Aristotele latino", in *Phantasia – imaginatio, V. Colloqui internazionale Roma 9–11 gennaio 1986*, Rome, Edizioni dell'Ateneo Roma, 1988, pp. 117–133.

Spruit, L., "Introduction" in Agostino Nifo, L. Spruit (ed.), *De intellectu*, Leiden-Boston: Brill, 2011, pp. 1–32.

Spruit, L., "Ludovico Boccadiferro" in M. Sgarbi (ed.), *Encyclopedia of Renaissance Philosophy*, Cham, Springer International Publishing Switzerland, 2022.

Spruit, L., "Marco Antonio Genua" in M. Sgarbi (ed.), *Encyclopedia of Renaissance Philosophy*, Cham, Springer International Publishing Switzerland, 2022.

Spruit, L., *Species Intelligibilis: From Perception to Knowledge: 2. Renaissance Controversies, Later Scholasticism, and the Elimina*, Leiden, Brill, 1995.

Spruit, L., "The Pomponazzi Affair. The Controversy over the Immortality of the Soul" H. Lagerlund, B. Hill (eds), *The Routledge Companion to Sixteenth Century Philosophy*, New York, Routledge, 2017, pp. 225–246.

Steel, C., *The Changing Self: A Study on the Soul in Later Neoplatonism: Iamblichus, Damascius and Priscianus*, Brussels, Paleis der Academiën, 1878.

Steel, C., Griffin, M., Sorabji, R. (eds), *Simplicius' On Aristotle's On the Soul 3.6–13*, London, Bloomsbury Academic, 2014.

Steel, C., "Breathing Thought: Proclus on the Innate Knowledge of the Soul", in J. J. Cleary (ed.), *The Perennial Tradition of Neoplatonism*, Leuven, Leuven University Press, 1997, pp. 293–309.

Steel, C., "Proclus et les arguments pour et contre l'hypothèse des ideés", in *Revue de Philosophie Ancienne* 2 (1984), pp. 3–27.

Steinbauer, D., "Proculus", in H. Cancik, H. Schneider (eds), *Brill's New Pauly: Encyclopedia of the Ancient World*, Leiden, Brill, 2007.

Steiris, G., "John Argyropoulos", in M. Sgarbi (ed.), *Encyclopedia of Renaissance Philosophy*, Cham, Springer International Publishing Switzerland, 2015.

Stewart, I., "Mathematics as philosophy: Barrow and Proclus", in *Dionysius* 18 (2000), pp. 151–181.

Streeter, E. C., "Leoniceno and the School of Ferrara", in *Bulletin of the Society of Medical History of Chicago* 1 (1916), pp. 18–22.

Suter, R., "The Scientifc Work of Alessandro Piccolomini", in *Isis* 60 (1969), pp. 210–222.

Sweeney, E., "Three Notions of Analysis (*Resolutio*) and the Structure of Reasoning in Aquinas", in *The Thomist* 58/2 (1994), pp. 197–243.

Sylla, E. D., "Mathematical physics and imagination in the work of the Oxford Calculatores: Roger Swinehead's *On natural Motion*", in E. Grant, J. E. Murdoch (eds), *Mathematics and its applications to science and natural philosophy in the Middle Ages: essays in honor of Marshall Clagett*, Cambridge New York, Cambridge University Press, 1987, pp. 69–102.

Sylla, E. D., "Medieval Quantifications of Qualities: The "Merton School"", in *Archive for History of Exact Sciences* 30/12 (1971), pp. 9–39.

Sylla, E. D., "The Oxford calculators", in N. Kretzmann, A. Kenny, J. Pinborg, E. Stump (eds), *The Cambridge History of Later Medieval Philosophy*, Cambridge, Cambridge University Press, 1982, VII – Natural philosophy, pp. 540–563.

Taylor, R. C., "ʿAbd al-Laṭīf al-Baghdādī's Epitome of the Kalām fī maḥḍ al-ḫayr (*Liber de Causis*)", in M. E. Marmura (ed.), *Islamic Theology and Philosophy*, New York, State University of New York Press, 1984, pp. 236–248.

Taylor, R. C., "Introduction" in Averroes, in R. C. Taylor (ed.) *Long Commentary on the De Anima of Aristotle*, New Haven, Yale University Press, 2011, pp. xv–cix.

Taylor, R. C., "Averroes on psychology and the principles of metaphysics" in *Journal of the History of Philosophy* 36/4 (1998), pp. 507–523.

Taylor, R. C., "Contextualizing the Kalām fī maḥḍ al-khair / *Liber de causis*", in D. Calma (ed.), *Reading Proclus and the Book of Causes: Western Scholarly Networks and Debates*, Leiden Boston, Brill, 2019, vol. 2, pp. 211–232.

Thijssen, J., "The nature of change", in R. Pasnau, Ch. Van Dike eds, *The Cambridge History of Medieval Philosophy*, Cambridge, Cambridge University Press, 2011, pp. 279–292.

Tieleman, T., "Methodology", in R. J. Hankinson (ed.), *The Cambridge Companion to Galen* (Cambridge: Cambridge University Press, 2008), chapter 5, pp. 49–65.

Touwaide, A., "Galen and Galenism", in M. Sgarbi (ed.), *Encyclopedia of Renaissance Philosophy*, Cham, Springer International Publishing Switzerland, 2022.

Trifogli, C., *Oxford Physics in the Thirteenth Century (ca. 1250–1270) Motion, Infinity, Place and Time*, in A. Speer, W. Goris, K. Emery, G. Kapriev (eds), *Studien und Texte zur Geistesgeschichte des Mittelalters, 72*, Leiden Boston, Brill, 2000.

Trifogli, C., "Change, time, and place", in R. Pasnau, Ch. Van Dike eds, *The Cambridge History of Medieval Philosophy*, Cambridge, Cambridge University Press, 2011, pp. 267–278.

Troilo, E., *Averroismo e aristotelismo Padovano*, Padova, Olschki, 1939.

Van Hee, R., "Jérémie Trivère (1504–1554): médecin humaniste, né il y a 500 ans", in *Revue Médicale de Bruxelles* 26 (2005), pp. 475–478.

Velilla-Jiménez, H. E., "El debate sobre la certeza de las matemáticas en la filosofía natural de los siglos XVI y XVII (*De quaestio* [sic] *de certitudine mathematicarum*)", in *Estudios de Filosofía* 3 (2018), pp. 59–93.

Velilla-Jiménez, H. E., "Francisco Sánchez and the Quaestio de certitudine mathematicarum: A Sceptical approach", in *Endeavour* 46/4 (2022), pp. 1–10.

Vescovini, G. F., "Blaise de Parme et l'intelligibilité mathématique", in J. Biard, A. Robert (eds), *La philosophie de Blaise de Parme: Physique, psychologie, éthique*, Florence, Micrologus Library, 2019, pp. 107–128.

Vlad, M., "L'Être premie entre Proclus et Denys l'Aréopagite", in D. Calma (ed.), *Reading Proclus and the Book of Causes: Western Scholarly Networks and Debates*, Leiden Boston, Brill, 2019, vol. 3, pp. 89–116.

Wakelnig, E., "Les Chapitres sur les thèmes métaphysiques d'al-ʿĀmirī et l'anonyme Kitāb al-ḥaraka: deux interprétations du *Liber de causis* en arabe", in D. Calma (ed.), *Reading Proclus and the Book of Causes: Western Scholarly Networks and Debates*, Leiden Boston, Brill, 2019, vol. 2, pp. 183–198.

Wallace, W. A., *Causality and Scientific Explanation*, vol. 1–2 (Ann Arbor: University of Michigan Press, 1972–1974).

Wallace, W. A., *Domingo de Soto and the early Galileo Essays on intellectual history*, (New York: Routledge 2018).

Wallace, W. A., *Prelude to Galileo: Essays on Medieval and Sixteenth-Century Sources of Galileo's Thought*, Dordrecht Boston, Springer, 1981.

Wallace, W. A., "Albertus on Suppositional Necessity on the Natural Sciences", in J. A. Weisheipl (ed.), *Albertus Magnus and the Sciences: Commemorative Essays*, Toronto, Pontifical Institute of Mediaeval Studies, 1980, pp. 103–128.

Wallace, W. A., "Aquinas on the Temporal Relation Between Cause and Effect", in *The Review of Metaphysics* 27/3 (1974), pp. 277–318.

Wallace, W. A., "Circularity and the Paduan *Regressus*: From Pietro d'Abano to Galileo Galilei", in *Vivarium* 33/1 (1995), pp. 76–97.

Wallace, W. A., "Duhem and Koyré on Domingo de Soto", in *Pierre Duhem: Historian and Philosopher of Science. Part I: Duhem as Historian of Science*, in *Synthese* 83 (1990), pp. 239–260.

Wallace, W. A., "St. Thomas's Conception of Natural Philosophy and its Method", in L. Elders (ed.) *Studi Tomistici, La philosophie de la nature de saint Thomas d'Aquin*, Città del Vaticano, Libreria Editrice Vaticana, 1983, pp. 7–27.

Wallace, W. A., "The "Calculatores" in Early Sixteenth-Century Physics", in *The British Journal for the History of Science* 4/3 (1969), pp. 221–232.

Wallies, M., *Themistii Analyticorum posteriorium paraphrasis*, Berlin, Academiae litterarum regiae Borusicae, 1900.

Walton, S. A., *An Introduction to the Mechanical Arts in the Middle Ages*, Toronto, Association Villard de Honnecourt for Interdisciplinary Study of Medieval Technology, University of Toronto, 2003.

Watson, G., *Phantasia in classical thought*, Galway, Galway University Press, 1988.

Weidemann, H., "Aristotle on the Reducibility of All Valid Syllogistic Moods to the Two Universal Moods of the First Figure (Apr A7, 29b1–25)", in *History and Philosophy of Logic* 25 (2004), pp. 73–78.

Weijersm, O., "Les gloses sur le *Liber de causis* dans les manuscrits parisiens", in D. Calma (ed.), *Reading Proclus and the Book of Causes: Western Scholarly Networks and Debates*, Leiden Boston, Brill, 2019, vol. 1, pp. 121–151.

White, K., "The Meaning of Phantasia in Aristotle's De Anima, III, 3–8", in *Dialogue: Canadian Philosophical Review*, 24/3 (1985), pp. 483–505.

Whitney, E., *Paradise Restored: The Mechanical Arts from Antiquity through the Thirteenth Century*, Philadelphia, American Philosophical Society, 1990.

Wilson, C., "Pomponazzi's Criticism of Calculator", in *Isis* 44/4 (1953), pp. 355–362.

Wilson, C., "William Heytesbury: Medieval Logic and the Rise of Mathematical Physics", in *British Journal for the Philosophy of Science* 8/31 (1957), pp. 254–256.

Wolfe, J. L., *Humanism, Machinery, and Renaissance Literature*, Cambridge and New York: Cambridge University Press, 2004.

Index of Manuscripts

Index of Ancient Authors

Index of Modern Authors